INDUSTRIAL WAXES
Volume 1

BY THE SAME AUTHOR

Practical Emulsions
Trademarks (Chemical)
Substitutes (Chemical)
Commercial Waxes
Chemical Specialties
New Cosmetic Formulary

EDITOR OF

Concise Chemical and Technical Dictionary
The Chemical Formulary Series

INDUSTRIAL WAXES

Volume I

NATURAL & SYNTHETIC WAXES

by

H. BENNETT, F.A.I.C.

Director, B. R. Laboratory
Miami Beach, Florida 33140

CHEMICAL PUBLISHING COMPANY, INC.
New York, N.Y. 10003
1975

CHEMICAL PUBLISHING CO., INC.
NEW YORK N.Y.

Printed in The United States of America

PREFACE TO FIRST EDITION

Waxes are used in industry for many purposes. Their principal use is in waterproofing. Some other important uses are in candles, polishes, electrical insulation, coatings, and carbon paper. There are myriad other uses.

Natural waxes, particularly beeswax, have been used for centuries for protective and decorative purposes. Petroleum waxes are used in greater quantity than the total of all other waxes. Synthetic waxes are increasing in use and have already displaced some of the natural waxes in certain applications. Combinations of different waxes and of waxes with other materials are commonly used to obtain special effects.

The purpose of this work is to present a comprehensive survey of all the types of waxes found in industry—their sources, their preparation, their properties, and their uses. Information is included on mixtures of various waxes and of waxes with resins and other materials. All this will be useful to the chemist, engineer, technician, and student.

The best of the information given in two editions of *Commercial Waxes* has been included in this work, and each section has been supplemented by new material, especially on proprietary products. As developments in the industry have expanded so vastly, it has been necessary to publish the available information in two volumes. The first volume comprises all natural waxes and synthetic products; the second includes compounded waxes, tests and techniques, and commercial applications. Each volume has its index, so that it can stand as a unit.

Grateful acknowledgment is made to those individuals and companies who supplied trade literature and other useful data reproduced here.

H. Bennett

PREFACE TO SECOND EDITION

Since 1963, when the first edition was published, waxes have increased in use in many fields of industry. Many synthetic waxes have been made and, because of their unique properties have replaced the natural and mineral waxes in many uses.

The basic data, in the first edition, has been retained with some deletions, additions and corrections.

In the section of wax formulations entirely new and modern practical formulas are given.

Grateful acknowledgment is made to those individuals and companies who have supplied data and formulas.

H. BENNETT

TABLE OF CONTENTS

PART 1 : NATURAL WAXES

PART 2 : SYNTHETIC WAXES

INTRODUCTION

In industry and commerce, a wax is considered to be an unctuous solid with varying degrees of gloss, slipperiness, and plasticity, which melts readily. Natural waxes may be of vegetable, animal, or mineral origin. Modified waxes are natural waxes that have been treated chemically to change their nature and properties. Synthetic waxes are made by the reaction of two or more chemicals. Compounded waxes are physical mixtures of various waxes or of waxes with resins or other compounds.

Natural waxes show variations in properties which may be due to soil, climate, age, oxidation, impurities, or adulteration. Thus different values for certain properties of a particular wax are often given. Different methods used in testing also produce different data. For most industrial uses, small variations in constants are not critical.

There is considerable misunderstanding as to the nature and classification of fats, waxes, gums, and resins. For example cocoa butter, although it is actually a fat, is commonly considered as a wax because of certain characteristic properties. On the other hand, wool wax, derived from lanolin, is technically a wax but, because of its properties, it is usually considered with fats. Kauri gum is commonly called a wax; and so on. Because of this overlapping and confusion, this presentation has been broadened to include some of these borderline substances.

PROPERTIES OF WAXES

The following list of properties of waxes is fairly complete. It must be understood that the significance or unimportance of any of these properties depends upon the use for which a given wax is intended.

Absorption and adsorption
Adhesiveness or stickiness
Bacterial content
Blocking (self-adhering)
Bursting strength
Candle power
Clarity
Coefficient of expansion
Cold flow
Color
Compatibility with other ingredients
Corrosiveness
Crystalline structure
Crystallization from solution
Density or specific gravity
Dielectric strength
Ductility
Edibility
Elasticity

Electrical state
Emulsifiability or dispersibility
Feel or hand
Film-surface continuity and topography
Flash point
Flexibility
Fluorescence
Form (amorphous or crystalline)
Fracture
Gelling or thixotropy
Hardness
Heating value
Homogeneity
Hygroscopicity
Iridescence
Length (when melted)
Lubricity
Melting point or range
Migration from composition
Odor
Opacity
Overprinting or coating
Penetration into adjacent materials
pH
Plasticity
Polishing properties
Polymerization
Power factor
Pressure (behavior under)
Purity
Reactivity
Refractive index
Resistance to bacteria and fungi
  to radiation
  to shock
  to temperature
Saponifiability
Setting time
Sheen and reflectivity
Slipperiness
Solubility
Stability to electrical discharges
  to light
  to oxidation
Sublimation
Surface tension
Taste
Tenacity or coherence
Tensile strength
Thermal conductivity
Toxicity
Transparency
Ultraviolet screening

Viscosity (molten in solution)
Volume change with temperature
Water repellency
Water-vapor transmission

No one wax always displays all of the properties needed for a specific application. Waxes are usually compounded with other waxes and/or with resins, elastomers, fillers, and other substances to modify their properties.

In selecting a wax for a certain industrial application, in addition to the properties listed, the following factors may also have to be considered:

Availability
Price
Ease of handling
(bulk, slab, flakes, or powder)
Legal restrictions
(health, safety, or patents).

ABBREVIATIONS

A.	Angstrom unit
AMP	American Melting Point
amp.	Ampere
A.P.I.	American Petroleum Institute
ASTM	American Society for Testing Materials
Av.	Average
Bé	Baumé (degrees)
bp	Boiling Point
C	Centigrade
c	Cold
cal	Calorie
calcd	Calculated
cc	Cubic centimeter
cm	Centimeter
cm^3	Cubic centimeter
D	Dispersible
d	Density
d	Dextrorotary
D H	Dispersible hot
distn	Distillation
dr	Dram
E	Engler
EMP	English melting point
F	Fahrenheit
FFA	Free fatty acid
fp	Freezing point
g	Gram
G	Gels
gr	Grain
h	Hot
hr	Hour
i	Insoluble
in.	Inch
inf	Infinite
kg	Kilogram
l.	Liter
l	Laevorotary
lb	Pound
m	Meter
max	Maximum
min	Minimum
mg	Milligram
ml	Milliliter
mm	Millimeter
mol wt	Molecular weight
mp	Melting point
n	Refractive index
n	Normal
No.	Number
oz	Ounce

P	Partially
PSH	Partly soluble hot
psi	Pounds per square inch
pt	Pint, point
qs	A quantity sufficient to make
rd	Specific optical rotation
ref	Refined
s	Soluble cold
SAE	Society of Automotive Engineers
Sap	Saponification
sec	Second
SH	Soluble hot
Sl	Slightly
sol	Soluble
soln	Solution
sp gr	Specific gravity
Suv	Saybolt universal viscosity
SX	Clearly miscible in certain proportions
tech	Technical
temp	Temperature
unsap	Unsaponifiable matter
visc	Viscosity
vol	Volume
wt	weight

SYMBOLS

$\sim$	Approximately	$>$	More than
$\pm$	Plus or minus	$<$	Less than

Part 1

NATURAL WAXES

Chapter 1

PARAFFIN WAX

The United States Pharmacopeia defines paraffin wax as a purified mixture of solid hydrocarbons obtained from petroleum, a colorless or white, more or less translucent mass, showing a crystalline structure, without odor or taste, and slightly greasy to the touch. Its melting point varies from 110–150°F (43.3–65.5°C). Industrially, Paraffin wax appears in many different grades with varying physical properties.

Waxes derived from petroleum are produced and used in quantities many times greater than the total of all other waxes. Paraffin wax is found in many kinds of crude petroleum and is extracted from the high-boiling fractions during the refining process. The annual production in the United States is about 1,732,800,000 lb (1961). As they are by-products of petroleum refining the prices of paraffin waxes usually are relativel low.

Occurrence in Nature

Petroleum is a natural resource found in various parts of the world. It is now generally conceded that petroleum was formed during a vast period of geologic time by the decomposition of organic refuse, largely of marine origin, aided by conditions of low temperatures and enormous pressures. An analysis of petroleum, freed from extraneous matter, shows that it is composed mainly of hydrogen (11–14%) and carbon (83–87%). Small quantities of sulfur, oxygen, and nitrogen are frequently present. The elements contained in petroleum are in chemical combination, as are most of the elements in nature. It is a peculiarity of carbon that it can enter into a vast number of chemical combinations with itself and other elements. It is a part of all living organisms. Because of this fact, early chemists distinguished between carbon compounds and all others, classifying the former as "organic" and the latter as "inorganic."

Most of the compounds contained in petroleum are hydrocarbons. They range from gases to solids with a predominance of liquid members. Relatively few pure chemical compounds have been isolated from petroleum, but thousands actually exist. This is due to the multiplicity of combinations that are possible between atoms of hydrogen and carbon.

COMPOSITION OF PETROLEUM

Petroleum is composed essentially of four distinct types of hydrocarbons which differ from one another in chemical structure. These are known as paraffins, olefins, naphthenes and aromatics.

Paraffins, or saturated, open-chain hydrocarbons, are those in which the carbon atoms are linked together with single bonds, the remaining valences of the carbon atoms being satisfied with hydrogen atoms. Methane (CH_4) is the simplest paraffin hydrocarbon, followed by ethane ($CH_3.CH_3$), and propane ($CH_3.CH_2.CH_3$, or C_3H_8).

The next compound in the series is *butane*, $CH_3.CH_2.CH_2.CH_3$, or C_4H_{10}. All these compounds are called *straight-chain* hydrocarbons because in their formulas the carbons are linked together in a succession, with no branches or side-chains.

Beginning with butane, however, compounds occur which contain the same numbers of atoms of carbon and hydrogen but have entirely different properties. Such compounds, called *isomers*, are represented by formulas with branched- or side-chains. The formula for *iso*-butane, for example, is

$$\begin{array}{c}CH_3{-}\overset{}{CH}{-}CH_3.\\ \quad|\quad\\ CH_3\end{array}$$

Obviously, normal butane ($CH_3.CH_2.CH_2.CH_3$) has the same composition as isobutane, but it is an entirely different compound because of a distinct difference in the arrangement of the carbon atoms. It can be seen that as the number of carbon atoms increases, the number of possible arrangements also increases, and at a more rapid rate. This accounts for the complexity of petroleum products and the difficulty in isolating pure compounds from them.

Olefins are closely related to the paraffins in that they have an open chain. They differ however, in that at some point in the chain two carbon atoms are joined together by means of a double (or more rarely a triple) bond. Such products are said to be "unsaturated," meaning that under certain conditions they will take up hydrogen or other elements. Olefins can be converted into paraffins by a reaction known as *hydrogenation*. This consists of the introduction of two hydrogen atoms into the molecule at the double bond and its conversion to a single bond.

Naphthenic and aromatic hydrocarbons are *cyclic*, i.e., their formulas show the carbon atoms in ring structures, Simple members of these series are benzene (C_6H_6) and cyclohexane (C_6H_{12}).

Naphthenes are saturated hydrocarbons (no double bonds) containing one or more ring nuclei. In some naphthenes paraffinic side chains are attached to the carbon rings.

Aromatic hydrocarbons are characterized by the presence of a benzene ring in the molecule. These too may have paraffinic side chains attached to the benzene ring. The relationship between naphthenes and aromatics is about the same as between paraffins and olefins.

Petroleum crudes, occurring in various parts of the world, differ from one another in the percentage of the four classes of hydrocarbons that they contain. In Pennsylvania crudes, for example, paraffinic hydrocarbons predominate, whereas in California crudes, naphthenes and aromatics constitute the greater portion. Crudes from mid-continent sources, in general, are of mixed base and contain substantial quantities of all four types of hydrocarbons.

GRADES OF PARAFFIN WAX

Paraffin wax is sold in various grades, which differ from one another chiefly in melting point. Generally, each grade covers a range of melting points of 2°F, for example 125–127°F. Melting points are given by either of two accepted standards: American melting point (AMP) and ASTM melting point, which is often called English melting point (EMP). The American melting point is a calculated value based on the ASTM melting point, the first being arbitrarily taken as 3°F higher than the second. The method for determining this value, which has the ASTM designation D87-37, involves the cooling of molten wax in a cooling bath at a constant rate of drop in temperature. The melting point of the wax is taken as the temperature at which the wax first shows the minimum change in temperature. This method is based on the fact that, on solidification, the wax gives up its latent heat of fusion and the rate of drop in temperature is considerably reduced.

The melting point of a wax, in itself, is no measure of quality or chemical composition because it is derived from the melting points of the constituents which influence one another in various ways. Other characteristics of the wax must also be considered in determining quality. These include oil content, stability to light, tensile strength, etc. Their relationship to the applicability of the wax will be discussed later in greater detail.

The most familiar types of paraffin wax on the market are *refined*, *semi-refined*, and *crude scale waxes*.

Refined Wax

Refined paraffin wax, which is the most commonly used grade of paraffin wax, has the following properties:

Melting Point (ASTM)	122–124°F
Color (Saybolt)	25–30
Flash Point	395°F
Penetration at 32°F	16

Table 1:1

Composition and Properties of Petroleum Waxes and Petroleum Wax Fractions

Wax identification	*Paraffin wax A*	*Urea-fractionated wax A*		*Micro. wax B*	*Urea-fractionated wax B*		*Semi-micro. wax C*	*Solvent-fractionated wax C*		
		Ad-ductible	*Non-adductible*		*Ad-ductible*	*Non-adductible*		*Cut 1*	*Cut 2*	*Cut 3*
% By wt. of parent wax	100	97	3	100	42	58	100	42	28	30
Composition, vol. %										
Normal paraffins	93.0	{91.9	{28.0	17.6	21.9	18.1	52.6	37.6	36.0	30.9
Isoparaffins	0.5			26.4	40.9	3.6	5.3	35.8	18.5	5.7
Noncondensed cycloparaffins	6.5	6.0	64.9	46.6	35.5	59.0	39.5	25.7	42.5	53.9
Condensed cycloparaffins										
2 rings	0.0	0.3	3.5	4.9	0.3	10.1	1.6	0.2	1.9	5.8
3 or more rings	0.0	0.1	0.2	1.7	0.0	4.2	0.0	0.0	0.1	0.8
Benzenes	0.0	1.3	3.4	2.7	1.4	4.7	1.0	0.7	1.0	2.7
Naphthalenes	0.0	0.4	0.0	0.1	0.0	0.3	0.0	0.0	0.0	0.2
Inspection properties										
Melting point (D87), °F.	131.4	131.9	107.8	...	...	...	158.4	168.0	150.6	136.8
Melting point (D127), °F.	133.6	...	...	168.5	179.0	141.5	161.0	171.0	153.0	140.0
Congealing point (D938), °F.	133	134	112	162	173	134	156	172	153	138
Oil content (D721), wt. %	0.1	...	...	3.4	0.7	5.4	1.9	0.8	2.0	6.2
Viscosity (D446), SUS @ 210°F.	37.8	...	...	78.5	59.2	103.0	51.8	55.4	50.5	52.8
Density @ 210°F., g./cc.	0.7529	0.7530	0.7760	0.7960	0.7785	0.8086	0.7753	0.7713	0.7751	0.7820
Refractive index @ 176°F.	1.4301	1.4297	1.4400	1.4509	1.4450	1.4566	1.4393	1.4375	1.4411	1.4410
Refractive index @ 212°F.	1.4253	...	...	...	...	...	1.4351	1.4342	1.4349	1.4382
Molecular weight	381	380	405	685	645	780	494	512	483	450
Physical properties (from expansion measurements)										
Crystal point, °C.	56.2	56.3	45.0	76.9	81.5	59.9	71.6	77.1	66.8	59.2
Transition point, °C.	39.5	39.1	...	...	...	...	...	...	58.6	48.1
ΔT for complete phase change, °C.										
Fusion	7.9	8.5	21.2	65.0	70.5	48.0	47.6	56.6	8.2	11.1
Transition	11.0	11.1	...	...	...	...	...	...	33.6	24.6
Liquid expansion, cc./g./°C.	0.0011	0.0011	0.0010	0.0009	0.0010	0.0009	0.0010	0.0010	0.0010	0.0009
Solid expansion, cc./g./°C.										
Form A (above transition)	0.0013	0.0011	...	...	...	...	...	...	0.0007	0.0034
Form B (below transition)	0.0007	0.0008	0.0015	0.0006	0.0005	0.0007	0.0007	0.0005	0.0006	0.0007
Expansion at phase change, cc./g.										
Fusion	0.1272	0.1267	0.1134	0.1611	0.2018	0.1564	0.1805	0.2023	0.1087	0.0998
Transition	0.0392	0.0385	...	...	...	...	...	...	0.0657	0.0485
Specific volume, cc./g.										
Crystal point	1.2795	1.2824	1.2347	1.2357	1.2668	1.2027	1.2618	1.2750	1.2584	1.2406
Transition point	1.1405	1.1463	...	...	...	...	...	...	1.1497	1.1408
Expansion, %										
Fusion	11.0	11.0	10.1	14.9	18.9	14.5	16.7	18.9	9.5	8.7
Transition	3.6	3.5	...	...	...	...	...	...	6.1	4.4
0–100°C.	22.6	22.4	18.6	17.8	21.3	15.4	21.0	22.1	20.8	18.9

STAP No. 6 A.S.T.M.

Table 1:1 *(continued)*

Composition and Properties of Petroleum Waxes and Petroleum Wax Fractions

Wax identification	*Micro. wax D*	*Solvent-fractionated wax D* *Cut 1*	*Cut 2*	*Cut 3*	*Micro. wax E*	*F*	*G*	*L*	*M*	*N*	*P*	*R*	*S*
% By wt. of parent wax	100	34	15	51	100	100	100	100	100	100	100	100	100
Composition, vol. %													
Normal paraffins	25.9	18.4	19.4	20.9	20.5	59.0	61.3	94.1	89.4	88.1	83.2	73.4	67.5
Isoparaffins	15.4	33.9	21.2	12.7	27.7	21.3	10.0	4.3	0.0	3.4	0.0	2.4	0.0
Noncondensed cycloparaffins	45.1	39.6	42.2	48.2	38.2	19.2	26.7	1.5	9.3	8.5	16.3	23.8	31.5
Condensed cycloparaffins													
2 rings	7.1	4.5	5.7	8.5	4.8	0.4	1.3	0.0	0.1	0.0	0.2	0.0	0.3
3 or more rings	2.6	1.2	2.7	3.6	2.7			0.0	0.0	0.0	0.0	0.0	0.0
Benzenes	3.7	2.3	6.4	5.6	5.6	0.1	0.7	0.0	1.0	0.0	0.2	0.4	0.7
Naphthalenes	0.2	0.1	2.4	0.5	0.5	0.0	0.0	0.1	0.2	0.0	0.1	0.0	0.0
Inspection properties													
Melting point (D87), °F.	...	...	...	...	...	129.4	129.8	118.5	126.6	126.1	133.7	140.9	144.6
Melting point (D127), °F.	170.0	177.0	167.0	160.5	170.5	135.5	133.5	120.9	130.0	129.3	135.0	144.0	148.0
Congealing point (D938), °F.	160	175	161	149	163	133	133	118.5	127	128	133	142	147
Oil content (D721), wt. %	1.2	0.7	0.6	1.4	3.8	0.4	0.3	0.9	0.4	0.2	0.3	0.2	0.2
Viscosity (D446), SUS @ 210°F.	77.2	74.6	75.3	82.4	93.3	39.7	40.5	35.7	37.4	37.3	39.0	42.2	43.6
Density @ 210°F., g./cc.	0.7929	0.7884	0.7912	0.7964	0.7972	0.7568	0.7606	0.7465	0.7518	0.7506	0.7556	0.7619	0.7648
Refractive index @ 176°F.	1.4481	1.4487	1.4489	1.4515	1.4520	1.4313	1.4330	1.4258	1.4289	1.4280	1.4302	1.4332	1.4349
Refractive index @ 212°F.	...	...	...	...	...	1.4260	1.4266	...	...	...	1.4261	1.4285	1.4295
Molecular weight	703	730	725	690	730	400	400	313	355	346	400	414	445
Physical properties (from expansion measurements)													
Crystal point, °C.	76.9	79.8	77.4	72.5	77.5	55.8	54.9	48.9	53.4	53.1	57.6	61.4	63.8
Transition point, °C.	...	...	...	...	...	39.5	39.0	34.8	36.3	37.8	42.2	47.4	...
ΔT for complete phase change, °C.													
Fusion	58.2	60.4	59.2	56.0	63.5	10.4	9.0	8.4	9.4	7.2	9.6	9.3	32.1
Transition	...	...	...	...	...	16.0	16.0	14.1	10.8	12.5	17.2	13.1	...
Liquid expansion, cc./g./°C.	0.0009	0.0009	0.0009	0.0009	0.0009	0.0011	0.0011	0.0011	0.0011	0.0011	0.0011	0.0010	0.0010
Solid expansion, cc./g./°C.													
Form A (above transition)	...	...	...	...	...	0.0015	0.0017	0.0013	0.0013	0.0013	0.0015	0.0020	...
Form B (below transition)	0.0007	0.0007	0.0007	0.0005	0.0006	0.0008	0.0007	0.0008	0.0008	0.0008	0.0007	0.0008	0.0007
Expansion at phase change, cc./g.													
Fusion	0.1605	0.1752	0.1664	0.1470	0.1535	0.1212	0.1177	0.1270	0.1267	0.1261	0.1285	0.1218	0.1744
Transition	...	...	...	...	...	0.0404	0.0395	0.0437	0.0374	0.0407	0.0437	0.0395	...
Specific volume, cc./g.													
Crystal point	1.2395	1.2505	1.2440	1.2310	1.2340	1.2728	1.2662	1.2834	1.2807	1.2813	1.2783	1.2726	1.2708
Transition point	...	...	...	...	...	1.1425	1.1369	1.1490	1.1442	1.1445	1.1412	1.1414	...
Expansion, %													
Fusion	14.9	16.3	15.4	13.6	14.2	10.5	10.2	11.0	11.0	11.0	11.2	10.6	15.9
Transition	...	...	...	...	...	3.6	3.6	4.0	3.4	3.7	4.0	3.6	...
0–100°C.	18.4	19.5	18.8	16.8	17.1	21.7	21.3	23.1	22.6	23.0	22.5	22.2	21.9

Penetration at 77°F	22
115°F	221
Saponification Value	0
Acid Value	0
Viscosity at 212°F	39
Structure	Crystalline
Specific Gravity	0.880–0.915

Soluble in: Benzene, ether, chloroform, carbon disulfide, carbon tetrachloride, turpentine, petroleum, fixed oils.

Insoluble in: Water, cold alcohol.

Uses: Candles; waterproofing, sealing; lubricating; protection of foods, plants, fruits, and vegetables; paper; polishes; cosmetics; crayons; pharmaceutical ointments and salves; electrical insulation, etc.

In general, the oil content of refined waxes is 0.5% or less. Usually they are reasonably stable to darkening when exposed to light, and their tensile strength is 160–400 lb psi when measured on the Perkins Tensile Tester. These methods for evaluating and testing the uniformity of waxes are conventional. Unfortunately, many users of waxes consider these characteristics as the sole measure of their quality for all applications. As existing relationship between quality and applicability is by no means so clear cut as is frequently assumed, erroneous conclusions regarding applicability are drawn far too often.

The largest percentage of the paraffin wax on the market is of the *refined* type. As mentioned previously, this type of wax contains a small percentage of oil, it is hard, and it has been refined to a point where it is tasteless, odorless, and stable to light. The bulk of the remainder is crude scale wax.

Crude Scale Wax

Crude scale wax differs from refined wax in several important details. First, it contains appreciably more oil (average 3%). For this and other reasons, its quality is poorer than that of refined waxes of comparable melting points. The high oil content of crude scale wax is evident in low tensile strength (40–100 lb per sq in.) and also in a softer and more crumbly structure.

Because of the lesser degree of refining of these waxes, they are not entirely tasteless and odorless, and their stability of color is questionable. There are wide variations in the over-all quality of crude scale waxes, largely because no standards or specifications (except melting point) are established for them. This condition is also due to the large number of petroleum refiners that produce crude scale waxes, and to the variations in the crudes from which they are derived.

Crude scale wax is sold in two forms: *White* and *yellow*. They are similar except for one step in the refining procedure for bleaching the wax. Most of crude scale wax sold in the market is the white type.

Semirefined Wax

Between refined waxes and crude scale waxes there is a third type, called *semirefined* wax. As far as quality is concerned, this designation has even less meaning than has "white crude scale wax." Just as crude scale waxes may vary in quality, so may semirefined waxes. In spite of a relatively uniform price, these grades differ appreciably, depending on the supplier and the source. Frequently, semirefined waxes approach fully refined waxes in quality, but just as frequently, they may be little, if any, better than crude scale waxes. Consequently, a purchaser of these grades, unless dealing directly with a producer, will have no real guarantee of uniformity.

Semirefined waxes are produced by a group of small operators whose main business is the refining of wax. These companies purchase crude scale wax on the open market and sweat it to a more highly refined grade. Depending entirely on the basic crude scale wax, and on their ability to conduct this operation, they offer waxes that vary widely in quality. As a group, semirefined waxes will have melting points in the range 120–135°F, oil contents of about 0.6–3%, and tensile strength of 40–200 lb per sq in.

Match wax is a low-melting (104–116°F) refined paraffin wax. Reclaimed paraffin wax is often available at a lower price. This refers to wax that has been scraped off or melted out of paper, textiles, etc. It is usually darker in color and may have different kinds of foreign impurities and decomposition products.

Black Wax is obtained by blowing air through heated petrolatum or petroleum waxes. This gives a viscous black product of melting point ±129°F; flash point (open cup), 465°F; and viscosity of 401 (210°F) Saybolt.

MANUFACTURE OF PARAFFIN WAXES

As previously noted, petroleum fractions are complex mixtures of many ingredients. Their properties depend, in general, on two factors: the average molecular weight, and the average configuration. The differences in properties are used for the separation of one fraction from another.

Crude petroleum is first subjected to distillation to break it up into fractions varying in boiling point and viscosity. Gaseous light ends, gasoline, naphtha, kerosene, and gas-oil fractions are removed in this order. From paraffinic crudes, there is then obtained a fraction known as wax distillate, or paraffin distillate, which consists of a mixture of light-bodied oils ranging from 40–300 seconds S.u.v.[1] at 100°F, and various waxy fractions of about the same boiling range. Generally, fractions having boiling points slightly higher than that of the wax distillate cut are not

[1] Saybolt universal viscosity.

distilled, but remain as a residuum and are further processed to prepare heavier bodied lubricating oils.

In the wax distillate, wax and oil must be separated for use. The wax distillate will have a pour point of about 80°F, which indicates that below this temperature it is, at best, a semifluid mixture of wax and oil with poor lubricating properties. Most oils are dewaxed to a pour point of 25°F, or lower, for use as lubricants. Similarly, satisfactory waxes for most uses must be deoiled to relatively low oil contents. Separation is based either on the differences in melting points of the wax and oil or on differences in the solubilities in a solvent. It should be emphasized, at this point, that there is a very close relationship between paraffin wax and a paraffin oil having, for example, a viscosity of 100 seconds S.u.v. at 100°F. Both have approximately the same boiling range and volatility and, when heated to above the melting point of the wax, have comparable viscosities. Care is essential, therefore, in the handling of molten paraffin wax, for at elevated temperatures its tendency to be oxidized and the loss due to evaporation are little less than those of paraffin oil. Paraffin wax differs from the oil primarily in that it has a nearly straight-chain structure, whereas the oil is more branched.

The first step in the separation of the wax distillate, pressing, consists essentially of filtering solid wax from liquid oil at low temperatures. The products of this operation are called *dewaxed oil* and *slack wax*. Often the paraffin distillate is diluted with dewaxed oil to facilitate handling. This is particularly necessary if the wax content is high because, otherwise, the chilled mass will be thick and difficult to pump from the chillers to the filter presses.

Frequently, the distillate is treated with a mild acid to lighten its color and remove highly polymerized impurities. After the acid treatment, the distillate is washed with water, neutralized with soda, and washed again. Finally, it is chilled at 13–15°F to precipitate the higher-melting, waxy hydrocarbons.

The temperature to which the distillate is chilled depends on the pour point that is desired in the dewaxed oil. For a 20°F pour point, a temperature of 13–15°F is necessary. For lower pour points, the pressing temperature must be lowered accordingly. The chilled distillate is pumped to a bank of filter presses in which crystalline wax is filtered from the dewaxed oil. This yields slack wax that contains about 35% oil. The dewaxed oil is finally resolved into various paraffin oils and other light lubricating stocks by distillation and conventional refining procedures.

Wax is separated from oil by either of two methods. The older procedure involves the so-called "sweating" operation; the more modern one, known as "solvent pressing" utilizes crystallization from a solvent mixture. Wax distillates vary in wax content over a range of 6–25%; 15% average content may be expected in the distillate.

Sweating Process

The slack wax, which melts around 107°F must be subjected to further processing, the first state of which may be a second treatment with mild acid. The molten slack wax is charged into *sweaters* which are generally shaped like pans. A conventional sweater comprises a room or oven which houses a stack of ten to fifteen or more horizontal trays, each equipped with a gravity drain. The slack wax is allowed to solidify over water, whose surface has been adjusted to be on a level with a wire mesh in each tray. When the wax is completely solid (at temperature around 80°F), the water is withdrawn and the slack cake, 5–10 in. thick, remains suspended on the screen. The temperature of the sweater is then raised at a rate of 2°F per hour by circulating warm water through banks of tubes embedded in the wax along the walls of the oven. Heat causes beads of oil to form on the surface of the wax, which resemble sweating (hence the name of the process). The oil and lower-melting wax are allowed to drain away as the temperature increases. Finally, a cake of high-melting waxes remains which is melted and separated. This is known as *yellow crude scale wax*; it averages 5% oil and has a melting point of 122–124°F. The oil and low-melting waxes constitute the *foots oil* which is used as cracking feed stock for the manufacture of gasoline.

The crude scale wax is given another treatment with mild acid, washed, neutralized, and again washed. The subsequent resweatings are frequently highly complicated, and generally involve the recycling of the lower-melting fractions (intermediates) to produce a series of three or four grades of wax. For example, waxes melting at 130–132°F, 126–128°F, and 122–124°F may be produced. Some refiners make a higher-melting grade and others one that melts as low as 118–120°F. These waxes which contain only 0.4% of oil, or even less, are used for blending to intermediate-melting waxes. Finished waxes are finally prepared by blending to melting point and filtering through bone char, bauxite, or some other medium to produce water-white products.

Many of the white crude scale waxes on the market, were treated in a similar manner, but because of their nature, they could not make them meet the specification for refined wax. In other words, in spite of resweating from yellow crude scale wax, they still contain 2% oil, and are low in tensile strength. As they are not entirely color-stable, they must be sold as *white crude scale wax*. Some refiners merely bleach yellow crude scale wax to obtain the same result. Such white crude scale waxes are inferior to those that have been resweated.

Sweat wax differs from slack wax in that it is a mixture of oil and paraffins of a limited range of melting points in the lower melting point series. It solidifies around 60°F and it is used where extreme penetration by wax is desired.

Solvent-Pressing Process

In the modern method of producing paraffin wax, by pressing it from solvents, the raw material is the same wax distillate as used in the sweating process. In general, it is pressed to slack wax by conventional procedures. In some cases, a solvent-pressed slack is used, whose oil content is much lower than usual. The slack wax is dissolved in a suitable solvent, which may be petroleum naphtha or a mixture of benzene and methyl ethyl ketone.

In one method, three parts of slack wax and two parts of solvent are chilled to 45°F and filtered in a press, as described before, to yield a crude scale wax that contains only 3% oil. This crude scale wax is further processed by repeated solvent-pressing or recrystallization to yield suitable base waxes. The quantity of solvent and the temperature employed are by no means fixed, but depend on the yield and on the melting point of the desired wax. Intermediate fractions are recycled, but this step is not so complicated as in the sweating process because of the better fractionation obtained. For the same reason, only two or three blending waxes are produced, these being drier, less ductile and, in general, more free of oil than sweated waxes.

COMPOSITION OF PARAFFIN WAX

Paraffin wax, as stated, is a mixture of a great number of hydrocarbons. Small amounts of liquid "impurities" are present in the form of oil. The solid constituents are saturated paraffinic hydrocarbons.

Frequently it is said that these hydrocarbons are all of the straight-chain type and that no branched-chain products are present. Considering that the oils from which they are separated are highly complex and contain a wide variety of branched-chain hydrocarbons, this would be indeed surprising, if true. Much exploratory work was done by numerous investigators. Much of the information in this section is derived from the publication of Ferris, Cowles, and Henderson,[1] although reference will be made to the work of other investigators as well.

As mentioned previously, paraffin wax is derived from the wax-distillate fraction of crude petroleum and is separated from other lower-melting waxes lost in foots oil and sweater oil. In addition, higher-melting waxes, which are sold as *microcrystalline*, *amorphous*, or *petrolatum* waxes, are derived from heavier lubricating-oil fractions. A definite relationship exists among all these waxes. Buchler and Graves[2] found that by isolating pure, oil-free fractions from waxes of these types and plotting a curve for melting-point and refractive index, a substantially straight line was obtained. Certain deviations from the straight line occurred at higher melting points,

[1] *Ind. Eng. Chem.* **23**, 681 (1931).

[2] *Ibid.* **19**, 718 (1927).

but by recrystallization from a solvent, small amounts of an impurity were lost and then the line for melting point and refractive index for the purified waxes became straight.

Phenomena of Crystallization

It has also been observed that paraffin wax crystallizes in both plates and needles, depending on the conditions of crystallization. One theory explained this as being due to differences in viscosity of the medium and also to the rate of cooling. This explanation discounted the idea that differences in composition had any bearing on the type of crystals formed. Buchler and Graves attributed the differences in the crystal system entirely to the presence of the impurity mentioned before, which they called "soft wax."

In the later work of Ferris, Cowles, and Henderson, a slack wax from a mid-continent crude oil was selected for study. It was completely deoiled by successive recrystallizations at about 5°F and fractionally distilled into narrow cuts. These were purified into twenty-six separate waxes by repeated recrystallization and recombination of fractions having the same melting points. When the purification was completed, waxes were obtained which, on further crystallization, were practically unchanged in character. It was, therefore, assumed that practically pure components had been produced.

Then types of crystals, which were formed when these pure waxes were crystallized from a solvent, were carefully studied. Three different types were found. Most of these waxes formed plate-type crystals; another group crystallized as needles, and the remainder assumed forms that could not be identified. The last group was called "malcrystalline," for lack of a better name. Further studies of these fractions showed that no matter what solvent was used each individual fraction invariably yielded the same type of crystals. These observations effectively disproved the theory that differences in crystal system are due to the crystallization medium.

Study of Pure Waxes

In order to study the significance of these three types of crystals, the physical characteristics of all the pure waxes were determined. By plotting these characteristics, the relationships between melting point and refractive index, melting point and boiling point, melting point and density, and melting point and molecular weight were found. In each case, three curves were obtained that corresponded exactly to the three types of crystals. All these observations show conclusively that paraffin wax contains three series of compounds, each of which has a characteristic crystal form, irrespective of the medium from which it is precipitated. It is now generally recognized that the rate of cooling affects merely the size of the crystals formed and not their type. Storage of wax that contains small crystals may result in the growth of larger crystals, but it can never cause

Table 1:2

DATA ON PURIFIED, OIL-FREE PETROLEUM WAXES[1]

Melting Point °F.	Melting Point °C.	n_D^{84}	d^{84}	Observed Mol. Wt.	Observed Mol. Refractivity	Observed % C	Observed % H	Formula	Calculated Mol. Wt.	Calculated % C	Calculated % H	Calculated Mol. Refractivity
Paraffin Wax												
80.5	27.0	1.4133	0.7347	255	86.5	84.55	15.04	$C_{18}H_{38}$	254	85.04	14.96	86.9
88.5	31.4	1.4155	0.7397	271	92.0			$C_{19}H_{40}$	268			89.6
97.5	36.4	1.4173	0.7417	287	97.6			$C_{20}H_{42}$	282			94.1
101.0	38.3	1.4183	0.7440	292	99.1			$C_{21}H_{44}$	296			98.8
105.5	40.8	1.4191	0.7458	294	99.6	84.86	14.98	$C_{21}H_{44}$	296	85.10	14.90	98.8
107.0	41.7	1.4191	0.7466	298	100.8			$C_{21}H_{44}$	296			98.8
114.2	45.7	1.4210										
116.7	47.1	1.4217	0.7513	330	111.2	84.75	14.91	$C_{23}H_{48}$	324	85.18	14.82	107.9
118.7	48.2	1.4221										
121.0	49.4	1.4226	0.7526	337	113.5			$C_{24}H_{50}$	338			112.6
126.5	52.5	1.4240										
127.5	53.0	1.4242										
128.0	53.3	1.4244	0.7558	357	121.2	85.31	14.75	$C_{25}H_{52}$	352	85.21	14.79	117.2
131.5	55.3	1.4252										
133.5	56.4	1.4257	0.7580	367	123.9			$C_{26}H_{54}$	366			121.8
137.2	58.4	1.4266	0.7607									
138.5	59.1	1.4269	0.7609	385	129.9			$C_{27}H_{56}$	380			126.3
144.2	62.3	1.4285	0.7642	410	137.7			$C_{29}H_{60}$	408			135.5
149.5	65.3	1.4297	0.7660	430	144.7	85.22	14.74	$C_{31}H_{64}$	436	85.32	14.68	144.7
156.0	68.9	1.4315	0.7687	456	153.7			$C_{32}H_{66}$	450			149.3
Petrolatum Wax												
159.7	71.0	1.4323	0.7706	474	159.6			$C_{34}H_{70}$	478			159.6
164.5	73.6	1.4333	0.7727	492	164.4	85.45	14.78	$C_{35}H_{72}$	492	85.36	14.64	163.2
166.7	74.8	1.4339	0.7739	502	168.9			$C_{36}H_{74}$	506			169.9
169.2	76.2	1.4347	0.7753	520	174.9			$C_{37}H_{76}$	520			172.4
170.7	77.1	1.4351	0.7759	539	181.3			$C_{38}H_{78}$	534			177.0
173.7	78.8	1.4359	0.7771	555	186.7	85.13	14.62	$C_{39}H_{80}$	548	85.40	14.60	181.6
177.2	80.7	1.4366	0.7784	572	192.3			$C_{41}H_{84}$	576			190.8
181.2	82.9	1.4375	0.7803	586	194.9			$C_{42}H_{86}$	590			195.4
182.7	83.8	1.4378		603		85.54	14.47	$C_{43}H_{88}$	604	85.43	14.57	200.0

[1] Buehler and Graves, *Ind. Eng. Chem.* **19**, 725 (1927).

Table 1 : 2 (continued)

Melting Point				Observed					Calculated			
°F.	°C.	n_D^{84}	d^{84}	Mol. Wt.	Mol. Refractivity	% C	% H	Formula	Mol. Wt.	% C	% H	Mol. Refractivity
					Slop Wax							
132.2	55.7	1.4255		360		84.87	14.83	$C_{26}H_{54}$	366	85.23	14.77	
136.2	57.9	1.4263	0.7600	385	129.7			$C_{27}H_{56}$	380			126.3
138.5	59.2	1.4270										
139.0	59.4	1.4270	0.7614	394	133.0			$C_{28}H_{58}$	394			130.9
140.5	60.3	1.4275										
142.5	61.4	1.4280										
144.5	62.5	1.4284										
145.5	63.1	1.4287	0.7644	405	136.5	85.18	14.81	$C_{29}H_{60}$	408	85.29	14.71	135.5
148.5	64.7	1.4294										
150.5	65.9	1.4300	0.7677	431	145.1	85.19	14.74	$C_{31}H_{64}$	436	85.32	14.68	144.7
152.0	66.7	1.4303										
153.5	67.5	1.4306	0.7677	437	147.5			$C_{31}H_{64}$	436			144.7
156.0	68.5	1.4312	0.7684	451	152.1			$C_{32}H_{66}$	450			149.3
157.0	69.4	1.4316	0.7695	460	155.0			$C_{33}H_{68}$	464			153.9
160.0	71.1	1.4323	0.7706	470	158.1			$C_{33}H_{68}$	464			153.9
161.5	72.0	1.4327										
162.2	72.3	1.4329	0.7715	483	162.7	85.22	14.65	$C_{34}H_{70}$	478	85.35	14.65	163.1
168.0	75.6	1.4342	0.7741	508	171.6			$C_{36}H_{74}$	506			167.7
176.5	80.3	1.4365	0.7785	563	189.2			$C_{40}H_{82}$	562			186.1
182.0	83.3	1.4379	0.7810	600	201.8	85.21	14.65	$C_{43}H_{88}$	604	85.43	14.57	199.9
					Rod Wax							
165.0	73.9	1.4334	0.7728	493	165.9			$C_{35}H_{72}$	492			163.2
167.5	75.3	1.4340	0.7739	516	173.6	85.40	14.55	$C_{37}H_{76}$	520	85.38	14.62	172.4
169.5	76.4	1.4345	0.7743	524	176.4			$C_{37}H_{76}$	520			172.4
171.0	77.2	1.4349	0.7757	527	177.2			$C_{38}H_{78}$	534			177.0
172.7	78.2	1.4353	0.7761	545	183.3	85.38	14.68	$C_{39}H_{80}$	548	85.40	14.60	181.6
174.2	79.0	1.4358	0.7771	550	185.0			$C_{39}H_{80}$	548			181.6
176.7	80.5	1.4363	0.7780	569	191.3			$C_{40}H_{82}$	562			186.2
180.5	82.5	1.4372		575		85.39	14.60	$C_{41}H_{84}$	576	85.41	14.59	190.8

a change in their type. This can be accomplished only by removing from the wax, or adding to it, an impurity which is capable of impressing its form on the basic type.

It was concluded that the plate-type crystals represent straight-chain hydrocarbons and constitute the basic fraction of paraffin waxes. The needle and malcrystalline types of crystals were probably the "soft wax," of Buchler and Graves, and these were assumed to be branched-chain hydrocarbons. Microcrystalline waxes are composed largely of these types, although high-melting paraffin waxes will sometimes be found to contain small amounts of needles. These conclusions were substantiated by further work in the laboratory on certain selected fractions believed to be rich in "soft wax."

Crystalline Types

As pure components are rarely isolated in practice, additional work was done to determine the effect of mixtures of the three components on behavior in crystallization. One phase of this investigation involved the determination of relative solubility of the three types in various solvents. They found that:

1. Needle-type waxes are much more soluble in any solvent than plate-type waxes of the same molecular weight; the malcrystalline type was intermediate between the two.
2. For all solvents investigated, waxes of the same melting point, regardless of type, have about the same degree of solubility.

Mixtures of the three types, each having the same melting point, were prepared and their behavior in crystallization was noted. It was found that the needle type was strong enough to impress its form on the plates if both types were allowed to crystallize simultaneously from the same solution. However, if crystallization was such that the needle type was kept in solution, either by high temperature or by excess of solvent, the plate type formed its characteristic crystals.

Similar work showed that the malcrystalline wax is even more powerful than the needle type in impressing its form on plates and on mixtures of plates and needles. Thus, any blends with the malcrystalline type will yield malcrystalline formations, unless this impurity is kept from the sphere of activity by higher solubility.

One practical utilization of these differences in crystallization is found in the manufacture of paraffin wax by the sweating process. It is well known that a wax distillate with good pressing characteristics contains plate-type crystals, and that a slack wax with good sweating characteristics will crystallize as large interlacing needles. In both cases, apparently the same wax is crystallized, but the conditions of crystallization are far different. In paraffin distillates, the percentage of oil is relatively high, thus holding the needles entirely in solution and allowing plates to form. However, when slack wax, which contains the needles dissolved in the oil, is chilled to a solid cake, both types are precipitated simultaneously and the needles impress their form on the plates.

It has also been recognized by refiners that if the wax distillate contains too much of high-boiling constituents it will not press satisfactorily. It was also found that as the boiling point increases the percentage of the malcrystalline-type wax also increases very rapidly. A long-cut paraffin distillate will contain excessive quantities of malcrystalline waxes which precipitate simultaneously with the plates and impress their form on them, making the distillate nonpressable.

From this early work it was concluded that paraffin wax contains only paraffinic hydrocarbons having the generic formula C_nH_{2n+2} and that cyclic hydrocarbons are absent. It was desirable to know what ranges of molecular weights, or lengths of carbon chains, are present. Buchler and Graves isolated many fractions believed to be relatively pure chemical compounds. In paraffin wax, these ranged from $C_{18}H_{38}$ to $C_{32}H_{66}$. Work with petrolatum waxes (microcrystalline waxes) disclosed the presence of hydrocarbons from $C_{34}H_{70}$ to $C_{43}H_{88}$. Clark and Smith[1] reported hydrocarbons ranging from $C_{21}H_{44}$ to $C_{57}H_{116}$, but it appears that they were including high-melting paraffin waxes (up to 150°F and higher) in their discussion. These investigators also claimed, as a result of X-ray studies, that the number of different straight-chain hydrocarbons in the wax is small and seldom exceeds ten.

There was soon considerable conflicting evidence through the work of later investigators. A definitive study was undertaken in the research department of the Sun Oil Company[2] to determine:

1. Relationships among paraffin wax, microcrystalline wax, and oil.
2. Chemical composition of waxes.
3. Relationship between physical properties of waxes and composition.

For the purposes of the study petroleum wax was split into a number of fractions.

The stock for these wax fractions was prepared by blending "slack waxes" from the commercial dewaxing of distillate and residual lubricating oils. The proportions used were such that the blend represented the entire wax content of the lubricating oil fraction of the mixed-base crude charged to the lubricating oil plant, together with the oil normally remaining in slack waxes.

This composite was distilled under vacuum, taking six distillates and a residuum. Each of these seven distillation fractions was subjected to fractional crystallization from methylethylketone-benzene at successively lower temperatures, and each crystallized fraction of wax was washed with

---

[1] *Ind. Eng. Chem.* **23**, 697 (1931).

[2] *Petroleum Waxes*, S. W. Ferris, W. T. Harvey, and J. F. Ritter, Research & Development Division, Sun Oil Company, 1960.

Table 1:3 PROPERTIES OF WAX FRACTIONS

Fraction Number	*1*	*2*	*3*	*4*	*5*	*6*	*7*	*8*	*9*	*10*
Yield, % of total wax	3.6	4.1	2.8	0.8	3.3	1.5	4.3	3.2	3.2	2.2
Urea adductable, %	43	74	25	70	99+	66	99+	99+	55	97
Average carbon atoms	52.7	40.6	53.0	38.6	34.5	33.9	29.8	29.5	33.1	28.6
"x" in C_nH_{2n+x}	1.2	1.8	0.8	1.4	1.9	1.7	2	1.9	1.6	1.8
Refractive index at 212°F	1.4382	1.4318	1.4413	1.4316	1.4272	1.4281	1.4235	1.4233	1.4283	1.4230
Melting point, °F (CC)[1]	191	174	169	164	162	152	149	147	145	142
Viscosity at 210°F, centistokes	19.9	9.3	20.6	8.7	6.5	6.0	4.8	4.6	4.6	3.7
Penetration at 40°F	3	6	5	5	3	9	3	3	8	6
Penetration at 77°F	7	16	14	15	11	24	12	12	23	13
Penetration at 100°F	11	32	26	24	19	40	18	18	41	21
Penetration at 120°F	19	47	38	35	27	56	31	36	137	71
Penetration at 140°F	29	120	200+	68	44	—	200+	—	—	—
Penetration at 160°F	46	—	—	—	—	—	—	—	—	—
Penetration at (MP−50)°F	29	50	38	31	24	40	17	18	36	16
Modulus of rupture at 73°F	399	146	Bends	260	278	206	397	401	196	425
Density at 210°F	0.790	0.779	0.799	0.780	0.771	0.772	0.764	0.764	0.773	0.764
Density at 77°F	0.934	0.922	0.929	0.922	0.927	0.920	0.918	0.918	0.916	0.917
Volume increase upon melting, %	14.0	15.3	11.8	14.1	15.0	13.5	14.0	14.5	13.0	14.5

Fraction Number	*11*	*12*	*13*	*14*	*15*	*16*	*17*	*18*	*19*	*20*
Yield, % of total wax	1.0	2.9	2.8	3.8	1.9	2.1	1.8	3.5	1.3	1.7
Urea adductable, %	99+	99+	18	13	92	99+	99+	99+	95	99+
Average carbon atoms	27.7	27.2	40.4	53.0	28.3	26.4	25.2	25.1	26.0	24.8
"x" in C_nH_{2n+x}	2	2	1	0	1.7	2	2	2	2	2
Refractive index at 212°F	1.4218	1.4213	1.4360	1.4444	1.4232	1.4207	1.4191	1.4189	1.4207	1.4187
Melting point, °F (CC)[1]	141	140	139	139	137	135	131	130	129	128
Viscosity at 210°F, centistokes	4.1	3.9	9.6	22.1	4.3	4.6	3.4	3.3	3.6	3.3
Penetration at 40°F	3	3	12	11	6	8	4	4	9	3
Penetration at 77°F	12	11	38	23	13	14	9	10	14	11
Penetration at 90°F	15	13	57	31	16	17	15	16	24	17
Penetration at 100°F	20	17	88	46	28	24	27	26	51	37
Penetration at 120°F	87	163	—	200+	200+	167	—	—	—	—
Penetration at (MP−50)°F	16	13	56	31	15	16	10	11	16	11
Modulus of rupture at 73°F	316	407	106	Bends	365	449	156	259	348	322
Density at 210°F	0.753	0.761	0.789	0.806	0.762	0.762	0.755	0.755	0.760	0.755
Density at 77°F	0.911	0.916	0.912	0.919	0.915	0.913	0.912	0.907	0.913	0.909
Volume increase upon melting, %	15.4	14.7	10.3	9.9	15.2	14.1	14.9	14.7	14.1	15.0

Table 1:3 (continued)

Fraction Number	21	22	23	24	25	26	27	28	29	30
Yield, % of total wax	2.0	1.6	3.5	1.5	1.6	3.3	2.4	2.5	1.3	1.5
Urea adductable, %	28	98	71	99+	6	0	99+	99+	83	45
Average carbon atoms	32.8	24.4	28.4	23.7	39.5	52.5	23.5	23.3	26.0	24.1
"x" in C_nH_{2n+x}	1.1	2	1.4	2	0.4	−0.7	2	2	1.6	1.8
Refractive index at 212°F	1.4303	1.4182	1.4253	1.4166	1.4369	1.4456	1.4162	1.4158	1.4224	1.4188
Melting point, °F (CC)[1]	126	125	122	121	120	120	119	117	116	114
Viscosity at 210 °F, centistokes	5.2	3.1	4.4	2.9	10.0	22.3	2.7	2.7	3.7	2.6
Penetration at 40°F	14	3	11	3	14	10	3	4	12	9
Penetration at 77°F	39	15	36	11	45	36	11	16	49	42
Penetration at 90°F	71	29	85	19	93	57	26	62	200+	109
Penetration at 100°F	200+	84	200+	48	200+	190	57	162	—	200+
Penetration at (MP−50)°F	38	14	22	8	33	27	8	10	28	20
Modulus of rupture at 73°F	124	251	160	105	Bends	Bends	104	102	300	Bends
Density at 210°F	0.780	0.754	0.764	0.750	0.794	0.808	0.749	0.750	0.764	0.754
Density at 77°F	0.912	0.908	0.898	0.904	0.911	0.913	0.897	0.899	0.893	0.880
Volume increase upon melting, %	12.7	14.7	10.7	14.4	10.8	10.6	13.9	14.1	13.9	10.6

Fraction Number	31	32	33	34	35	36	37	38	39
Yield, % of total wax	0.7	2.8	3.8	6.0	1.3	1.9	0.8	2.4	6.8
Urea adductable, %	88	9	0	0	39	0	45	0	0
Average carbon atoms	22.7	33.3	39.2	53.4	28.5	33.5	26.3	39.7	54.3
"x" in C_nH_{2n+x}	2	0.7	−0.2	−1.3	0.6	0.2	1.0	−1.0	−2.0
Refractive index at 212°F	1.4153	1.4321	1.4383	1.4472	1.4280	1.4360	1.4252	1.4397	1.4500
Melting point, °F (CC)[1]	110	110	101	101	100	98	96	87	83
Viscosity at 210°F, centistokes	2.6	6.2	10.2	22.2	4.9	7.2	3.9	10.3	24.4
Penetration at 40°F	8	20	19	20	32	75	43	32	41
Penetration at 77°F	55	117	42	117	60	200+	152	107	200+
Penetration at 90°F	200+	—	137	—	200+	—	—	200+	—
Penetration at (MP−50)°F	18	52	30	28	35	95	50	23	31
Modulus of rupture at 73°F	71	Bends	Bends	Bends	Bends	Bends	Bends	Bends	Bends
Density at 210°F	0.756	0.783	0.794	0.812	0.783	0.784	0.778	0.798	0.819
Density at 77°F	0.872	0.902	0.891	0.899	0.872	0.884	0.871	0.876	0.863
Volume increase upon melting, %	7.7	9.4	7.1	5.7	7.3	6.9	7.1	5.1	3.9

[1] The melting point of paraffin waxes is customarily determined by ASTM D87-57, but this method is inapplicable to many microcrystalline waxes because no "flat" of the prescribed length is obtained. D127-49 or D938-49 is normally applied to microcrystalline waxes. In this work, "cooling-curve melting point" was used for all fractions, by a procedure similar to D87-57, except that the first marked deflection in the cooling curve was taken as the melting point.

fresh solvent to reduce oil content (ASTM D721-56T) to approximately 0.5%. The fractions were decolorized by contact with clay: the distillate fractions to about +25 Saybolt, and the residual fractions to yellow colors in the range (liquid) of 2–4 NPA (ASTM D155-45 T). The crystallization temperatures selected resulted in a total of 39 fractions, whose properties are given in Table 1. The numbers assigned were in the order of melting point, with the fraction having the highest point being 1.

Limitations of Melting Point

The most frequently mentioned property of petroleum waxes is melting point (the temperature at which the wax changes from a solid to a liquid), for many years considered to be the most important property, with higher melting points commanding higher prices. From the standpoint of performance, however, melting point is of only minor significance.

The properties of waxes depend, in the final analysis, upon their chemical composition. As waxes are composed of hydrocarbons, the important variable becomes the arrangement of the carbon atoms within the molecule; whether they lie in straight chains, branched chains, or in cyclic structures. Melting point itself cannot indicate which types of molecules predominate, because, with a given type of structure, for example straight-chain, the melting point increases as the length of the chain (and therefore molecular weight) increases.

Refractive Index

On the other hand, the relationship between the melting point of a wax and another quite different physical property which is also sensitive to changes in molecular configuration may be highly significant. The most useful second property is the refractive index of the liquid wax. All the other properties of the 39 wax fractions, given in Table 1:2, will be studied in the light of the relationship of melting point to refractive index.

Charting the Fractions

Figure 1:1 plots melting point vs. refractive index of the wax fractions, using the fraction numbers used in Table 1:3. Succeeding charts will show the same points, with other properties recorded. Points on the later charts may readily be identified by their position within the pattern and their numbers may be found by reference to Figure 1:1.

The diagonal line at the left of Figure 1:1 is that for the normal paraffin hydrocarbons, using the data of American Petroleum Institute, Project 44. Although hydrocarbons that fall above this line are known, they do not occur in recoverable amounts in petroleum.

The diagonal line at the right of Figure 1:1 would pass through those wax components that exhibit the highest values of refractive index for any given melting point. It is shown as a broken line because it has not been

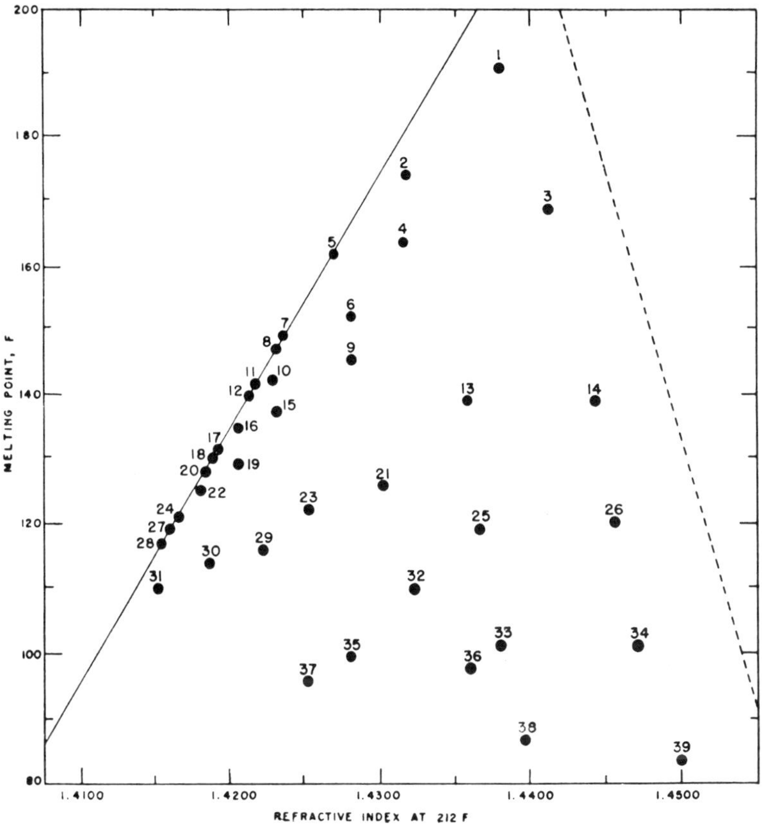

FIG. 1:1. NUMBERS OF PETROLEUM WAX FRACTIONS.[1]

determined exactly how far to the right that line falls. Further fractionation of fractions 1, 3, 14, 26, 34 and 39 would be required to establish this point.

The two lines just discussed, together with a horizontal line set arbitrarily at 80°F melting point, form the "wax triangle." The significance of the position of the base of the triangle will be discussed later in connection with *cerols*.

Eleven of the thirty-nine fractions fall on the line for normal paraffin hydrocarbons, with the others distributed rather uniformly over a triangular area. This is the "wax triangle" discussed here.

An empty area at the lower left corner of the triangle of Figure 1:1 will be noted. This does not mean that such compounds do not exist in petroleum. Fractions within this area could readily have been prepared, but only by dewaxing crude distillates of lower molecular weight than those normally included in lubricating oil distillates.

---

[1] Figures 1:1 to 1:15 from *Petroleum Waxes*, Sun Oil Co., 1960.

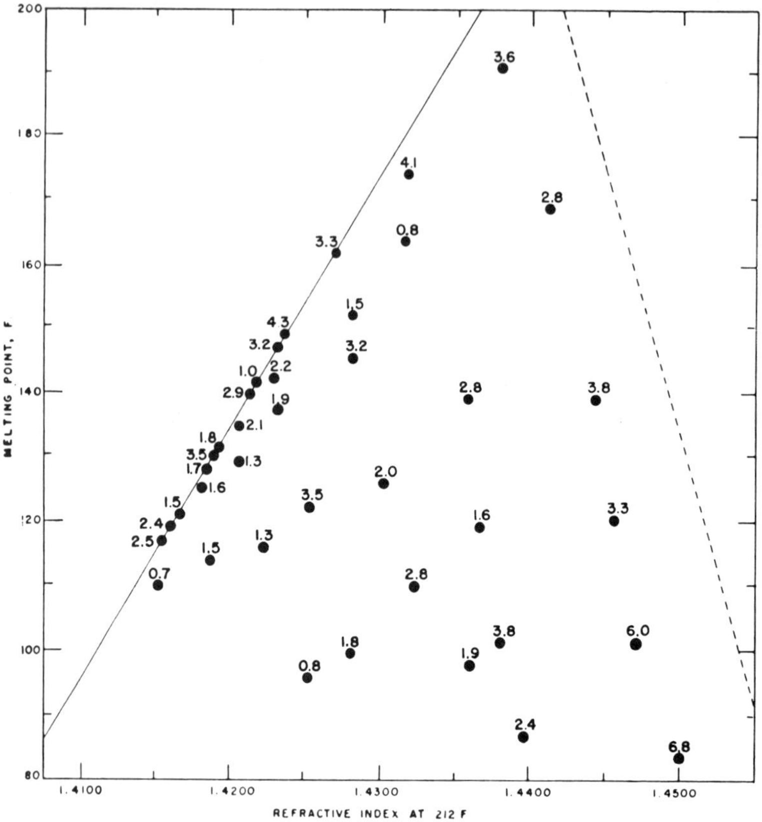

Fig. 1:2. Percentage Yield of Each Wax Fraction.

Before proceeding to use the wax triangle as an aid in analyzing other properties of the fractions, it is important to emphasize that, by varying the conditions of distillation and crystallization, it is possible to vary the positions of the fractions obtained. If Fractions 1 and 3 were each recrystallized, and the lower melting portion of Fraction 1 was combined with the higher melting portion of Fraction 3, the new fraction would fall between the points for 1 and 3. It is possible, in fact, to prepare a fraction falling at any desired point within the triangle.

The wax triangle, then, is literally full of compounds. If a fraction falls on the line for normal paraffins, it may be assumed to represent a relatively simple mixture of such hydrocarbons; but a point falling to the right of the line indicates a more complex mixture, composed of those hydrocarbons whose individual points lie in the area surrounding the point for the mixture. In Figure 1:2, the yield of each fraction is recorded at the point for that fraction. The yields add to 100% of the total wax recovered.

The combined yield of those fractions that fall on the normal paraffin line is only 28.1%. The rest of the wax covers the triangle fairly evenly, except for some concentration of yield at the lower right corner.

Fractions on the normal paraffin line are completely adductable with urea. The further from that line, the lower the adductibility, with some fractions showing none. Adductibility with urea was at one time held to be proof of straight-chain structure. This is no longer accepted. Fetterly,[1] for example, found that a compound with a methyl side group could be adducted if a minimum of 10–13 straight-chain carbon atoms were present. Adduction of an ethyl group required over 24 straight-chain carbons.

If a hydrocarbon or hydrocarbon mixture is completely adductable its content of straight-chain compounds is probably high, and if a non-straight-chain material is adducted, it may be assumed that the adducted hydrocarbons are either slightly branched chains, or, if cyclic, that they contain long-chain substituents.

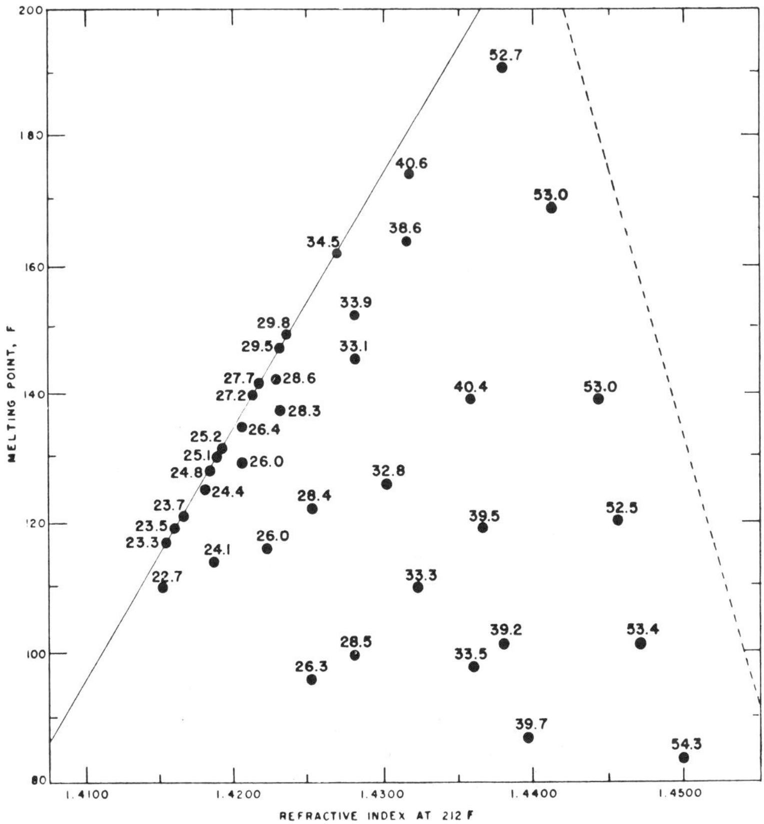

FIG. 1:3. AVERAGE NUMBER OF CARBON ATOMS.

[1] L. C. Fetterly, *Petrol. Refin.*, **31**, 110 (1952).

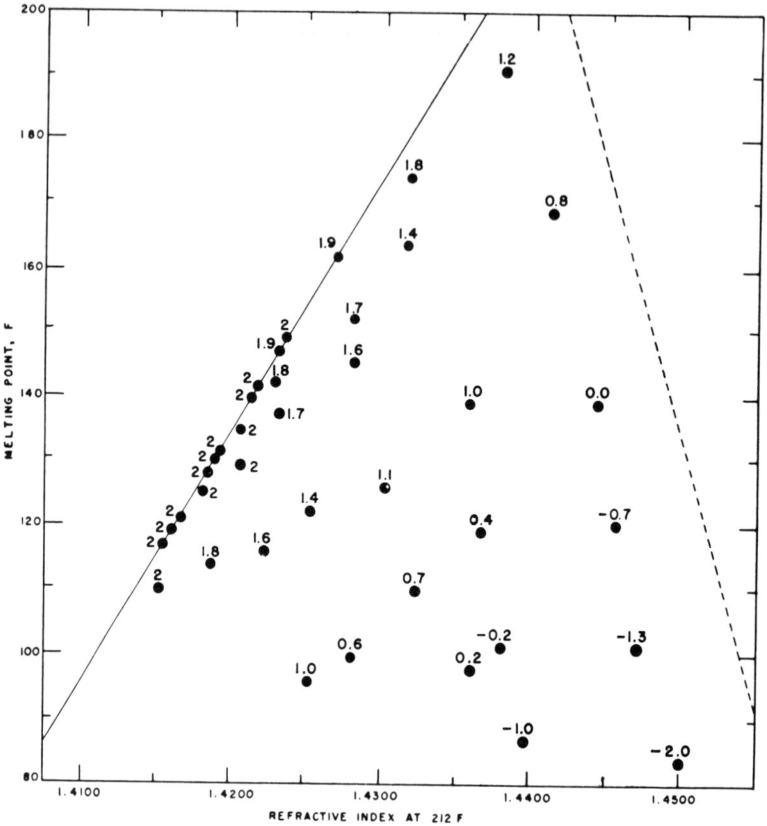

FIG. 1:4. VALUE OF x IN $C_nH_{2n}+x$.

Figure 1:3 shows that fractions of the same melting point vary widely in the number of carbon atoms. A fraction with a given number of carbon atoms may have a melting point anywhere over a long range. Note the fractions with 33, for example, which emphasize the diversity of hydrocarbon types.

As shown in Figure 1:4, on the normal paraffin line, $x = 2$ as it should. Simple branching of the saturated hydrocarbons does not change the figure, and therefore some of the fractions lying off the line still show the same value. When x is equal to 0, the presence of one saturated ring is indicated; as x becomes -2, two rings are indicated. Intermediate values presumably indicate mixtures; with $x = 1$, one may postulate that half the molecules contain a single saturated ring. (Olefinic unsaturation is usually not found in petroleum waxes, and only very small proportions of aromatics are present.)

Composition of Wax Fractions

Figures 1:3 and 1:4 present some of the data used to reach an approximation (Figure 1:5) of the types of hydrocarbons present at various positions within the wax triangle.

The figures for yields given on Figure 1:5 are for the particular crude oil used in this work; they will be different for other wax-bearing crudes. The evidence is, however, that the same hydrocarbon *types* occur in all such crudes. Just as paraffinic lubricating oil may be recovered by solvent-extraction from most lubricating distillates, but with different yields, so the wax fractions from different crudes contain varying proportions of straight-chain hydrocarbons, with the remaining hydrocarbons distributed over the wax triangle. Relative proportions of the types other than normal straight-chain paraffins probably also vary, but a given type from one crude can probably be matched by a fraction from another crude. In other words,

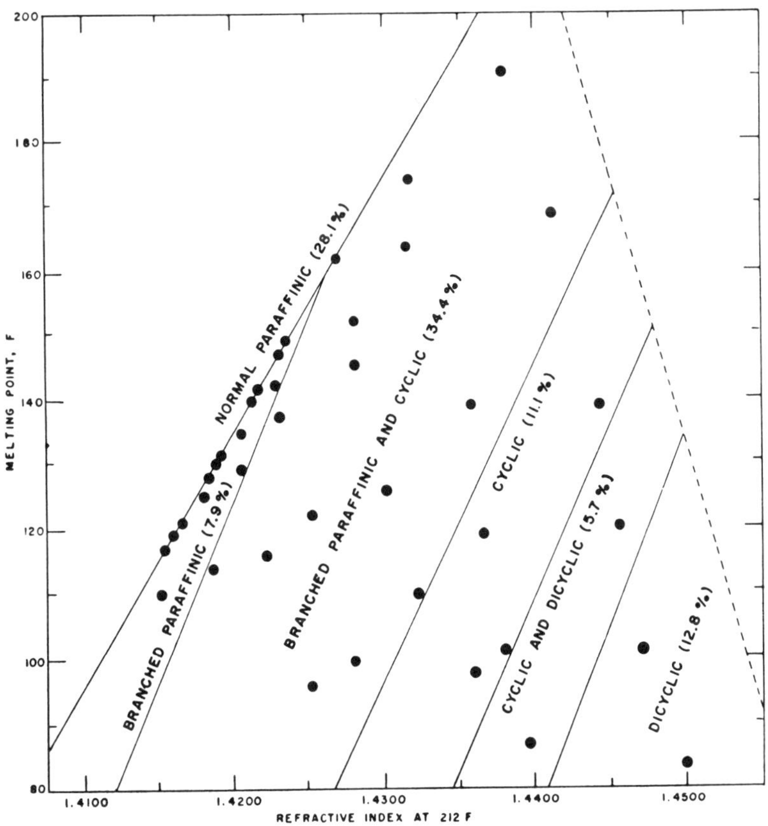

(Figures in parenthesis give percentage of total wax yields falling in each category.)

FIG. 1:5. TYPES OF HYDROCARBONS.

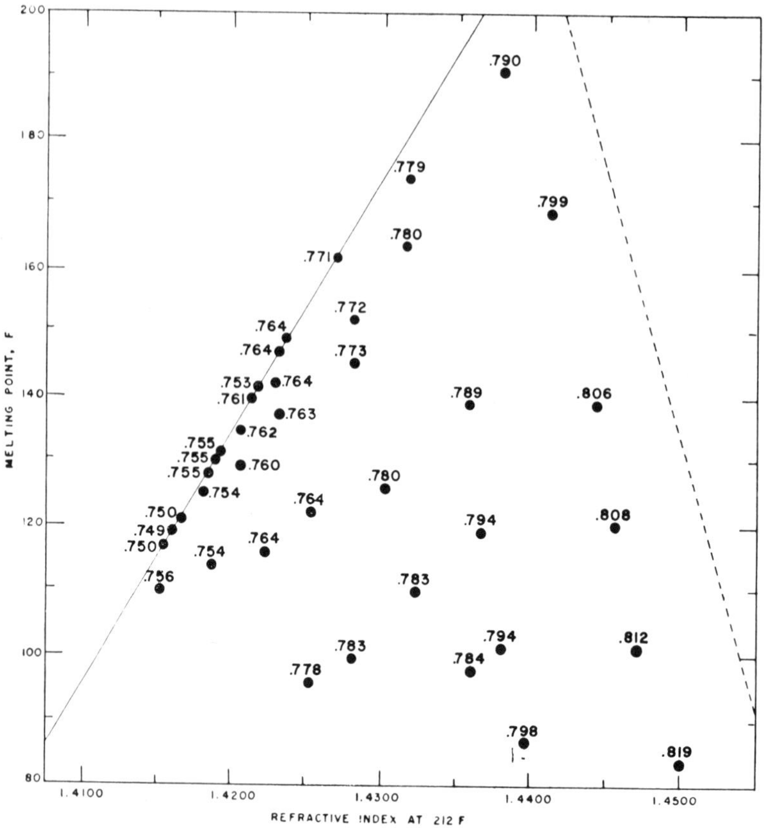

FIG. 1:6. DENSITY OF WAX FRACTIONS AT 210°F.

it appears that any wax-bearing crude will yield a fraction—in some yield or other—falling at any given position within the wax triangle.

As can be seen in Figure 1:5, proceeding to the right from the line of the normal paraffins, the chains progressively branch, then cyclic compounds are encountered and finally dicyclic. On the basis of evidence not presented here, the individual rings of the cyclic structure probably contain either 5 or 6 carbon atoms.

Physical Properties of Wax Fractions

Some physical properties of the fractions used in the preceding study are recorded on Figures 1:6 to 1:12.

First examination of several of these figures indicates that "constant property" lines (constant density, viscosity, etc.) could be drawn. Closer scrutiny reveals that any such lines would be tortuous. This is one more indication of the complex composition of waxes. Departing from the straight-chain configuration, composition becomes complex, not only because of branching, but also because of the introduction of cyclic or dicyclic structures. The substituents on the cycles may be few and large, or many and small, and either straight or branched.

Figure 1:6 shows that higher densities in the liquid state are encountered as one proceeds to the right from the line of normal paraffins toward the cyclic and dicyclic regions.

Tailored and By-Product Waxes

Petroleum waxes were first commercially produced because it was necessary to remove the waxes from lubricating oils in order to make the oils fluid at low temperatures. The de-waxing operation produced, in addition to the low-pour oil, a "slack wax" still containing substantial proportions of oil and cerol from certain slack waxes of low-molecular weight, yielding an "oil-free wax." Much later, solvent methods were developed to produce oil-free waxes from the slack waxes obtained from distillates and residua of higher molecular weight.

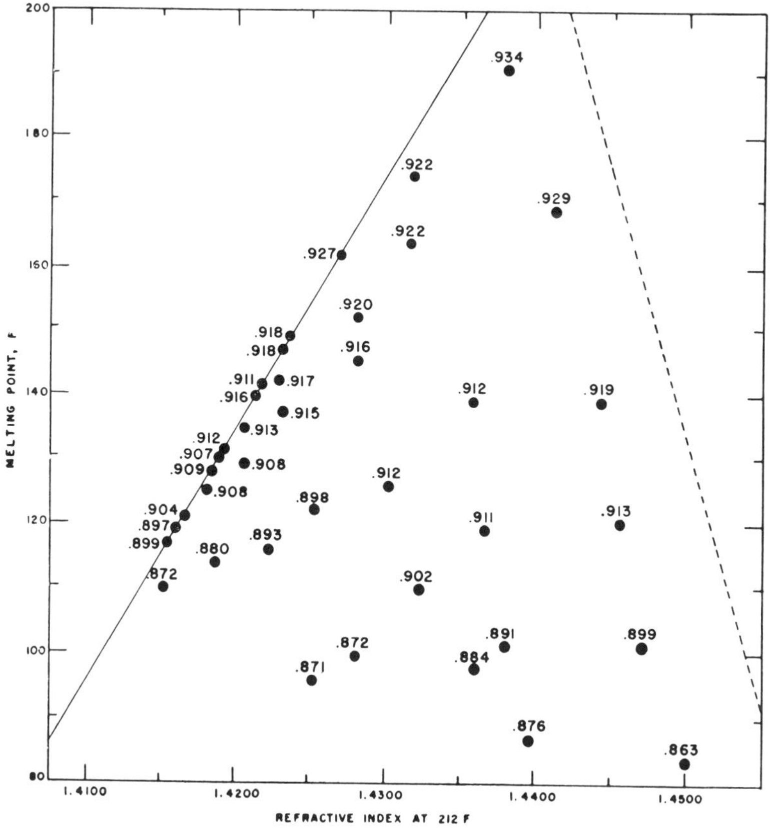

FIG. 1:7. DENSITY OF WAX FRACTIONS AT 77°F.

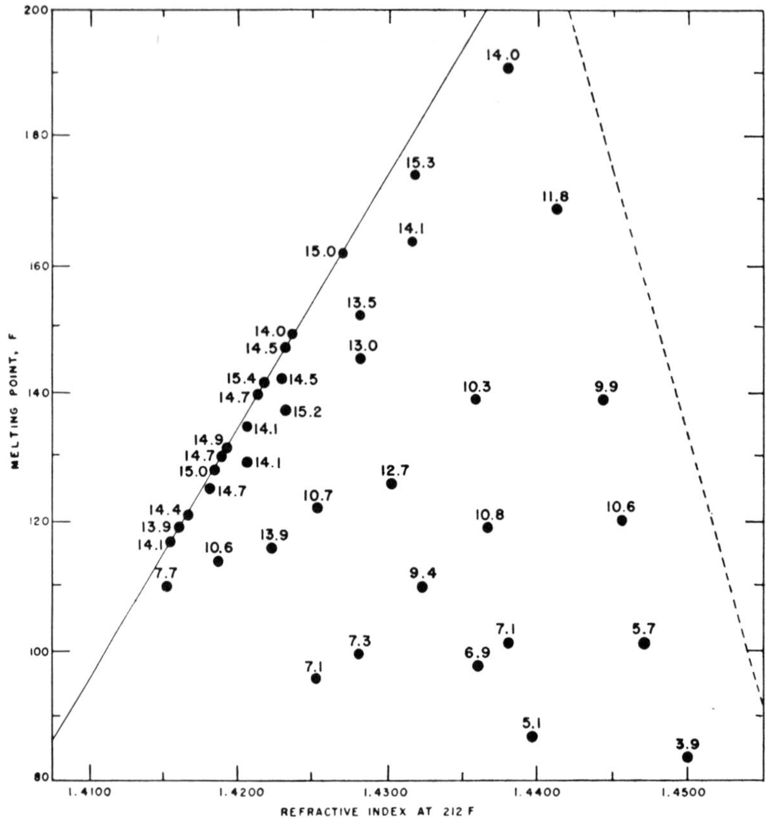

FIG. 1:8. PERCENTAGE OF INCREASE IN VOLUME ON MELTING.

Waxes, then, were merely by-products in the manufacture of lubricating oil. As already noted, melting point was considered the most important property by which paraffin waxes could be distinguished. In the case of the early microcrystalline waxes, their sticky nature was recognized and utilized.

Table 1:4

PROPERTIES OF SELECTED WAX FRACTIONS

Fraction Number	*22*	*23*	*24*	*25*	*26*
Melting point, °F (CC)	125	122	121	120	120
Refractive index at 212°F	1.4182	1.4253	1.4166	1.4369	1.4456
Viscosity at 210°F, centistokes	3.1	4.4	2.9	10.0	22.3
Molecular weight	356	400	335	564	727
Per Cent adductable by urea	98	71	99+	6	0
Modulus of rupture at 73°F	251	160	105	Bends	Bends

How little melting point does mean is shown in Table 1:4, which compares some of the properties of five wax fractions that melt in the range of 120–125°F. Other properties vary markedly, as shown both in the table and in Figures 1:13 and 1:14. There is no temperature at which the five waxes exhibit the same penetration (Figure 1:13), and there are differences in the rate at which the waxes harden as temperature decreases.

Similarly (Figure 1:14) the changes in density with change of temperature are widely different. The upper solid portions of the lines show different rates of expansion as the solid is heated. The length of the dotted portions of the lines shows whether the expansion on melting is large or small. Once the waxes are liquid, the density-temperature lines are roughly parallel.

Petroleum waxes can be tailored to specific applications by carefully splitting the broad spectrum of compounds into fractions, frequently

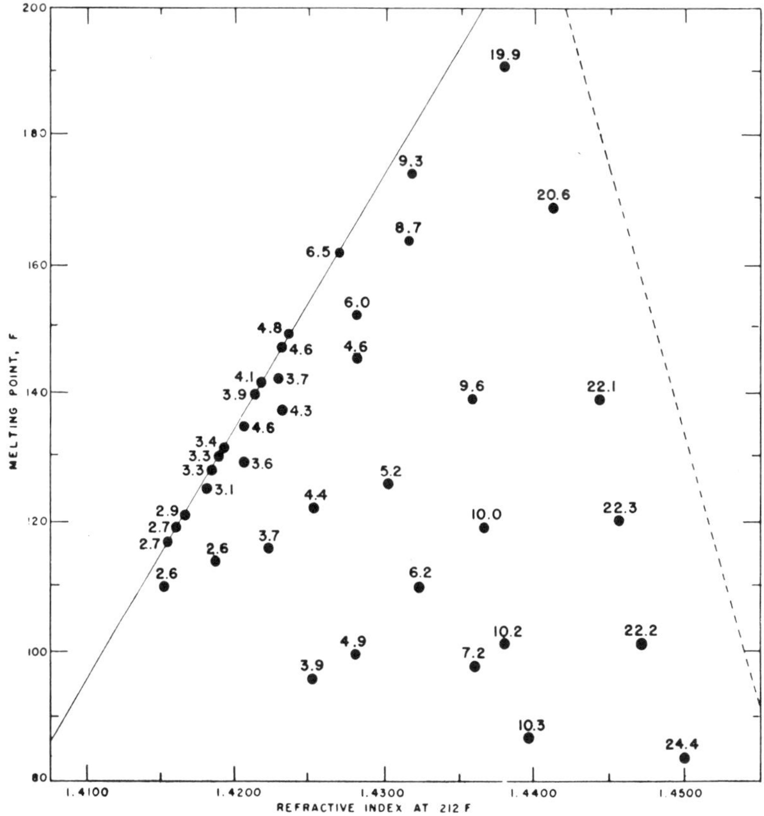

FIG. 1:9. VISCOSITY OF WAX FRACTIONS (CENTISTOKES).

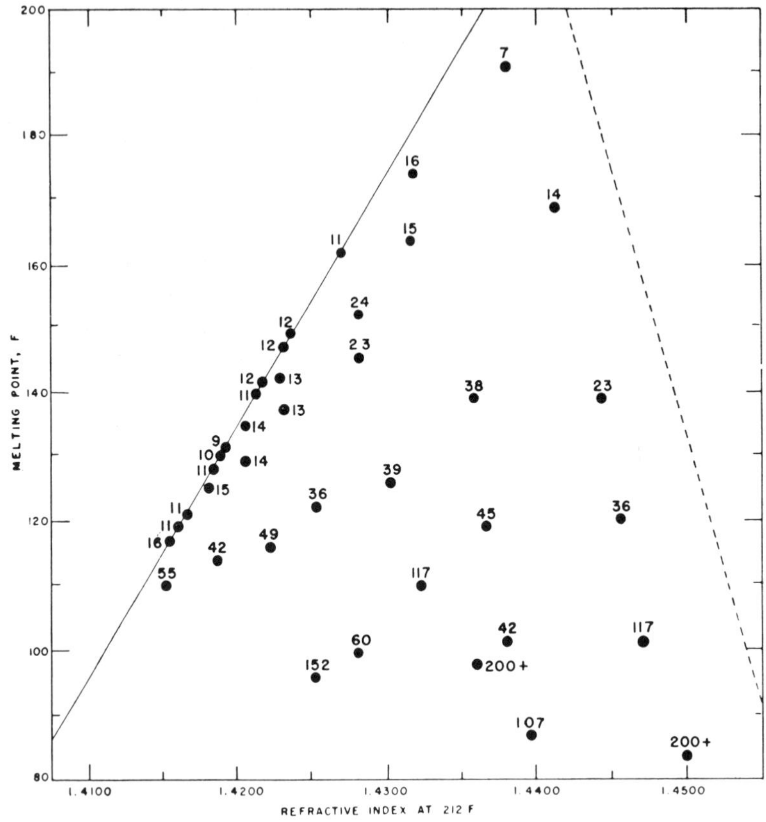

FIG. 1:10. PENETRATION OF WAX FRACTIONS AT 77°F.

followed by judicious blending of widely differing fractions to produce the final wax.

Most petroleum waxes still are by-products of the manufacture of lubricating oil, in that one of the principal means of tailoring—distillation to control molecular weight—is used to control the viscosity of the lubricating oil. The wax recovered and purified from a particular oil is a quality product, and it may be completely satisfactory for some applications. If it is, that is largely a matter of chance; if it is not, it is usually impossible to correct the deficiencies without interfering with the properties of the primary product, lubricating oil.

Waxes of controlled characteristics, tailored for specific applications, must be produced as primary products, with all available separation and control methods applied for the sole purpose of segregating the group of hydrocarbons that exhibit the desired properties. (See chapter on Compounded Waxes.)

Paraffin and Microcrystalline Waxes

The distinction between paraffin and microcrystalline wax was at one time easy and straightforward. The paraffin waxes were produced (and many still are) by pressing a paraffin *distillate* and sweating the resulting slack wax for final removal of oil. As sweating is applicable only to waxes of relatively low melting point and molecular weight, all the waxes so produced were of similar characteristics; i.e., they formed *large crystals*, and were relatively hard, white, of low viscosity, and confined to a melting range of about 120° to 140°F. *Microcrystalline* waxes, on the other hand, were obtained by solvent crystallization methods from wax distillation residua. They were of high molecular weight; usually melted at 150°F or above; were yellow, soft and of high viscosity.

Many now prefer, in accordance with the above, to designate *distillate* (low molecular weight) waxes as paraffins, and residual (high molecular weight) waxes as microcrystallines. Because the original paraffin waxes

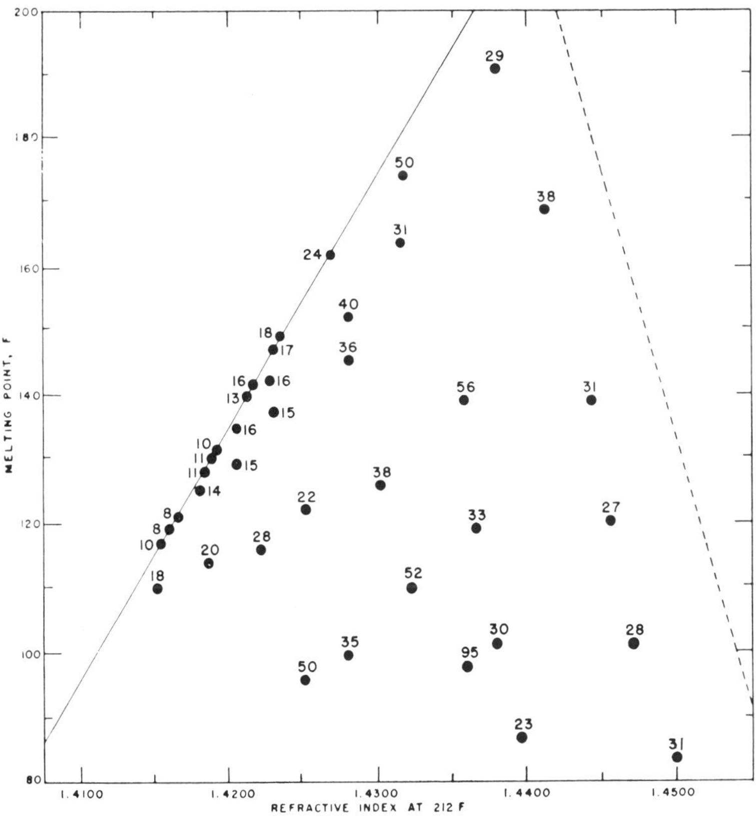

FIG. 1:11. PENETRATION OF WAX FRACTIONS AT (M.P.-50)°F.

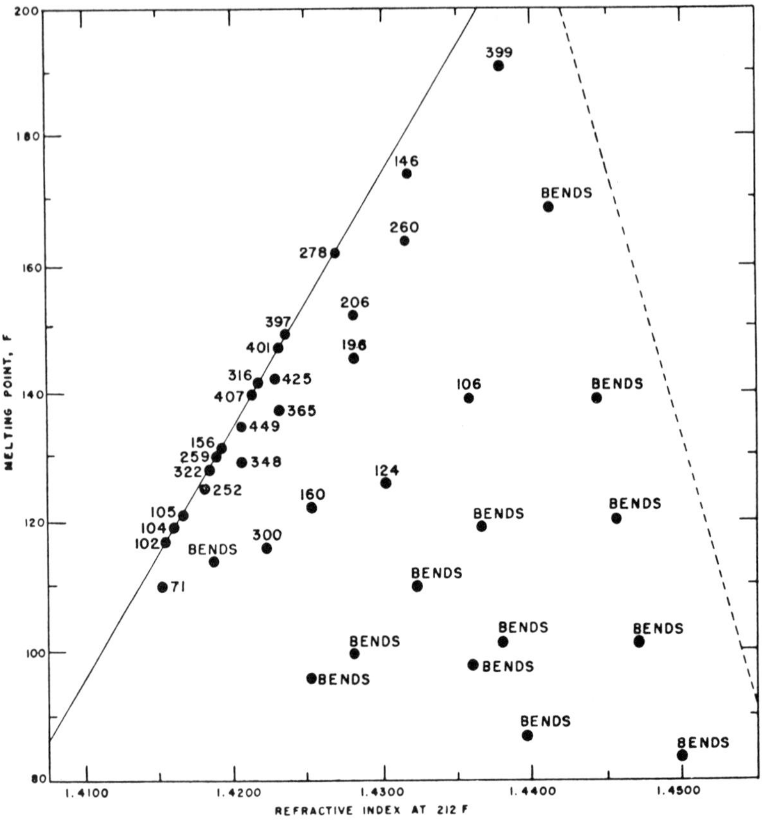

Fig. 1:12. Modulus of Rupture of Wax Fractions at 73°F.

were relatively high in straight-chain hydrocarbons (as indicated, for example, by adduction with urea), normal paraffin content has been used as a means of distinction. But this distinction is also questionable, for waxes rich in normal paraffin are now being made from residual stocks. Some users of wax call such waxes *microcrystalline* (being residual), while others call them *paraffins*.

At present, then, there is no fully satisfactory way of so defining paraffin wax and microcrystalline wax that any given wax may unequivocally be assigned to one group or the other. A section of the TAPPI-ASTM Committee on Petroleum Wax has been attempting for some time to agree upon definitions for paraffin wax and microcrystalline wax. They have now concluded that another category, tentatively designated *semi-microcrystalline* will be needed. Under the system of designation now being considered, paraffin waxes are those on an RI-MP chart falling on or close to the line of the normal paraffins. Those falling farther away from the line (of higher refractive index) would be called semi-microcrystalline, or microcrystalline, depending upon their viscosities, with waxes of higher viscosities called micrystalline. If the plan is accepted, it will be necessary to agree on the limits—how far from the normal line a wax may lie and still be termed paraffin; and what viscosity separates semi-microcrystalline from microcrystalline waxes.

Waxes, Oils, and Cerols

The difficulties of differentiation would not, however, cease even if paraffin and microcrystalline waxes were satisfactorily defined, for even the distinction between wax and oil is unclear.

One definition of a petroleum wax is: a substantially oil-free, *solid* mixture of hydrocarbons. If one were to attempt to specify what is meant by solid, he might say—"melting at 100°F or higher."

Similarly, one might define a petroleum lubricating oil—at least one that is to function properly at low temperatures—as a mixture of hydrocarbons which is sufficiently free of wax to be *liquid* at 0°F.

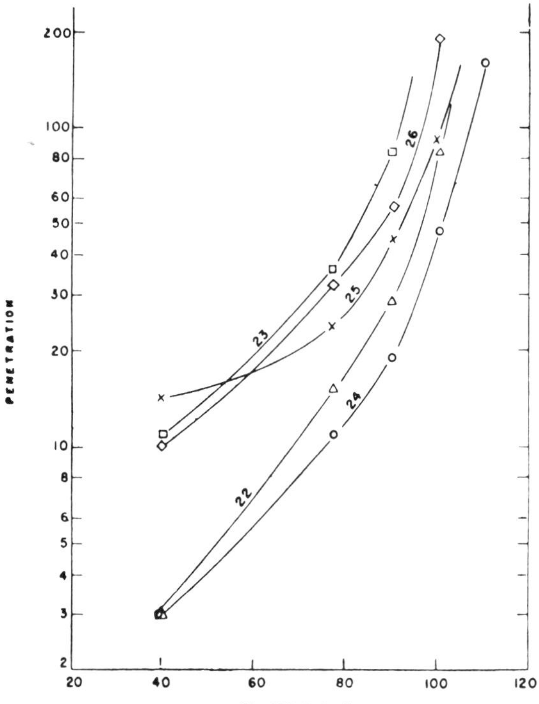

(The numbers, 22 to 26, refer to wax fractions in Table 1:2. There is no temperature at which the five waxes exhibit the same penetration.)

FIG. 1:13. PENETRATION OF SELECTED WAX FRACTIONS AT DIFFERENT TEMPERATURES.

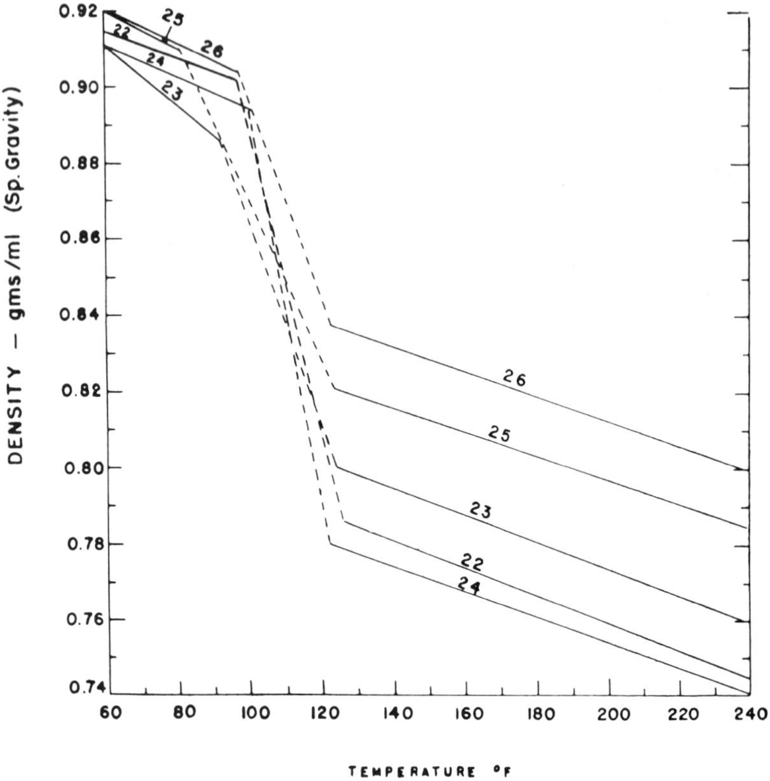

(Numbers refer to fractions listed in Table 1 : 2.)

FIG. 1 : 14. DENSITY OF SELECTED WAX FRACTIONS.

Making somewhat free use of the term *melting point*, and using it as the equivalent of the pour point of a lubricating oil, the two definitions just suggested for waxes and oils could be represented on the RI-MP chart as in Figure 1 : 15 with the wax triangle discussed previously at the top.

But what of the range between "100°F meltıng point" and "0°F melting point"? That area of the larger triangle is just as full of hydrocarbons as is the wax triangle at the top. In other words, there may readily be recovered, by fractional crystallization and distillation, a series of fractions, each free of "low pour oil," and melting at 60°F, or at any other level between 100°F and 0°F. If such a fraction were examined at 20°F it would behave like a wax; at 90°F it would behave like an oil. These compounds have been called *Cerols* (wax-oils).

In examining the properties of the 39 wax fractions by means of the RI-MP chart, the base line of the wax triangle was taken at MP = 80°F. Figure 1 : 15 shows that, if waxes melt at 100°F or higher, four of those "wax" fractions are not waxes but cerols.

Summarizing, petroleum waxes and petroleum lubricating oils lie within the same ranges of molecular weight, and therefore occur together when the crude oil is separated into fractions by distillation. The waxes are merely those hydrocarbons which, by reason of the relatively high symmetry of their configuration, exhibit high melting points; the oils are those hydrocarbons which by reason of their lack of symmetry are liquid at low temperature. Between these lie the cerols, with melting points too low to qualify as waxes and too high to qualify as oils.

Oil Content of Waxes

In view of the factors just discussed, it is not surprising that considerable difficulty has been experienced in obtaining trustworthy results in determining the oil content of waxes.

Oil content is customarily determined by crystallizing the wax from a solvent under specified conditions, and reporting as oil the material remaining in solution.

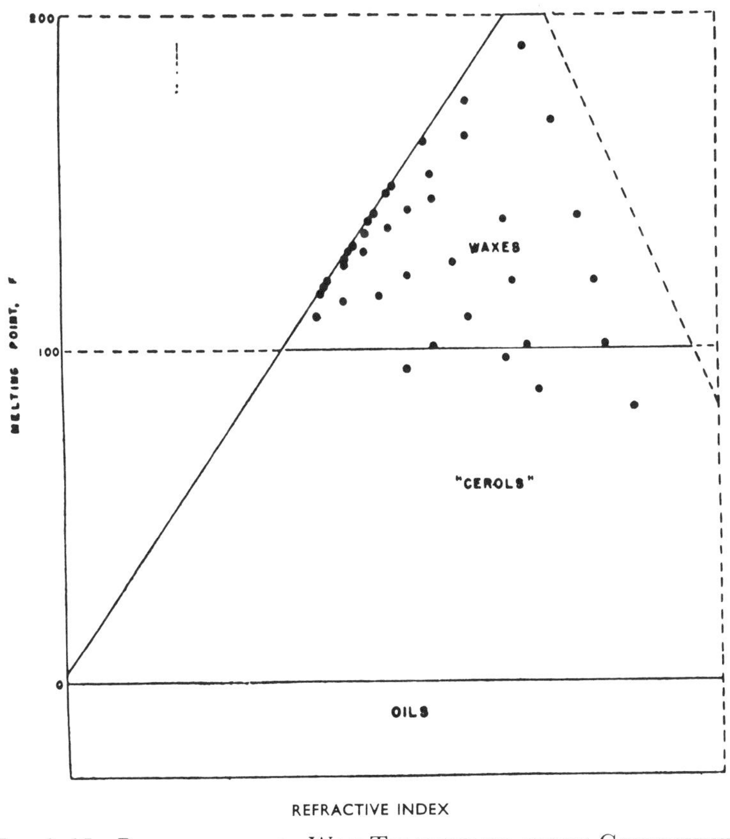

FIG. 1:15. RELATIONSHIP OF WAX TRIANGLE TO OTHER CONSTITUENTS OF PETROLEUM OILS.

Table 1:5

PROPERTIES OF PETROLEUM WAXES

	Paraffin Waxes								*Crude Scale*		*Slack Waxes*		
	125–127	128–130	133–135	138–140	140–142	143–145	147–150	157–167	White	Yellow	4% Oil	10% Oil	15% Oil
MP D-87	125–127	128–130	133–135	138–140	140–142	143–145	147–150		124–126	122–126	134		118–128
MP D-127								157–167				120–123	
Oil D-721	0.5	0.5	0.5	0.4	0.2	0.4	0.4	0.5	3.0	3.0	4%	8–10	15%
Penetr. 77°	14	15	15	12	8.5	15	8	11	—	—	—	—	—
90°	127	43	15	20	13.5	22	10	15	—	—	—	—	—
100°	279	239	48	52	18.0	37	16	21	—	—	—	—	—
105°	292	260	172	80	25	47	21	32	—	—	—	—	—
110°	320	275	179	119	105	60	35	35–50	—	—	—	—	—
Block	87	92	105	103	109	112	122	130	—	—	—	—	—
Tensile	172	220	368	260	350	260	260	288	—	—	—	—	—
Color	29	29	30	30	+30	30	30	30	21	8	Good	Dark brown	Good
Odor (max)	2.5	2.5	2.5	2.5	2.0	2.5	2.5	2.5	3.4	3.4	Good	Good	Good
Visc. 210°	37.50	40	37.40	41	39	41	41.2	53.89	—	—	—	38	—

The trouble lies not in the experimental procedure, but rather in the fact that there is no clear demarcation between the two groups of substances that the methods attempt to separate. The waxes merge into the cerols and the cerols into the oils. The higher the melting point of a petroleum hydrocarbon in the wax-lubricating oil range, the lower its solubility in a given solvent. At present there is a trend to include lower-melting components in waxes for certain applications. The further this trend goes, the less meaningful will be the determination of oil in wax, because the lower-melting waxes are relatively soluble in the solvents used, and therefore part of such waxes appears to be oil.

In other words, if one were to select a strictly oil-free, high-melting (relatively insoluble) wax and a low-pour oil, and were to prepare known mixtures, there would be no difficulty in developing a method which would determine the oil content with good precision. But in the waxes of commerce, "oil content" may represent anything from low-pour oil to high-melting cerols or low-melting waxes, and the latter two groups may be significantly contributing to some desirable property of the wax.

(See further discussion in the section on *Wax-Oil Mixtures.*)

PROPERTIES OF PARAFFIN WAXES

Density and Tensile Strength[1]

The tensile strength of paraffin wax increases rapidly from 30 to about 50°F when it begins to fall off; the maximum is reached around 0°. Cooling the wax from 15 to −6° over a period of 10 hours does not reduce the strain or flaws any more than does rapid cooling. This decrease of tensile strength is due to tiny cracks. Above 30°F the wax behaves like a plastic solid, because of liquefaction of some of the hydrocarbons. Below 25°F, the density of paraffin wax is a linear function of temperature. If

Table 1:6

TRUE DENSITY OF AND VOLUME PER CENT IN SOLID COMMERCIAL PARAFFIN WAXES

M.P. of Wax °F	*M.P. of Wax* °C	r_D *at* 20°C	*Temperature* °F	*Temperature* °C	*n* (*Obs.*)	d_1 (*Obs.*)	d_2 (*Calcd.*)	$d_2 - d_1$	*Volume* % *Air*
121	49.4	0.3343	50	10.0	1.5306	0.900	0.925	0.025	2.7
			60	15.6	1.5289	0.897	0.922	0.025	2.8
126	52.2	0.3342	50	10.0	1.5321	0.911	0.927	0.016	1.8
			60	15.6	1.5297	0.909	0.924	0.015	1.7
131	55.0	0.3339	50	10.0	1.5348	0.912	0.932	0.020	2.2
			60	15.6	1.5323	0.909	0.928	0.019	2.1
136	57.8	0.3341	50	10.0	1.5366	0.917	0.934	0.017	1.9
			60	15.6	1.5347	0.914	0.931	0.017	1.9
141	60.6	0.3337	60	15.6	1.5366	0.914	0.935	0.011	2.3
			70	21.1	1.5344	0.911	0.932	0.011	2.3

[1] W. F. Seyer and K. Inouye, *Ind. Eng. Chem.* **27**, 567 (1935).

density is plotted against tensile strength, the curve is not linear. It is evident that the tensile strength and the density are affected by the nature of the crystal matrix formed, because paraffin wax is known to contain three types of crystals: plates, needles, and macrocrystals. Occluded air may also be a contributing factor.

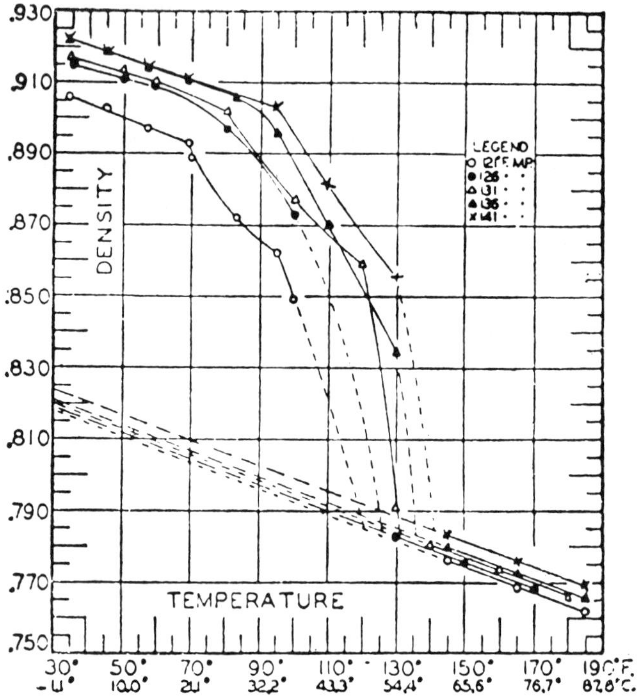

FIG. 1:16. DENSITY OF SOME COMMERCIAL WAXES.

Ductility

In general, refined waxes are described as being brittle and nonductile, especially when compared with microcrystalline or amorphous waxes. However, when this property is considered over a narrow range, it soon becomes evident that wide differences exist among various waxes. It is well known that, to a limited extent, refined waxes are thermoplastic below their melting point, and that changes in temperature have a pronounced effect on ductility. Conversely, at any given temperature, lower-melting commercial waxes are, in general, more ductile than those of higher melting point, probably because they are closer to the liquid state. Besides these variations, there is also a measurable difference in ductility between waxes of the same melting point when they are subjected to the same conditions of temperature and pressure.

The importance of this difference can be illustrated by the performance of two waxes used for coating milk cartons with the Excello machine. This machine immerses the carton in a bath of molten wax, coating it inside and out, and delivers the coated article to a conveyor which carries it through a chilling tunnel. As it cools, the coating drains from the side walls to the bottom, building up a thick cake of wax which is not completely congealed. Immediately after leaving the conveyors, the container is filled with milk at 38°F, the filling time being only a few seconds. When the less ductile of the two waxes is used, the impact of the cold stream on the plastic mass of wax sets up strains, which cause open cracks, and allow leakage of the milk to the paper board. In one test, 4% of all cartons filled were classed as "leakers." With a more ductile wax, the strains are counteracted by cold flow of the wax and cracks are practically nonexistent. In a similar controlled test, the more ductile wax produced only a fraction of one percent of "leakers."

Another practical application is the use of waxes for coating frozen-food containers which are subjected to temperatures below 0°F. Even papers creased at ordinary temperatures show that more ductile waxes cause less flaking than others.

A test has been developed to evaluate ductility. This is somewhat similar to a determination of tensile strength in that carefully prepared briquettes are broken, but has the important difference that constant loads are applied at controlled temperatures and are chosen so as to cause a measurable elongation of the briquette. Instead of obtaining a break in 10–15 seconds, as in measurements of tensile strength, times as long as several hours are required for breaking.

Attempts have been made to ascertain a correlation between ductility and other characteristics of refined waxes, such as tensile strength, oil content, sealing strength, etc., but apparently there is none. Instead, it appears that the most important factor is the distribution of the component parts with respect to melting point. Waxes with a wide range of melting points of the ingredients are much more ductile than those with narrower ranges. Also, it appears that the evenness of distribution has an important effect on ductility. Waxes having a very even distribution, as contrasted with those composed of a blend of narrow cuts, will be noticeably more ductile. It is possible, therefore, to prepare a highly ductile synthetic wax by blending together equal, or nearly equal, portions of several grades of wax.

The ductility of paraffin wax also depends on temperature, but different waxes are not always affected in the same manner. In general, ductility will increase as temperature is increased, reaching a maximum at some point below the melting point, and above that it will decrease as the temperature is raised further. The maximum ductility is apparently associated with the transition point of the wax, which is 40–50°F below its melting point and is the result of a change in crystal structure.

Chemical Composition

Whatever differences there are in various grades and types of petroleum wax, they trace back, in the final analysis, to differences in chemical composition. Petroleum waxes are composed almost entirely of hydrocarbons; the variations therefore become a matter of different configurations of carbon and hydrogen in molecules containing from about 18 to about 70 carbon atoms/molecule, and roughly twice as many hydrogen atoms.

Many of the proposed definitions emphasized that paraffins contained relatively high percentages of normal or slightly branched alkanes, whereas microcrystalline waxes contained substantial proportions of highly branched or cyclic hydrocarbons.

Analysis of waxes in terms of individual hydrocarbons is entirely out of the question. Normal alkanes of high purity can be separated from paraffin waxes, and they have also been synthesized. Their properties are therefore well known. Very few of the branched and cyclic isomers have been synthesized, and it is doubtful whether any have been recovered, in high purity, from waxes. The difficulty, of course, lies in the tremendous number of possible configurations in a complex molecule; someone has calculated that there may be 69,491,178,805,831 different isomers of tetracontane, $C_{40}H_{82}$.

Examination by spectrophotometers, and by the mass spectrometer, and correlation of various physical properties does, however, make it possible to make a reasonable estimate of the types of hydrocarbons present in a wax or wax fraction.

Table 1:7

Ductility of Wax–Oil Blends of 180 Cone Penetration

Wax component of blend	*TAPPI-ASTM classification*	*Normal paraffins, %*	*Ductility of blend*
Laminating P	Micro	25	30
Laminating R	Micro	33	25
Micro D	Micro	65	0
Coating L	Micro	70	0
Paraffin A	Paraffin	82	0
Paraffin C	Paraffin	88	0
Coating L adduct	Paraffin	100	0

Table 1:8

Heat Content and Heat of Solution of Paraffin Waxes (M.P. 110, 125, and 140°F., in Btu/lb)[1]

Temp. °F	Heat content of wax			Effective heat content of wax in solution			Heat of solution			Temp. °F.
	Melting point			Melting point			Melting point			
	110°	125°	140°	110°	125°	140°	110°	125°	140°	
0	−15	−13	−11	62	65	68	77	78	79	**0**
5	−13	−11	−9	65	68	71	78	79	80	5
10	−11	−9	−8	67	70	73	78	79	81	10
15	−8	−7	−6	70	73	76	78	80	82	15
20	−6	−5	−4	72	75	78	78	80	82	20
25	−3	−2	−2	75	78	81	78	80	83	**25**
32	0	0	0	78	81	84	78	81	84	32
35	+1	+1	+1	79	82	85	78	81	84	35
40	4	3	+3	82	85	88	78	82	85	40
45	7	6	5	85	88	91	78	82	86	45
50	10	8	7	87	90	93	77	82	86	**50**
55	13	11	9	90	93	96	77	82	87	55
60	16	13	11	92	95	98	76	82	87	60
65	19	16	13	95	98	101	76	82	88	65
70	23	19	16	98	101	104	75	82	88	70
75	27	22	19	100	103	107	73	81	88	**75**
80	32	27	22	103	106	109	71	79	87	80
85	38	31	25	105	108	111	67	77	86	85
90	44	36	29	108	111	114	64	75	85	90
95	51	42	33	111	114	117	60	72	84	95
100	60	49	38	113	116	119	53	67	81	**100**
105	73	57	43	116	119	122	43	62	79	105
110	119	65	49	119	122	125	0	57	76	110
115	121	76	55	121	124	127	0	48	72	115
120	124	90	62	124	127	130	0	37	68	120
125	127	130	71	127	130	133	0	0	62	**125**
130	130	133	81	130	133	136	0	0	55	130
135	132	135	94	132	135	138	0	0	44	135
140	135	138	141	135	138	141	0	0	0	140
145	138	141	144	138	141	144	0	0	0	145
150	141	144	147	141	144	147	0	0	0	**150**
160	146	149	152	146	149	152	0	0	0	160
170	152	155	158	152	155	158	0	0	0	170
180	157	160	163	157	160	163	0	0	0	180
190	163	166	169	163	166	169	0	0	0	190
200	169	172	175	169	172	175	0	0	0	**200**
220	181	184	187	181	184	187	0	0	0	220
240	193	196	199	193	196	199	0	0	0	240
260	205	208	211	205	208	211	0	0	0	260
280	217	220	223	217	220	223	0	0	0	280
300	230	233	236	230	233	236	0	0	0	**300**
320	243	246	249	243	246	249	0	0	0	320
340	256	259	262	256	259	262	0	0	0	340
360	269	272	275	269	272	275	0	0	0	360
380	282	285	288	282	285	288	0	0	0	380
400	295	298	301	295	298	301	0	0	0	**400**

[1] *Thermal Properties of Petroleum Products,* U. S. Bureau of Standards, Pbl. 97, 45 (1929).

Table 1:9

Variations of Physical Constants in Different Shipments of Paraffin Wax

M.P., °F.	ASTM Ball and Ring Softening Point, °F. Regular Min.	Regular Max.	Sloping Ring Min.	Sloping Ring Max.	Hardness Standard ASTM Needle Penetration 82°F. Min.	82°F. Max.	90°F. Min.	90°F. Max.	100°F. Min.	100°F. Max.	Tensile Strength lb per ¼ sq in. Area 90°F. Min.	Tensile 90°F. Max.	Elongation or Stretch (mm) 90°F. Min.	Elongation 90°F. Max.
Domestic 124–126	113.9	127.2	120.0	128.8	137	44	½ sec.	98	—	—	0	11.1	0	750
Domestic 125–127	121.1	129.1	127.6	136.2	90	21	2 sec.	236	½ sec.	2 sec.	20.2	45.0	½	2
Domestic 128–130	121.5	131.5	129.1	135.0	35	14	117	42	1 sec.	2 sec.	15.7	40.4	¼	45
Domestic 133–135	126.6	138.2	127.8	140.4	—	—	39	16	1 sec.	50	19.6	52.2	¼	¾
Domestic 135–137	127.1	136.2	136.2	139.5	—	—	42	18	3 sec.	140	29.8	63.3	¼	¾
Domestic 138–140	129.2	136.9	137.2	141.8	—	—	24	12	166	41	17.9	49.0	⅛	½
Imported 125–130	127.9	132.0	132.7	134.0	18	14	47	27	2 sec.	294	30.6	38.2	¼	¾
Imported 130–135	131.3	133.5	137.0	139.4	—	—	23	19	189	105	43.0	54.1	½	¼
Imported 133–135	130.8	137.2	136.4	141.0	—	—	26	15	207	32	35.9	59.7	¼	½
Imported 135–137	136.5	140.2	141.8	142.8	—	—	18	13	41	28	38.2	46.9	¼	½
Imported 138–140	136.6	142.1	142.2	147.4	—	—	15	10	29	19	37.8	61.8	¼	½
Imported 140–145	142.8	142.8	—	—	—	—	12	12	20	20	35.8	35.8	¼	¼

* The lowest and highest tests for each grade of paraffin wax are listed. The maximum and minimum tests were taken from a total of 426 tests made on different samples of paraffin waxes during 5 years.

Heat of Fusion

Paraffin wax of an average molecular weight of 400 has a molecular heat of fusion of about 21,000 cal.[1] Its latent heat of fusion is 38.9–43.9 cal, increasing with rise of specific gravity, solidifying point, and molecular weight.[2]

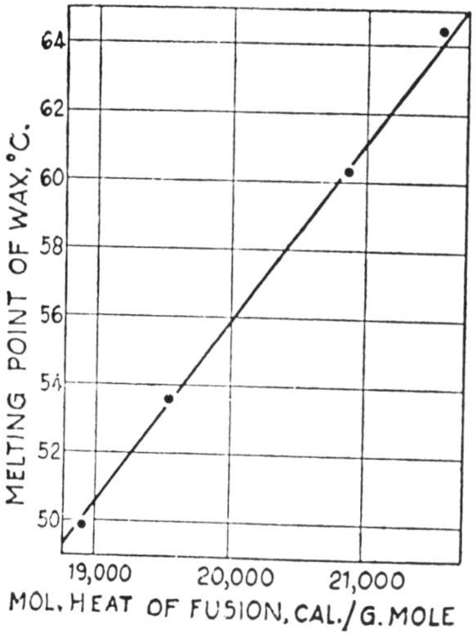

FIG. 1:17. HEAT OF FUSION OF PARAFFIN WAX.[3]

Expansion

The coefficient of expansion by heat of solid paraffin wax is about 0.005108. Paraffin wax obtained from lignite tar has approximately the same coefficient of expansion but, in the molten state, the value is almost double. The variation of the specific gravity of solutions of paraffin in solvents of high density can be expressed by the rule of mixtures, but, in the case of solvents of low density, one observes certain irregularities in the neighborhood of the point of fusion. The approximate density of solutions, at temperatures above the fusion point of paraffin wax, may be calculated by the rule of mixtures.[4]

The coefficient of volume expansion for petroleum waxes and *n*-paraffins are shown in the following tables and figures.[5]

There is good agreement among the freezing point values obtained in this study and those reported in the literature; however, the transition temperatures do not show as good general agreement (Table 1:12). This is probably due to differences in the purity of the samples used for these investigations, as it has been shown that the transition point is much more

---

[1] Sakhanov, *Neffyanoe Slantsevoe Khoz.* **6**, 820 (1924).
[2] G. Von Kozicki, *Chem. Ztg.* **42**, 147 (1918).
[3] A. Berne-Allen and L. T. Work, *Ind. Eng. Chem.*, **30**, 813 (1938).
[4] Anon., *J. petrole* **21**, 3 (1921).
[5] From study by P. R. Templin, *Ind. Eng. Chem.*, **48**, 154–161 (1956).

sensitive to impurity than is the freezing point. This was also clearly indicated in the present work by the freezing and transition point values for the pure and impure samples of *n*-octacosane (Table 1:14).

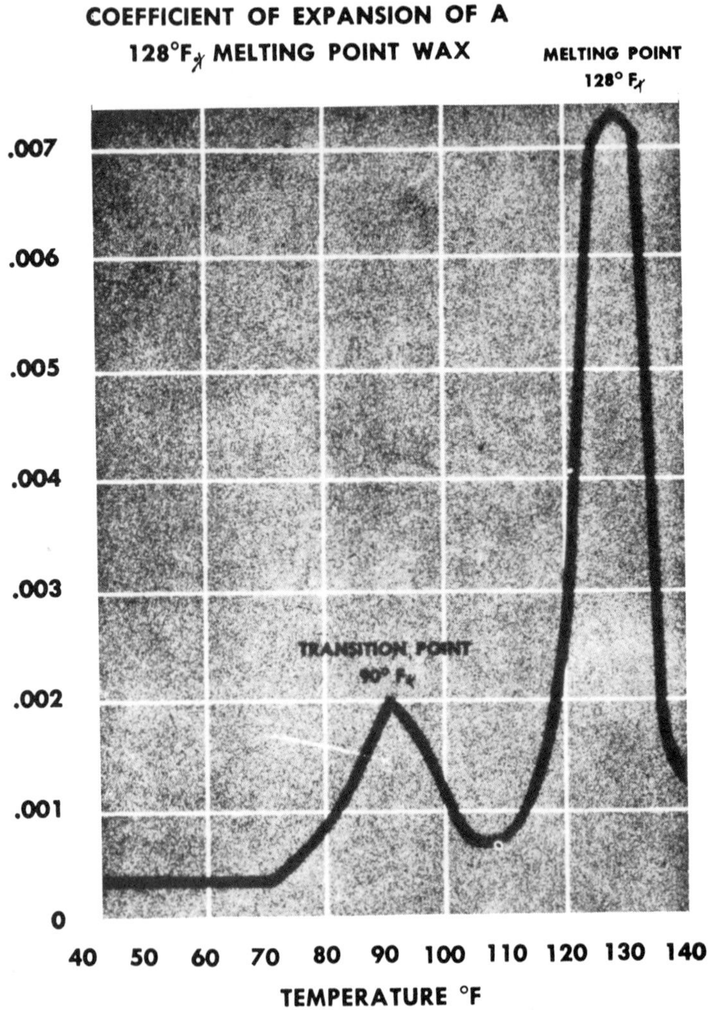

FIG. 1:18. COEFFICIENT OF EXPANSION OF A WAX, M.P. 128°F.

A comparison of the six curves in Figure 1 : 22 reveals that n-C_{24} is somewhat different from the other compounds with respect to the total volume change from liquid to the B solid form. This change is considerably less for n-C_{24} than for any of the other pure *n*-paraffins tested.

Although the expansion curves for the *n*-paraffins are very similar in character, several differences should be pointed out. n-C_{24}, n-C_{26}, n-C_{28}, and n-C_{32} indicate that a solid first-order transition takes place a few degrees below the freezing point. n-C_{20} gives no transition point of this nature, and n-C_{36} shows two of these solid-state transitions. In spite of these differences, however, the volume change from the liquid state to

Table 1:10

PHYSICAL PROPERTIES OF WAXES
(From expansion measurements)

Sample *Type grade*	125 Amp	128 Amp	133 Amp	136 Amp	*Microcrystalline Petrowax A*
	Refined Paraffin				
Crystal point, °C	51.6	53.2	55.9	56.8	74.0[1]
Transition point, °C	34.0	35.5	38.4	40.5	—
ΔT required for complete phase change, °C					
At fusion	8.6	8.5	8.8	7.7	35.0
At transition	10.5	10.3	11.4	12.0	—
Expansion					
Coefficient, cc/g/deg					
Liquid	0.0011	0.0011	0.0010	0.0010	0.0010
Solid form A	0.0016	0.0014	0.0013	0.0014	—
Solid form B	0.0010	0.0009	0.0008	0.0008	0.0009
Nonisothermal expansion[2]					
At fusion					
cc/g	0.1228	0.1237	0.1264	0.1262	0.1372
%	10.6	10.7	11.0	11.0	12.5
At transition					
cc/g	0.0356	0.0369	0.0387	0.0390	—
%	3.2	3.3	3.5	3.5	—
Solid form B at transition point to liquid at crystal point					
cc/g	0.1716	0.1735	0.1763	0.1770	0.1372
%	15.5	15.7	16.0	16.1	12.5
Isothermal expansion					
At fusion					
cc/g	0.1059	0.1119	0.1149	0.1150	0.1069
%	9.0	9.6	9.9	9.9	9.5
At transition					
cc/g	0.0252	0.0276	0.0300	0.0298	—
%	2.3	2.5	2.7	2.7	—
Solid form B to liquid at crystal point.					
cc/g	0.1436	0.1483	0.1536	0.1548	0.1069
%	12.7	13.1	13.7	13.3	9.5
Specific volume, cc/g					
At crystal point	1.2780	1.2796	1.2789	1.2782	1.2355[1]
At transition point	1.1420	1.1430	1.1413	1.1402	—

[1] Values taken at point of deviation on curve resulting from crystallization of liquid sample. Exact point at which crystals first appear is several degrees higher but extremely difficult to observe because of crystal size and high color of this microcrystalline wax.

[2] Expansion as it occurred over a temperature range (nonisothermal expansion) compared with expansion had it taken place entirely at one temperature (isothermal expansion).

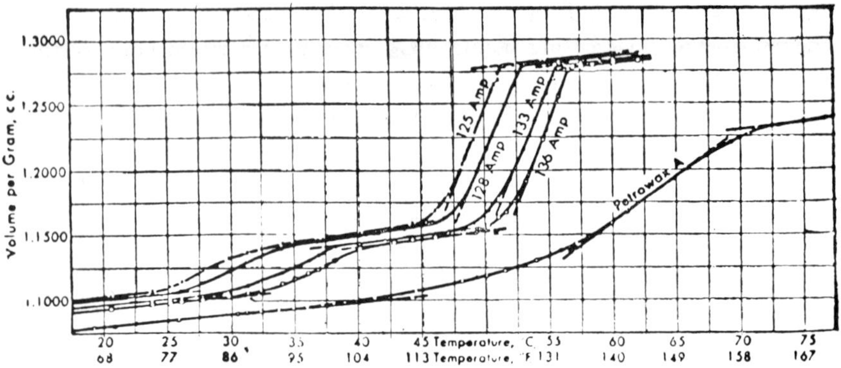

FIG. 1:19. VOLUME EXPANSION OF GULF WAXES.[1]

Table 1:11

PHYSICAL PROPERTIES OF *n*-PARAFFINS
(From expansion measurements)

PSC No.	540	541	106	176	197	190
Carbon No.	20	24	26	28	32	36
Crystal point, °C	36.4	50.3	56.0	61.1	69.3	75.9
First transition point, °C	—	48.1	53.2	58.1	65.5	73.8
ΔT, °C	—	2.2	2.8	3.0	3.8	2.1
Second transition point, °C	—	—	—	—	—	72.3
Liquid expansion, cc/g/°C	0.0011	0.0011	0.0011	0.0011	0.0011	0.0011
Solid expansion, cc/g/°C						
Form A	—	0.0033	0.0022	0.0020	0.0021	0.0020
Form A_2	—	—	—	—	—	0.0026
Form B	0.0005	0.0005	0.0005	0.0007	0.0005	0.0004
Expansion at phase change, cc/g						
Fusion	0.2077	0.1215	0.1254	0.1329	0.1346	0.1377
First transition	—	0.0662	0.0780	0.0774	0.0692	0.0454
Second transition	—	—	—	—	—	0.0200
Specific volume, cc/g						
Crystal point	1.2853	1.2833	1.2800	1.2837	1.2818	1.2778
First transition point	—	1.1547	1.1488	1.1450	1.1398	1.1361
Second transition point	—	—	—	—	—	1.0873
Expansion, %						
Fusion	19.3	10.5	10.9	11.5	11.7	12.1
First transition	—	6.1	7.3	7.2	6.5	4.2
Second transition	—	—	—	—	—	1.9
Low temp. solid at transition point to liquid at crystal point	19.3	17.9	19.5	20.2	19.7	19.7

[1] Figures 1:19 to 1:22 from Templin, *Ind. Eng. Chem.*, **48**, 154–161 (1956).

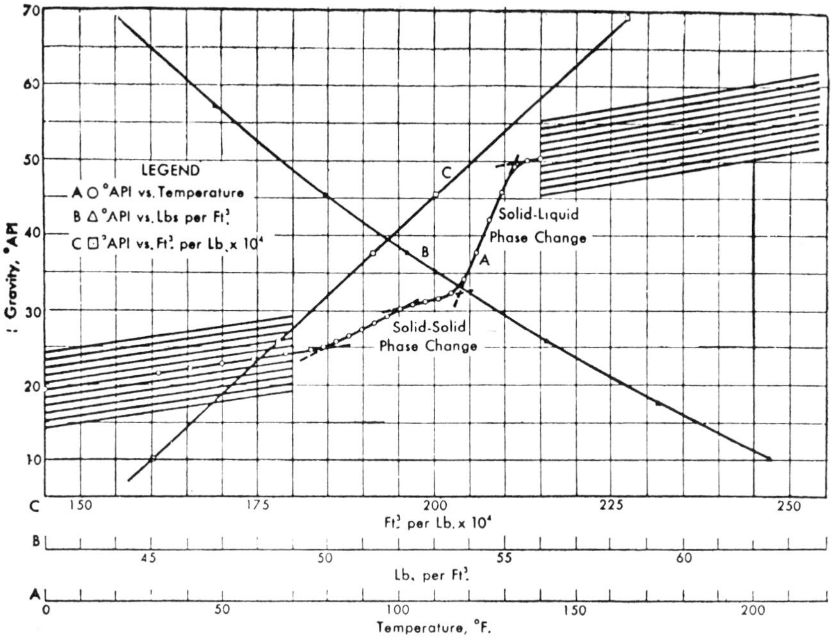

FIG. 1:20. CONVERSION CURVES FOR REFINED PARAFFIN WAXES.

Table 1:12

FREEZING AND TRANSITION POINTS OF *n*-PARAFFINS

Freezing (1) and Transition (2) Temperatures, °C

n-C_{20}		*n*-C_{24}		*n*-C_{26}		*n*-C_{28}		*n*-C_{32}		*n*-C_{36}	
1	2	1	2	1	2	1	2	1	2	1	2
36.4	—	50.3	48.1	56.0	53.2	61.1	58.1	69.3	65.5	75.9	73.8
36.2	32.0	50.7	47.9	55.8	48.8	61.2	54.0	69.5	63.5	...	—
—	—	—	—	56.1	51.5–52	61.1	57–57.4	69.2	65.2–65.4	75.6	73.9–74.1
36.6	—	50.6	—	56.2	—	—	—	—	—	—	—
36.6	—	50.68	48.11	56.2	52.54	61.20	56.90	—	—	—	—
—	—	51.2	40–41	58.0	45.5–46	—	—	—	—	—	—
—	—	50.7–8	47.0	—	—	61.2–3	54.2	—	—	75.8	73.5

the low-temperature solid form was very nearly the same regardless of whether there was one, two, or no solid transition at all.

Pure *n*-paraffins (C_{20}–C_{30}) have properties superior to those of commercial paraffin waxes. The addition of polyethylene gives a product of higher tensile strength with the former than with the latter.[1]

[1] K. W. Brooks, *Oil and Gas J.*, **58**, 89 (1960).

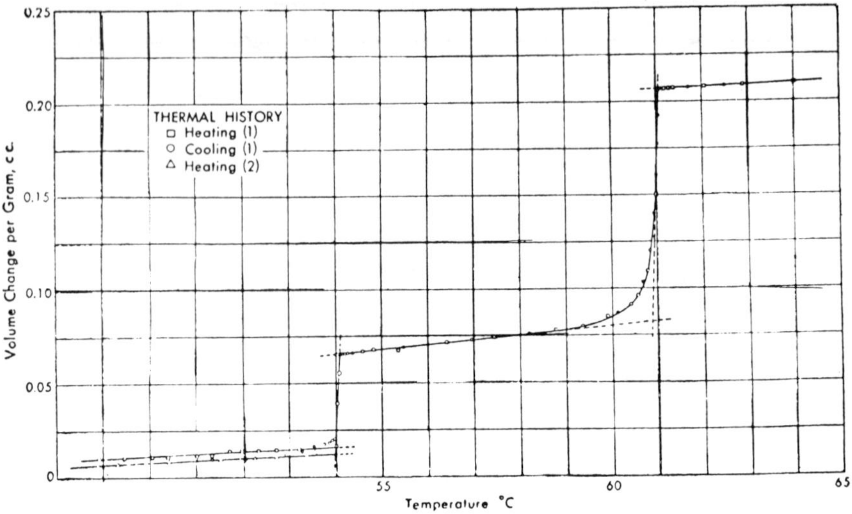

FIG. 1:21. VOLUME EXPANSION OF SEMI-PURE n-C_{28}.

Melting Point: Transition Point

Paraffin waxes, from different sources, have very similar compositions. Differences in the melting points of commercial paraffin waxes are due to differences in their oil content. Paraffin wax apparently retains oil as a solid solution of oil in wax. The melting points of the solid solutions lie between the melting points of the components. Due to the existence of solid solutions, paraffin wax does not have a sharp melting point.

Investigators of paraffin wax have noticed that, at temperatures below the apparent solidification point of the wax, changes in crystal structure frequently take place. The temperature at which this occurs is known as the *transition point*. Not all waxes exhibit transition points and most waxes, if quickly chilled to below the transition point, will show no change unless stored at slightly elevated temperatures where crystals may reform. Waxes, having uniform melting-point distribution, particularly over a wide range of temperature, will show no transition point, or at best an indistinct one. Waxes of uneven distribution invariably have the most pronounced transition points and show the greatest changes in ductility at these temperatures.

When waxes having transition points are cooled slowly, the following phenomena occur. When a large mass of wax is cooled and reaches the so-called melting point, latent heat of fusion is released and the temperature of the wax remains constant or increases slightly until solidification is complete. Below this temperature, cooling proceeds again at a regular rate. However, a large quantity of wax remains as an isotropic viscous fluid for some degrees below the soldification point. When this fluid

Table 1:13

VOLUME EXPANSION OF *n*-PARAFFINS

n-C_{20} (PSC No. 540)		n-C_{24} (PSC No. 541)		n-C_{26} (PSC No. 106)				n-C_{28} (PSC No. 176)		n-C_{32} (PSC No. 197)		n-C_{36} (PSC No. 190)	
Temp °C	Sp. vol. cc/g	Temp °C	Sp. vol cc/g	Temp °C	Sp. vol. cc/g	Temp °C	Sp. vol. cc/g	Temp °C	Sp. vol. cc/g	Temp °C	Sp. vol. cc/g	Temp °C	Sp. vol. cc/g
48.11	1.2987	59.61	1.2935	61.13	1.2858	51.99	1.0699	70.64	1.2936	85.59	1.2988	85.59	1.2880
47.49	1.2980	58.65	1.2925	61.11	1.2857	52.28	1.0700	69.49	1.2924	84.63	1.2978	84.63	1.2872
46.98	1.2974	57.12	1.2907	61.07	1.2857	52.54	1.0707	69.20	1.2921	83.79	1.2970	83.79	1.2863
46.72	1.2971	55.93	1.2885	61.04	1.2857	52.74	1.0815	69.07	1.2919	83.03	1.2965	83.03	1.2853
45.55	1.2958	55.48	1.2890	59.93	1.2846	52.84	1.0933	68.77	1.2917	82.36	1.2958	82.36	1.2845
44.51	1.2946	54.86	1.2886	59.16	1.2836	52.95	1.1083	68.22	1.2911	81.12	1.2944	81.12	1.2831
43.55	1.2934	53.97	1.2876	58.11	1.2820	53.07	1.1250	67.24	1.2899	79.04	1.2923	79.04	1.2812
45.04	1.2952	53.19	1.2866	57.61	1.2813	53.20	1.1487	66.18	1.2889	78.18	1.2911	78.18	1.2802
47.21	1.2977	52.92	1.2863	57.44	1.2810	53.37	1.1491	65.27	1.2881	77.13	1.2901	77.13	1.2792
47.64	1.2982	52.26	1.2857	56.86	1.2802	53.63	1.1500	64.73	1.2874	76.80	1.2896	76.80	1.2789
45.48	1.2958	51.54	1.2849	57.68	1.2816	53.83	1.1506	64.67	1.2871	76.11	1.2890	76.59	1.2785
43.75	1.2939	51.19	1.2846	57.88	1.2819	54.12	1.1518	63.59	1.2861	75.84	1.2890	76.39	1.2785
46.43	1.2968	50.34	1.2837	58.04	1.2821	54.61	1.1549	62.35	1.2848	75.47	1.2885	76.11	1.2781
47.27	1.2977	50.25	1.2186	58.60	1.2831	55.09	1.1608	61.44	1.2839	74.90	1.2877	75.93	1.2780
47.86	1.2985	50.22	1.2092	59.52	1.2842	55.21	1.1633	64.46	1.2873	74.20	1.2869	75.84	1.1965
47.97	1.2985	50.16	1.1974	60.21	1.2850	55.31	1.1665	64.70	1.2877	73.65	1.2864	75.81	1.1825
48.94	1.2997	50.05	1.1843	60.60	1.2855	55.42	1.1705	65.05	1.2879	73.05	1.2859	75.74	1.1665
49.89	1.3007	49.63	1.1675	61.79	1.2866	55.45	1.1738	65.39	1.2882	69.55	1.2823	75.66	1.1554
50.29	1.3012	48.76	1.1583	63.08	1.2881	59.61	1.2842	66.74	1.2899	75.84	1.2889	75.47	1.1450
50.77	1.3017	48.11	1.1553	59.94	1.2845	58.65	1.2831	69.32	1.2925	75.46	1.2887	74.90	1.1391
51.97	1.3032	47.49	1.1531	58.33	1.2825	57.12	1.2817	69.13	1.2922	72.18	1.2851	74.20	1.1370
43.57	1.2937	46.98	1.1516	56.50	1.2799	55.93	1.2804	65.83	1.2888	71.92	1.2845	73.65	1.1358
41.74	1.2915	46.72	1.0908	55.97	1.2790	55.48	1.1757	64.86	1.2875	70.64	1.2834	73.05	1.0889
40.28	1.2900	45.55	1.0899	55.94	1.2570	54.86	1.1580	65.03	1.2877	69.49	1.2820	74.52	1.1376
38.88	1.2882	44.51	1.0891	55.92	1.2421	53.97	1.1512	64.41	1.2870	69.30	1.2819	73.54	1.1353
37.63	1.2870	43.55	1.0886	55.89	1.2326	53.19	1.1486	61.13	1.2838	69.26	1.2581	73.46	1.0902
36.77	1.2860	45.04	1.0894	55.87	1.2244	52.92	1.1479	61.11	1.2837	69.20	1.2219	73.48	1.0905
36.28	1.2854	47.21	1.0906	55.84	1.2154	52.26	1.1463	61.07	1.2100	69.14	1.1843	73.50	1.0904
36.00	1.2850	47.64	1.0895	55.79	1.2056	51.54	1.1447	61.04	1.1923	69.07	1.1655	73.53	1.0905
35.95	1.1068	45.48	1.0875	55.69	1.1900	51.19	1.0776	60.93	1.1668	68.77	1.1507	72.41	1.0873
35.87	1.1035	43.75	1.0866	55.48	1.1747	50.34	1.0770	61.01	1.1799	68.22	1.1466	70.43	1.0667
35.39	1.0892	46.43	1.0881	54.90	1.1593	50.25	1.0761	60.98	1.1733	67.24	1.1438	69.55	1.0662
35.24	1.0872	47.27	1.0884	53.81	1.1519	50.22	1.0756	60.95	1.1685	66.18	1.1415	70.14	1.0663
34.81	1.0832	47.86	1.0887	52.11	1.1466	50.16	1.0753	60.92	1.1648	65.27	1.1395	70.90	1.0668
34.09	1.0799	47.97	1.1299	53.27	1.1493	50.05	1.0750	60.89	1.1629	64.73	1.1384	71.72	1.0668
33.75	1.0788	48.07	1.1550	52.70	1.1477	49.63	1.0747	60.86	1.1607	64.67	1.0729	71.96	1.0856

n-C_{20} (PSC No. 540)		n-C_{24} (PSC No. 541)		n-C_{26} (PSC No. 106)				n-C_{28} (PSC No. 176)		n-C_{32} (PSC No. 197)		n-C_{36} (PSC No. 190)	
Temp °C	Sp. vol. cc/g	Temp °C	Sp. vol. cc/g	Temp °C	Sp. vol. cc/g	Temp °C	Sp. vol. cc/g	Temp °C	Sp. vol. cc/g	Temp °C	Sp. vol. cc/g	Temp °C	Sp. vol. cc/g
32.11	1.0776	48.94	1.1593	51.89	1.0766	48.76	1.0743	60.83	1.1595	63.59	1.0700	71.86	1.0831
30.51	1.0767	49.89	1.1746	51.15	1.0757	46.98	1.0732	60.79	1.1581	62.35	1.0689	71.82	1.0819
31.35	1.0770	50.29	1.2380	50.30	1.0746	45.55	1.0725	60.46	1.1527	60.51	1.0677	72.69	1.0892
33.15	1.0766	50.77	1.2843	48.37	1.0733	44.51	1.0721	59.93	1.1496	61.44	1.0682	73.96	1.1363
30.16	1.0747	51.97	1.2856	46.64	1.0722	43.55	1.0714	59.16	1.1475	64.46	1.0701	75.84	1.2781
31.62	1.0754			47.68	1.0729	45.04	1.0722	58.11	1.1454	64.70	1.0703	76.40	1.2787
32.68	1.0758			49.39	1.0738	47.21	1.0733	57.95	1.1449	64.80	1.0704	77.12	1.2796
33.55	1.0776			50.57	1.0736	47.64	1.0734	57.79	1.1445	64.96	1.0713	77.70	1.2802
34.52	1.0811			51.31	1.0724	45.48	1.0725	57.67	1.1443	65.05	1.0720	72.18	1.0868
35.68	1.0947			50.19	1.0710	43.75	1.0714	57.61	1.0722	65.18	1.0752	71.92	1.0863
36.00	1.1099			45.66	1.0686	46.43	1.0727	57.44	1.0695	65.25	1.0790	76.26	1.2791
36.12	1.1225			48.75	1.0703	47.27	1.0732	56.86	1.0671	65.33	1.0875		
36.24	1.1471			50.75	1.0700	47.86	1.0735	55.52	1.0660	65.39	1.1072	75.57	1.1492
36.35	1.2078			47.69	1.0682	47.97	1.0734	53.04	1.0645	65.44	1.1200	73.54	1.0958
36.41	1.2855			45.73	1.0671	48.07	1.0731	54.39	1.0651	65.52	1.1360	72.34	1.0869
37.20	1.2864			46.77	1.0677	48.94	1.0736	56.20	1.0661	66.74	1.1422	70.95	1.0687
38.68	1.2881			48.24	1.0684	49.89	1.0742	57.68	1.0682	69.32	1.2820	69.83	1.0678
				50.18	1.0695	50.29	1.0742	57.88	1.0715	69.23	1.2819	71.62	1.0685
				51.55	1.0699	50.77	1.0743	57.99	1.0861	70.13	1.2829	71.87	1.0687
				51.76	1.0699	51.97	1.0748	58.04	1.1129	69.23	1.2819	72.16	1.0703
								58.11	1.1451	69.13	1.1786	72.24	1.0724
								58.60	1.1460	69.07	1.1651	72.44	1.0887
								59.52	1.1482	65.83	1.1406	73.02	1.0902
								60.21	1.1507	65.53	1.1400	73.24	1.0905
								60.60	1.1541	65.48	1.1397	73.65	1.0925
								61.01	1.1937	65.35	1.1395	73.76	1.1153
								61.06	1.2483	64.86	1.1384	73.80	1.1292
								61.79	1.2846	64.56	1.1378	73.83	1.1352
								63.08	1.2860	64.41	1.1375	74.48	1.1375
										64.96	1.1382	75.94	1.2785
										65.09	1.1276	75.85	1.2316
										65.03	1.1094	75.87	1.2657
										64.41	1.0870	75.88	1.2782
										61.13	1.0693		
										61.11	1.0692		
										61.07	1.0691		
										57.61	1.0669		
										54.39	1.0652		
										56.20	1.0661		
										58.04	1.0671		
										63.08	1.0697		

Table 1:14

EXPANSION OF n-OCTACOSANE

	Pure n-C_{28}	*Semipure* n-C_{28}
Crystal point, °C	61.1	61.1
Transition point, °C	58.1	54.1
ΔT, °C	3.0	7.0
Expansion coefficient, cc/deg/g		
Liquid	0.0011	0.0010
Crystalline form A	0.0020	0.0025
Crystalline form B	0.0007	0.0011
Volume change, cc/g		
At fusion	0.1329	0.1245
At transition	0.0774	0.0534

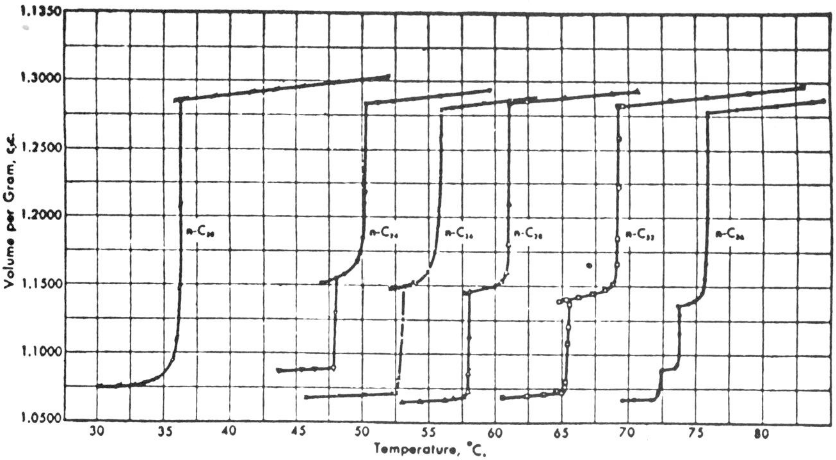

FIG. 1:22. VOLUME EXPANSION OF *n*-PARAFFINS.

congeals, contraction commences and air begins to enter the wax. This point probably represents the final solidification of all constituents. At a still lower temperature, transition takes place, which is accompanied by a definite change in volume caused by contraction of certain solid portions of the wax, an evolution of heat, and a release of the air that was dissolved during the prior cooling. The release of heat causes a change in the rate of cooling and it is used for determining the transition point. In some cases, the change in rate is pronounced, whereas in others, it is indefinite. Following transition, the wax again continues to cool at a regular rate. Frequently, a microscopic study of these cooling phenomena will show that a portion of the wax actually becomes liquefied and recrystallizes long after it appears to have become permanently solid. The quantity involved is rather small but it is spread uniformly throughout the mass of wax.

The practical aspects of the transition point in waxes are not definitely established yet. However, it appears that waxes showing indistinct transitions, or no transition points, are more translucent than those in which the transition is marked. It might also be reasoned that the absence of a transition point would indicate a more satisfactory wax for moisture-proofing, as there would be no fractures between crystal boundaries due to contraction. This is another way of saying that the more ductile waxes—those without transition points, but with a uniform distribution of components—are more satisfactory for this purpose, because they can be deformed without fracture.

An additional observation of interest is that molded articles, such as candles, can be made most readily from waxes with sharp transition points. It is found that these waxes show more contraction on cooling and thus have less tendency to adhere to the mold than do those with indistinct transition points.

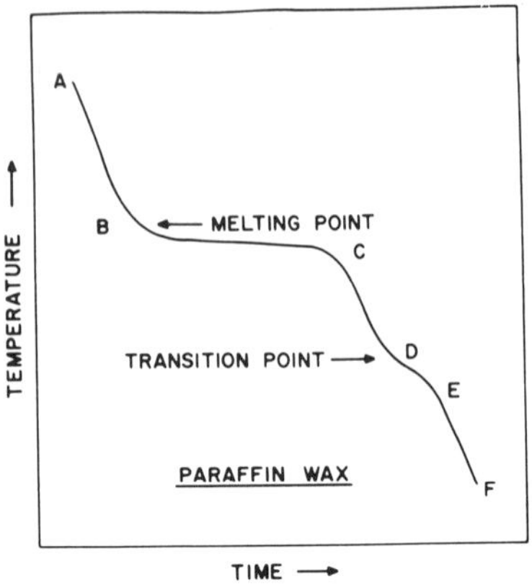

Cooling curve of a paraffin wax

FIG. 1:23

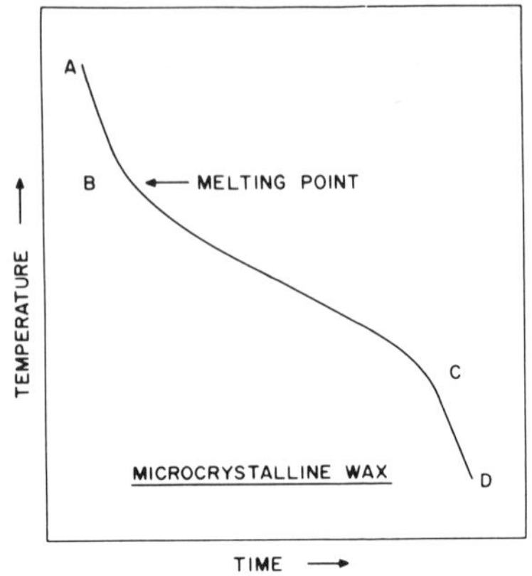

Cooling curve of a microcrystalline wax

FIG. 1:24

In order for a petroleum wax to have a solid–solid transition, it must contain a substantial proportion of paraffinic hydrocarbons in the carbon number range of about 20 to 40.

Thermal expansion of *n*-paraffins over the temperature range of 0 100°C. decreases as the chain length increases beyond 28 carbon atoms. The expansion of petroleum wax also decreases with increasing molecular weight.

As the molecular weight of petroleum wax increases, the wax becomes more complex with respect to molecular structure and hydrocarbon distribution. Paraffin wax is the simplest form of petroleum wax and its volume–temperature characteristics are similar to those of *n*-paraffins in the wax range. These characteristics include a well-defined solid-state transition, narrow melting and transition ranges, and relatively large volume changes at the melting point and transition point. As the complexity of the wax increases in going from paraffin wax to the higher molecular weight semimicrocrystalline and microcrystalline waxes, the melting and transition ranges broaden and the thermal expansion decreases.

The volume–temperature characteristics of a wax represent the composite effect of all of its hydrocarbon components in proportion to their concentration. Similarly, the volume–temperature characteristics of a blend are proportional to the concentration and volume–temperature characteristics of the blend components. The volume–temperature behavior of a wax blend is predictable from a knowledge of the volume–temperature behavior of the blend components.

Regardless of the method used to make a wax more complex, e.g., blending, isomerization, or concentrating complex structures by fractionation, the effects on volume–temperature behavior are similar.

The transition point of a paraffin wax can be eliminated by adding to it about 25% by weight of microcrystalline wax. Eliminating or reducing the sharp volume change associated with the transition point increases resistance to thermal shock.

Viscosity, specific volume, density, molecular weight, and refractive index are all interrelated properties of wax. With the exception of molecular weight these properties change with temperature. At any temperature where one of these properties changes abruptly, the others should also show an abrupt change. Wherever an abrupt change occurs in the volume–temperature curve for a wax, changes in other temperature-sensitive properties should be expected.

Melting Point–Refractive Index Relationship

It has been amply demonstrated that different types of hydrocarbons exhibit different relationships between melting point and refractive index, and this means of classification was exhaustively considered. In a group of hydrocarbons of the same melting point, the normal alkane exhibits the lowest refractive index; progressive branching raises the refractive index; and cyclic structures raise it even more, as shown in Fig. 1:25.

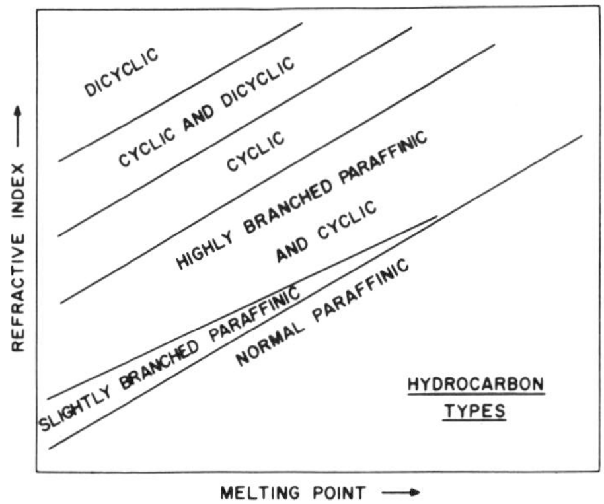

Relationship between melting point and refractive index for different types of hydrocarbons

FIG. 1:25

A chart of this type therefore offers an excellent means of determining the overall composition of a wax, and of differentiating paraffin waxes from microcrystalline waxes.

Transition Effects in Paraffin Wax[1]

When different temperatures are plotted against time, the minimum marks the melting point, the maximum, probably, the final solidification point, and a second low point (in some waxes), a transition point at which internal forces modify the crystalline structure, with release of energy. This second low point marks the maximum rate of change from the α- to the β-condition. Isoparaffins show no transition points. Some waxes exhibit no transition point unless cooling is rapid; others, none at all. Waxes that are blended for a particular melting point, tend to be nearly homogeneous in structure and more stable in temperature change if they have no transition point. Translucency accompanies a transition point. The transition change leads to contraction in crystalline boundaries and this is unfavorable where wax is used to prevent the passage of water vapor, as in electric condensers, or waxed papers. It is an advantage, however, when the wax is to be molded.

The transition from needle-shaped prisms to rhomboidal plates or leafy masses occurs at 10–15° below the melting point. The crystal form depends on the solvent used, the rate of cooling, and the wax used. Wax 7–15% of its own volume of air, occludes at ordinary temperature.[2]

Blocking

Blocking is a term used throughout the paper-converting trade which refers to the tendency of two sheets of waxed paper to adhere to each other. It is encountered where sheets of waxed paper are stored in stacks or piles and where a tightly wound roll is exposed to elevated temperatures. Various degrees of blocking are possible, all of which are primarily the result of either fusion of the wax from both sheets or localized adhesion of wax particles. In the extreme case, from which the term is derived, a practically solid cake, or block, is formed. Lesser degrees of blocking result in the formation of pock marks in the surface caused by localized blocking or a general marring of the waxy surface due to a slight over-all tendency to form a wax bond between two layers of paper. As such a difference in degree exists, and as a solid block practically precludes rerunning the paper, all forms of blocking are objectionable and should be avoided.

Blocking is due to complete or incipient fusion of wax from the surfaces of both layers of paper; it is influenced by temperature, time and pressure. As any of these three increases, so does the tendency to block. Because time and pressure are somewhat beyond control (except in the case of the user), the effect of temperature must be overcome. This is accomplished most simply by the use of higher-melting waxes when elevated temperatures of storage are anticipated. Carefully controlled laboratory data have

---

[1] C. R. Scott-Harley, *J. Inst. Petroleum* **25**, 238 (1939).

[2] J. A. Carpenter, *J. Inst. Petroleum Tech.* **12**, 288 (1926).

shown that, with waxes from the same source, blocking is almost entirely due to a low melting point.

In general, no safe temperature limits can be established because of the effect of pressure and time, but an increase of about 5° in melting point is equivalent to a 10° increase in permissible storage temperature. This approximation holds only over the range of 125–135°F, in which most domestic refined waxes fall. Practically all converters purchase higher-melting grades during summer months, and those that cater to the southern trade ought to make certain that lower-melting waxes are not used in paper stocks to be shipped to such points.

In addition to melting point, there are two other factors that influence blocking in a lesser degree. The first of these is oil content, which, if too high will tend to increase blocking. However, the quantity of oil required for so adverse an effect is excessive and would be objectionable for other reasons. Refined waxes usually contain less than 0.5% of oil.

The second factor is the ductility of the wax. As previously shown, two waxes of identical melting point may have entirely different melting point distribution. Thus, in one of them, there may be low-melting fractions that are softer than usual. The presence of these constituents will result in a wax of greater ductility which is more easily deformed by constant pressure. Such waxes tend to block at slightly lower temperatures than harder, more brittle waxes of equal melting point.

Fully refined paraffin waxes and blends with plastic microcrystalline wax (up to 40% of latter), and with or without 5% polyethylene of low molecular weight, show a direct relationship to hardness and blocking temperature.[1]

The use of wax fractions with a narrow range of molecular weight causes an increase in blocking temperature without raising the melting point. The presence of *n*-alkanes of low melting point harms blocking temperature.[2]

Blocking and Transition Point[3]

The rapid changes in density of waxes undergoing transition may explain the erratic nature of the results of tests on penetration obtained at temperatures in or near the transition region, particularly if the wax has very sharp transition characteristics. It would be expected that because densities change sharply when waxes undergo transition, hardness as measured by the needle penetration test would show a similar stepwise behavior. Such changes in hardness are more difficult to observe than the corresponding changes in density vs. refractive index because of the slowness with which phase equilibrium is attained in larger wax specimens. The longer time required for complete transition in bulky samples is probably

[1] K. G. Arabian, Tappi, **39**, 348 (1956).
[2] Tappi, **41**, 275–80 (1958).
[3] From study by D. V. Harrison *et al.* Chem. Eng. Data, Series 3, 352–358 (1958).

Table 1:15

Volume Changes in Petroleum Waxes

Test		Sample of Wax							
		A	B	C	D	E	F	G	H
Melting point, °F (ASTM D 87–42)		129.8	128.7	143.8	171.6[1]	143.4	143.1	143.3	124.2
Melting point, °F (By refractive index)		129.2	129.2	143.6	168.8	143.6	142.9	143.6	124.5
Oil content, % (ASTM D 721–56T)		1.1	0.5	0.7	1.0	0.6	0.8	1.7	0.3
Refractive index, at 176°F		1.4268	1.4282	1.4306	1.4481	1.4315	1.4351	1.4311	1.4286
Transition range, °F	Top	107.6	104.0	118.4	...[2]	119.3	122.0	120.2	96.8
	Mid-point	106.9	92.0	117.5	—	116.2	114.8	116.4	84.6
	Bottom	106.2	77.0	116.6	—	105.8	105.8	104.0	71.6
Penetration (ASTM D 1321									
77°F		12	14	11	13	10	9	14	21
90		17	38	15	20	13	12	19	58
100		25	96	19	25	17	16	22	100
110		76	130	31	41	26	26	36	144
120		226	—	110	95	111	75	142	—
25		—	—	165	118	148	105	202	—
Density, grams per cc (ASTM D 1217–54)									
130°F		—	—	—	—	—	—	—	0.7832
140		0.7774	0.7797	—	—	—	—	—	—
149		—	—	0.7819	—	0.7828	—	0.7825	—
150		—	—	—	—	—	—	—	0.7768
158		—	0.7736	0.7780	—	0.7800	—	—	—
169		0.7705	—	—	—	—	0.7856	—	—
176		—	—	0.7722	—	0.7734	0.7806	0.7724	—
185		—	—	—	0.8027	—	—	—	—
194		—	—	—	0.8017	—	—	—	—
Blocking, °F	Pick point	96	83	112	117	115	119	112	83
Tappi T652 sm 57)	Blocking point	105	93	119	119	117	121	122	88
Mass spectrometer analysis									
Normal paraffins, %		90.0	61.1	78.3	—	63.7	48.1	—	42.5
Isoparaffins, %		10.0	30.1	18.6	—	31.4	32.1	—	31.6
Cycloparaffins, %		—	8.8	2.1	—	1.2	9.1	—	25.9
Alkyl Benzenes, %		—	—	1.0	—	0.7	0.7	—	—
Total		100.0	100.0	100.0	—	97.0[3]	90.0[4]	—	100.0

[1] Drop melt point ASTM D127–49; [2] Indefinite;
[3] 3% heavier molecular weight material not volatilized; [4] 10% heavier molecular material not volatilized.

also important in its effect on repeatabilitity of penetration results in the transition region, which appears to be associated with penetration values between about 30 and 100 units.

The waxes used may be identified as follows:

A: Prepared by distilling commercial paraffin wax to obtain 60% heart cut of narrow-boiling-range material.

B, C, H: Commercial paraffin waxes.

D: Commercial microcrystalline wax.

E, F: From 5% and 25% respectively (wt) of D into C.

G: 1% (wt) medicinal white oil into C; visc., 340 at 210°F.

The data of Table 1:15 indicate that the blocking temperature of a petroleum wax falls in the transition range, or very close to it. With wide-cut waxes, such as B and H, blocking occurs at approximately 50% transition, while in the very close-cut waxes, such as A and C, it can be either just below or just above the narrow transition range. In blends of wax D with wax C the blocking point falls in the upper end of the transition range.

The lack of exact agreement between transition and blocking temperatures, as noted here, may be due in part to difficulties experienced in detecting the blocking point of certain waxes. Perhaps better agreement

Table 1:16

REFRACTIVE INDICES, DENSITIES, AND SPECIFIC VOLUMES FOR WAX A[1]

°F	n_D	n_o	n_e	d	1/d
149.0	1.4328	—	—	0.7744	1.2913
136.4	1.4356	—	—	0.7787	1.2842
131.0	1.4365	—	—	0.7801	1.2819
129.2	1.4370	—	—	0.7809	1.2806
128.3	1.4370	1.4736	1.5243	0.8625	1.1594
123.8	1.4379	1.4752	1.5254	0.8646	1.1566
113.0	—	1.4782	1.5285	0.8692	1.1505
108.5	—	1.4804	1.5306	0.8723	1.1464
107.6	—	1.4817	1.5318	0.8743	1.1438
106.2	—	1.4995	1.5484	0.9000	1.1111
102.2	—	1.5016	1.5515	0.9035	1.1068
95.0	—	1.5040	1.5539	0.9070	1.1025
89.6	—	1.5060	1.5560	0.9101	1.0988
82.4	—	1.5075	1.5577	0.9123	1.0961
77.0	—	1.5084	1.5585	0.9136	1.0946
66.2	—	1.5108	1.5607	0.9170	1.0905

°F Fahrenheit.
n_D Refractive index, liquid state.
n_o Refractive index, solid state, ordinary rays.
n_e Refractive index, solid state, extraordinary rays.
d Density (calculated).
1/d Specific volume (calculated).

[1] See Table 1:15.

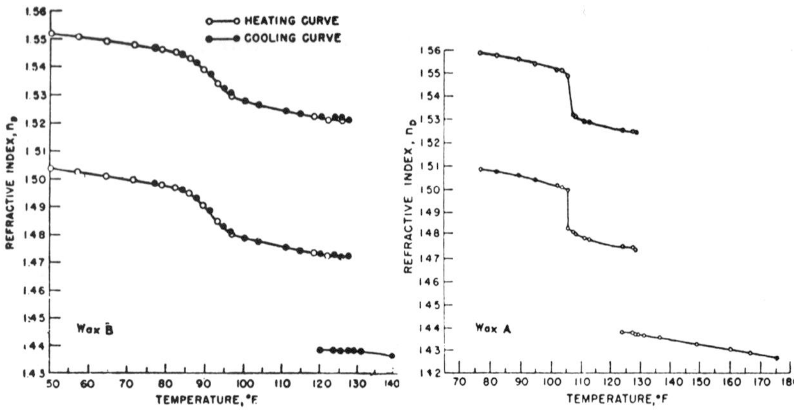

FIG. 1:26. REFRACTIVE INDICES VS. TEMPERATURES.[1]

would be obtained by use of the pressure method, which has shown certain waxes to be sensitive to pressure applied during the test.

It appears from this study that blocking takes place at a temperature at which transition is sufficiently advanced so that a significant amount of "soft wax" (form stable to high temperature) is present. This would lower the cohesive strength of a wax coating enough to allow transfer of wax from one sheet to another. The rapid expansion in volume during transition could help account for the severe blocking sometimes seen in tightly wound rolls of waxed paper.

In Figure 1:27 solid densities for waxes A, B, C, and H are plotted against temperature referred to the melting point. Irrespective of composition these waxes show very nearly the same densities in the high-temperature stable form. The transition region for wax A, which is the

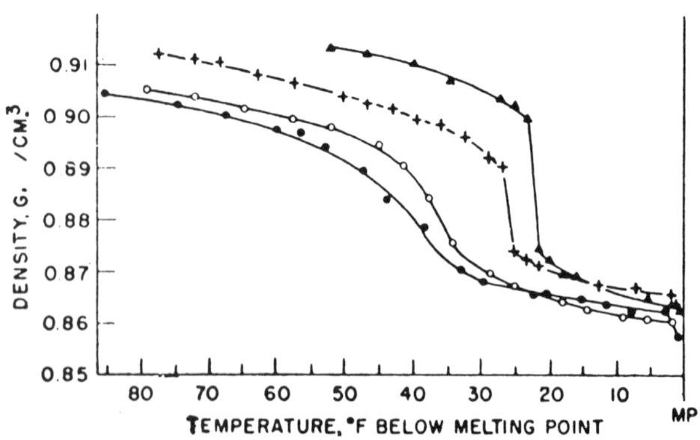

FIG. 1:27. DENSITY-TEMPERATURE CURVES SHOWING EFFECT OF COMPOSITION.

[1] Figures 1:26 to 1:28 from Harrison, *Chem. Eng. Data*, Series 3, 352–358 (1958).

simplest mixture of hydrocarbons, falls closest to the melting point, while the transition region for wax H, the most complex mixture, falls farthest below the melting point. The corresponding regions for the other two waxes fall between these curves in proper order with respect to increasing complexity of composition. Although the transitions of these waxes extend over temperature ranges varying from 1° to 27°F, the upper limits of these ranges fall between 22° and 29°F below the melting point.

The relationship between penetration and density for the eight waxes studied is indicated in Figure 1:28. Here the penetration, plotted on a logarithmic scale against density on a linear scale, gives nearly a straight line for each wax. This relationship is apparently invariant as the waxes undergo transitions, and suggests that the closeness of packing on a molecular scale is of prime importance as a factor affecting the hardness of a wax.

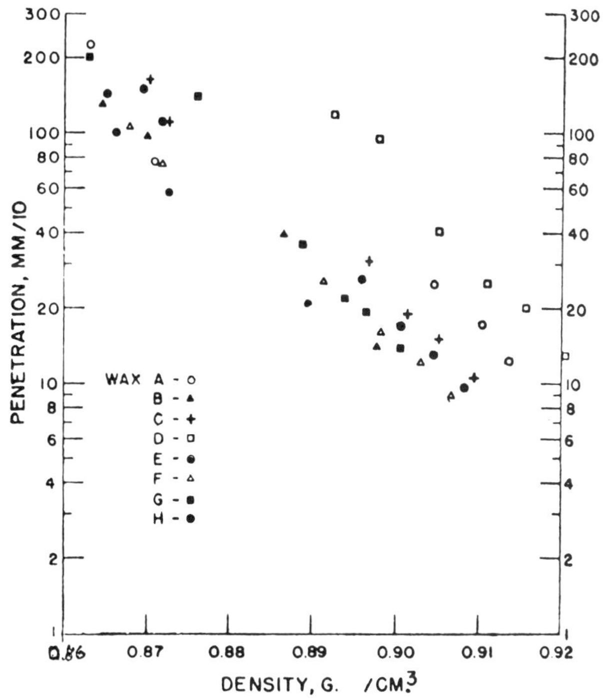

FIG. 1:28. PENETRATION VS. DENSITY.

Solubility of Petroleum Waxes

Frequently, it is desired to apply petroleum waxes in solution and data concerning their solubility at definite temperatures are essential. In addition, petroleum waxes are used extensively as coatings. These waxes are waterproof, but they are often expected to resist other solvents as well.

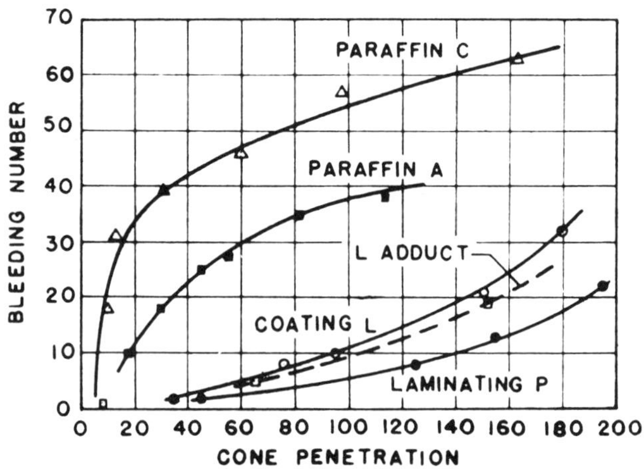

FIG. 1:29. Bleeding number versus penetration

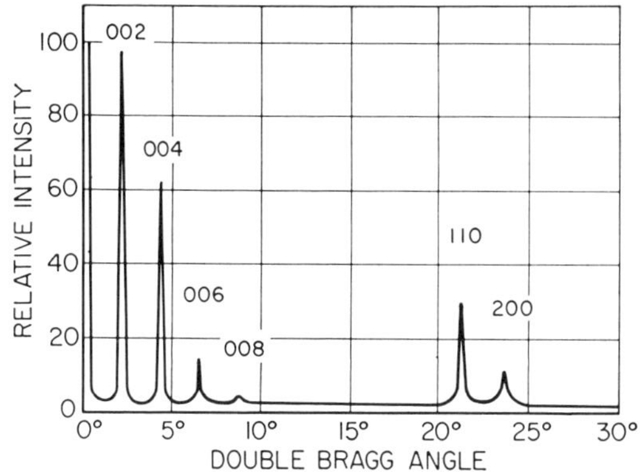

FIG. 1:30. X-ray pattern. Paraffin wax. Air-cooled. Reflection

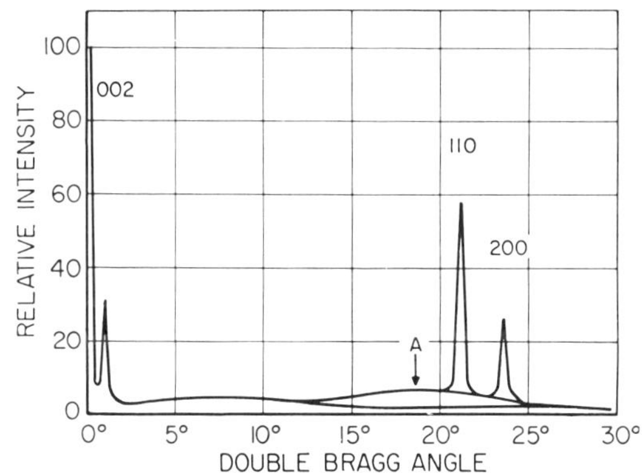

FIG. 1:31. X-ray pattern. Microwax. Air-cooled. Reflection

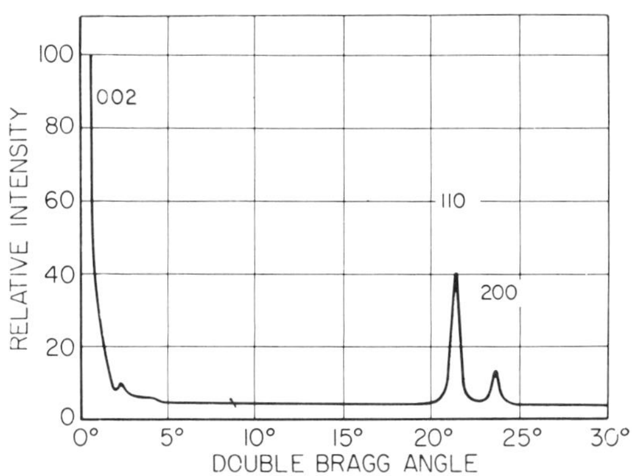

FIG. 1:32. X-ray pattern. Paraffin wax. Air-cooled. Transmission

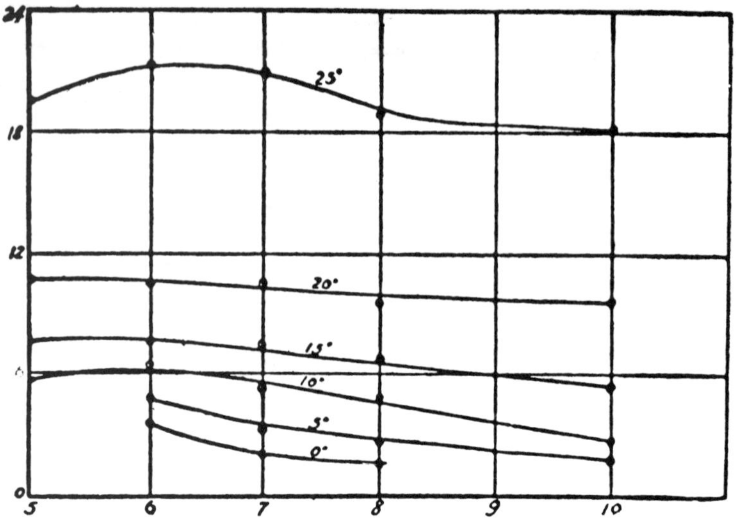

FIG. 1:33. SOLUBILITY OF PURE PARAFFIN WAX IN HYDROCARBONS.

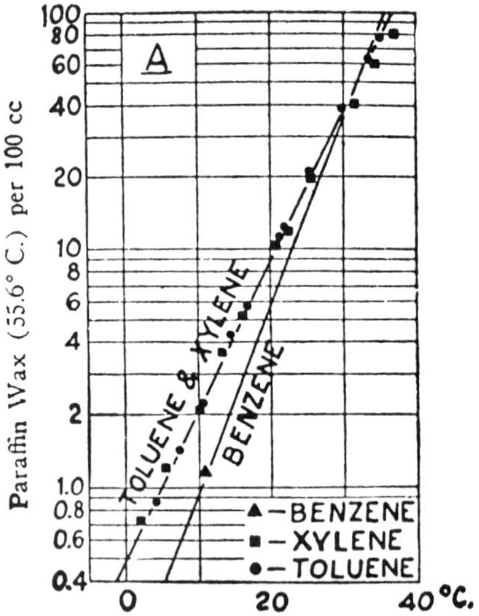

FIG. 1:34. SOLUBILITY OF PARAFFIN WAX IN AROMATIC HYDROCARBONS.

In such cases, waxes of minimum solubility are required. Data on solubility are also of value in separating waxes from other substances, because a solvent may be available which will preferentially dissolve one component.

The solubility of a solid paraffin in hydrocarbons of low molecular weight increases rapidly with a rise in temperature from that of isodecane to that of pentane. The increase in solubility with rise of temperature is more rapid with the higher-molecular-weight solvent. This can be seen by dividing the solubilities at 10°C into the solubility at 25°C, the ratio being 2.3, 3.8, 4.0, 3.9, and 6.6 for, respectively, pentane, hexane, heptane, octane, and isodecane.

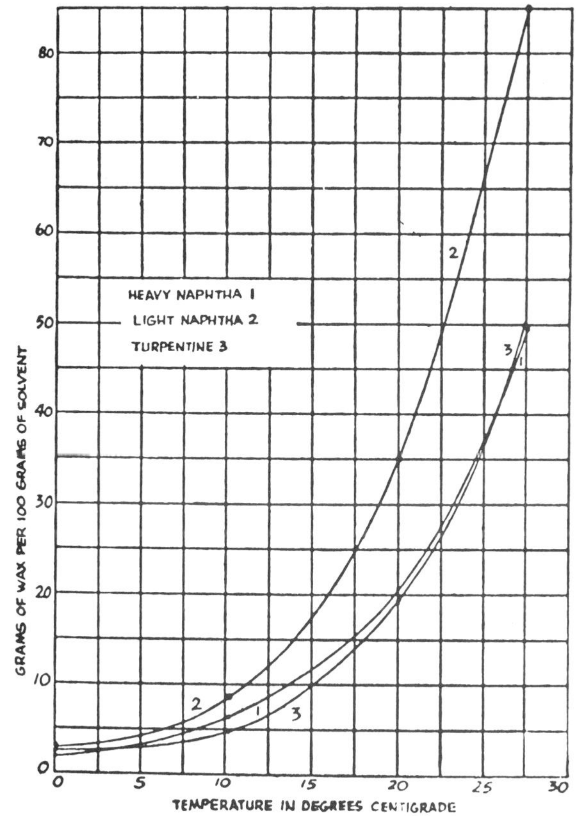

FIG. 1:35. TEMPERATURE-SOLUBILITY CURVES FOR PARAFFIN WAX IN NAPHTHAS AND TURPENTINE.[1]

Paraffin wax (M.P. 56°C) is dissolved in solvents at 19°C as follows: Benzene 9.34%; toluene 9.74%; xylene 9.90%; benzene (at 20°C) 11.82% (Fig. 1:33). The amount of paraffin wax required to cause formation of gels decreases with the viscosity of the solvent, as shown by the addition of paraffin to benzene, xylene, chloroform, turpentine, and carbon disulfide. The solubility varies with the isomeric form. Tests with various mineral

[1] Figures 1:35 to 1:41 are reprinted from J. W. Poole, *Ind. Eng. Chem.*, **23**, 171 (1951).

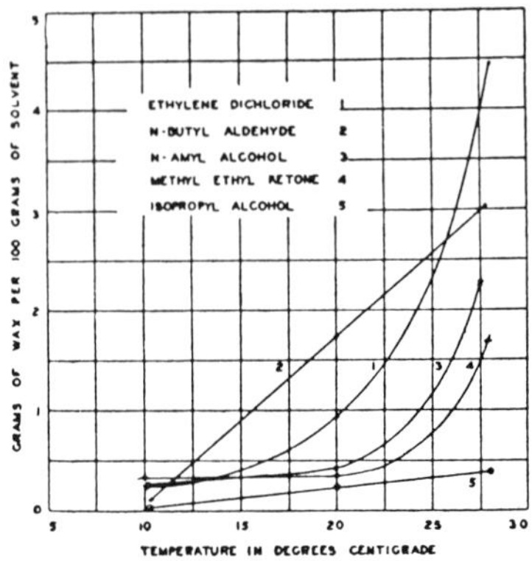

Fig. 1:36. Temperature-Solubility Curves for Paraffin Wax in Various Organic Solvents.

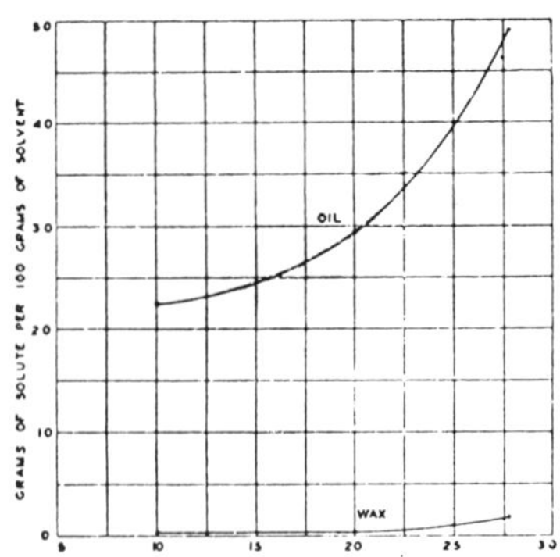

Fig. 1:37. Temperature-Solubility Curves for Paraffin Wax and Mineral Oil in Methyl Ethyl Ketone.

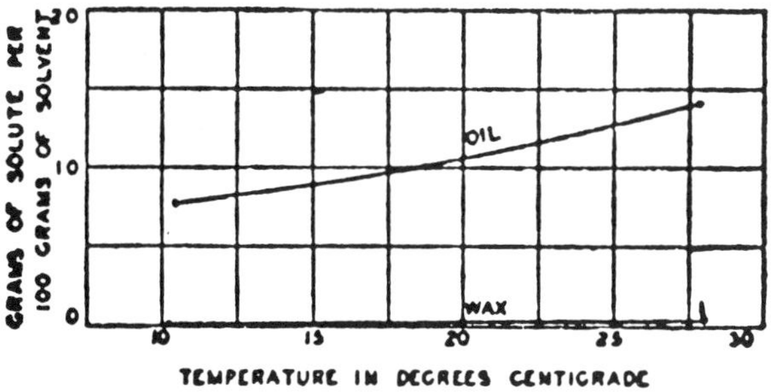

Fig. 1:38. Temperature-Solubility Curves for Paraffin Wax and Mineral Oil in Isopropyl Alcohol.

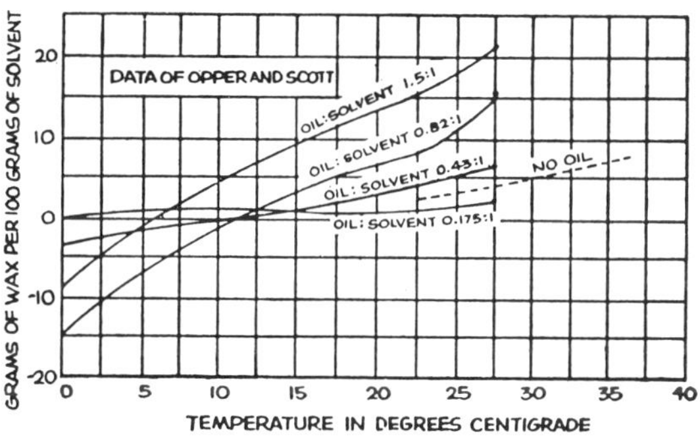

Fig. 1:39. Temperature-Solubility Curves for Paraffin Wax in Solutions of Mineral Oil and Butanol.

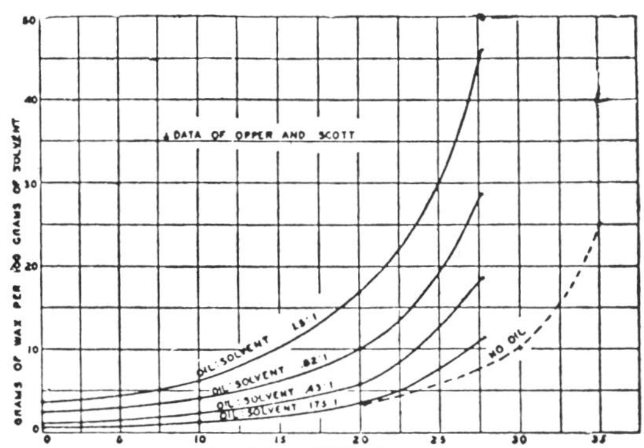

Fig. 1:40. Temperature-Solubility Curves for Paraffin Wax in Solutions of Mineral Oil and Butyl Acetate.

solvents show that minimum viscosity is associated with maximum adsorption.[1]

For any particular solvent, the solubility is inversely proportional to the melting point of the paraffin wax, and for any particular paraffin wax, the solubility decreases with increasing density of the solvent. Benzene, of 0.677 density, dissolves about ten times as much paraffin wax as does machine oil, of 0.908 density. In the liquid state, the paraffin waxes are miscible in all proportions with all petroleum fractions, benzene, and isoamyl alcohol. As the solubility of the paraffins increases more rapidly at higher temperatures, a given lowering of temperature will cause the deposition of more paraffin wax at a high temperature than at a low temperature. The difference between the temperature at which paraffins first separate when a solution is cooled and the solidifying point of the paraffins varies inversely with the temperature of deposition, that is, the lower the concentration of the solution. For solutions that contain more than 15 g per 100 g of solvent, the difference is 1–3°; from 15–5 g, 3–7°; and for less than 5 g, 7–20°.

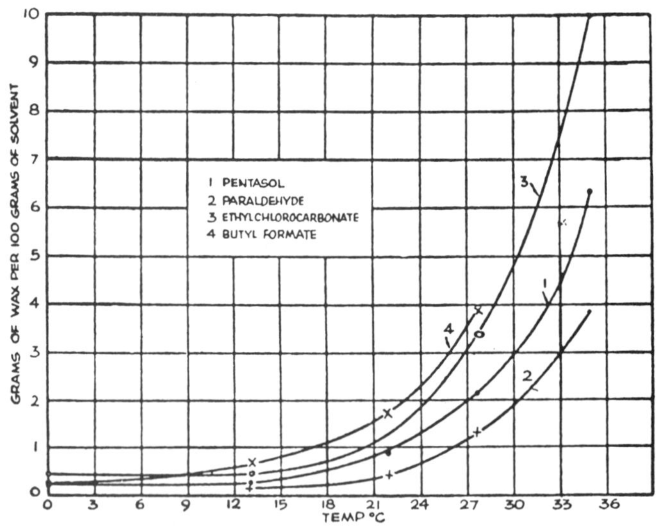

FIG. 1:41. TEMPERATURE-SOLUBILITY CURVES FOR PARAFFIN WAX IN SELECTED SOLVENTS.

The higher the freezing point of the solution, the greater is the proportion of wax that separates out when it is cooled to the freezing point. During solidification, all the solvent is absorbed by the solid paraffins. As this reduces the rise of temperature, a small separation of paraffins at low temperatures is sufficient to cause solidification. The quantity of

[1] M. A. Rakuzin, *Petroleum Z.* **21**, 1489 (1925).

Table 1:17

SOLUBILITY OF PETROLEUM WAXES IN VARIOUS SOLVENTS

Solvent	Wax in Solution, %						Pounds of Wax per 100 lb Solvent					
	60°F.	70°F.	80°F.	90°F.	100°F.	110°F.	60°F.	70°F.	80°F.	90°F.	100°F.	110°F.
133–135°F. AMP Refined Paraffin Wax												
Sovasol No. 6	4.4	8.5	16.0	27.5	43.0	61	4.6	9.2	19.0	38.0	75	156
Turpentine	4.6	9.5	19.2	31.8	46.0	63	4.8	10.6	24.7	47.0	85	170
Pine Oil	0.6	1.4	2.9	6.2	13.4	29	0.6	1.5	3.0	6.6	16	41
125–127°F. AMP Refined Paraffin Wax												
Sovasol No. 3	16.0	26.0	41.0	61.0	81.0	90	19.0	35.0	70.0	156.0	425	—
Sovasol No. 4	13.0	23.0	37.0	56.0	76.0	88	15.0	30.0	59.0	127.0	316	—
Sovasol No. 5	10.0	19.0	32.0	49.0	69.0	83	11.0	25.0	47.0	96.0	222	—
Sovasol No. 6	7.9	14.5	26.5	42.0	61.0	81	8.6	16.5	36.0	72.0	156	—
Sovasol No. 5-A	12.0	21.0	33.0	50.0	70.0	84	14.0	27.0	50.0	100.0	233	—
Benzene	10.0	18.0	35.0	56.0	77.0	90	11.0	21.0	54.0	127.0	335	—
Carbon Tetrachloride	10.0	15.0	31.0	48.0	68.0	83	11.0	18.0	45.0	92.0	210	—
Ethylene Dichloride	0.2	0.8	1.6	7.0	30.0	81	0.2	0.8	1.6	7.5	43	—
Perchloroethylene	10.0	15.0	26.0	42.0	63.0	82	11.0	18.0	35.0	72.0	170	—
Trichloroethylene	12.0	20.0	32.0	50.0	68.0	83	14.0	25.0	47.0	100.0	210	—
Halowax 1000	2.0	4.0	8.0	13.0	31.0	62	2.1	4.2	8.7	15.0	45	—
Turpentine	9.2	17.5	29.2	43.5	60.0	78	10.0	21.2	41.0	77.0	150	—
Pine Oil	1.1	2.5	5.4	11.6	25.2	55	1.1	2.6	5.7	14.0	34	—
S/V 907	2.7	5.1	9.5	17.5	33.0	55	2.8	5.4	10.5	20.0	50	—
Petrolatum, M.P. 120°F.												
Sovasol No.6	10.5	19.3	33.0	51	72	—	11.7	25.0	50.0	104	260	—
Turpentine	12.0	22.0	36.0	54	74	—	14.0	28.0	56.0	117	284	—
Pine Oil	1.8	4.0	8.5	19	40	85	1.8	4.2	9.2	25	67	—
S/V 907	4.5	8.2	16.0	30	55	89	4.8	8.9	19.0	43	122	—

absorbed solvent depends on the size of the crystals, and this, in turn, depends on the previous treatment of the solution.

The use of Figure 1:43 may be illustrated as follows: What is the solubility of paraffin wax in petroleum when the melting point of the wax is 45°C, the temperature of the solution is 26°C, and the average boiling point of the pure distillate is 260°C? The difference between the melting point and the temperature of the solution is 45 – 26, i.e., 19°C. Connect 19 on the right-hand scale with 260 on the left-hand scale and read the

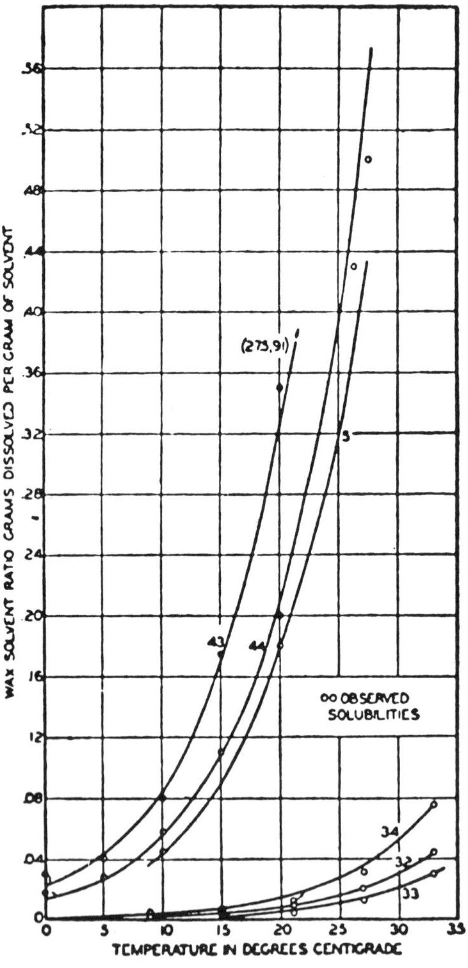

FIG. 1:42. TEMPERATURE-SOLUBILITY CURVES FOR PARAFFIN WAX IN VARIOUS SOLVENTS, SHOWING AGREEMENT BETWEEN OBSERVED AND CALCULATED VALUES.[1]

[1] J. W. Poole and T. A. Mangelsdorf, *Ind. Eng. Chem.* **24**, 1217 (1932).

solubility on the inclined axis as 31 g wax per 100 ml solvent. Similarly, the chart reveals that 100 ml of a distillate, boiling at 190°C, will dissolve 5 g of paraffin wax when the melting point of the wax is 60°C and the temperature of the solution 23°C. Note that one reads solubilities on the right-hand side of the inclined scale when temperature differences are read on the right-hand side of the temperature-difference scale and conversely.

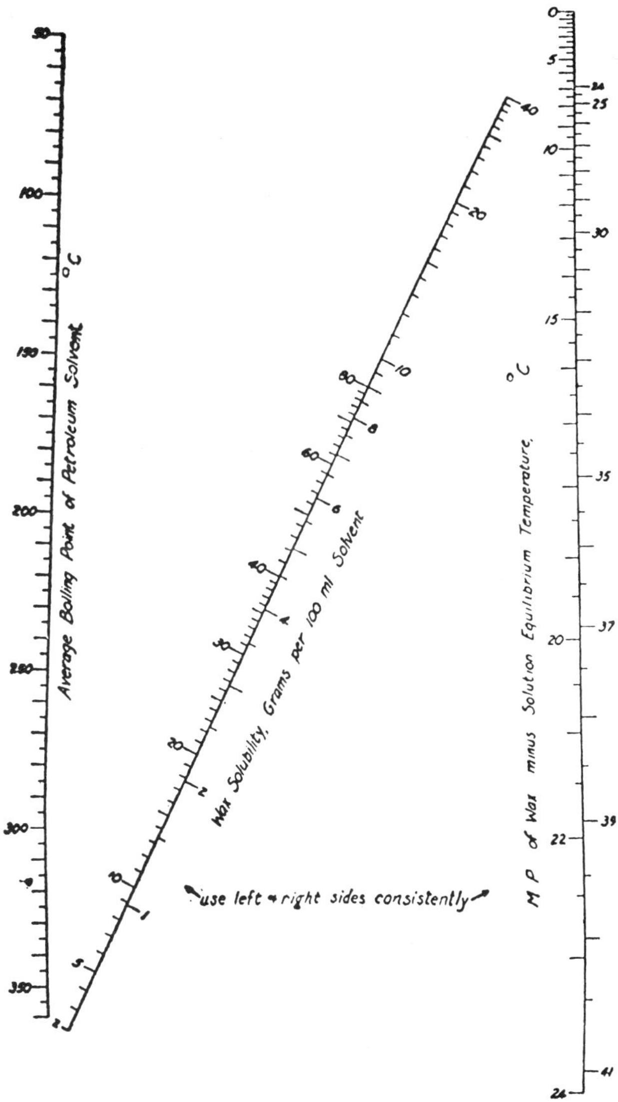

FIG. 1:43. SOLUBILITY OF PARAFFIN WAX IN PETROLEUM FRACTIONS.[1,2]

[1] D. S. Davis, *Ind. Eng. Chem.* **32,** 1293 (1940).
[2] A. Berne-Allen and L. T. Work, *Ibid.* **30,** 811 (1938).

It has been found that the melting point of the wax has a profound effect on its solubility in any given solvent. The higher-melting waxes are less soluble than those of lower melting point.

The solubility of paraffin wax in Pliotite increases with increase in temperature to the melting point of the wax. At the melting point, they become miscible in all proportions. These mixtures (Pliowax) are tougher and give more flexible and lustrous films than the wax alone.

Stability

A) To Light

Many grades of paraffin wax will darken appreciably if allowed to stand in direct, or even indirect, sunlight for any length of time. This property of a wax was often attributed to the crude oil that was its source and it was assumed to be characteristic of all waxes derived from the same crude. However, careful investigation shows that this is not the case. As the pure solid constituents of paraffin wax are remarkably stable to ultraviolet light, they will discolor only if subjected to it for prolonged periods or at elevated temperatures. However, the oil or liquid constituent is far from stable. It is found that, even in the presence of only a small (0.5) percentage of oil in a wax, this oil may be so unstable that its oxidation is sufficient to cause the entire mass of wax to become noticeably darker.

Stability to light is an obvious requirement in waxes. To be salable waxed paper must not yellow. Other articles, including candles, are even more degraded by discoloration and often deteriorate and become unsightly.

Two methods are available for improving light-stability of wax: removing the oil or refining until it is color-stable. As the first method is impracticable, wax producers have to resort to additional refining. After adequate treatment, even 0.5% of residual oil will not be harmful. Thus, the relation between color-stable and color-unstable waxes is the same as that between two theoretically oil-free waxes, to one of which a small percentage of white oil has been added, and to the other similar quantities of an unrefined paraffin distillate.

It follows from this that the quantity of oil present in a wax is not in itself a measure of color-stability. The type of oil must also be considered, for, if the oil is color-stable, even excessive quantities will not be detrimental to color-stability.

Paraffin wax that contains more than 0.1% oil is unstable to ultraviolet rays. Paraffin wax treated with Fuller's earth and exposed to ultraviolet light is light-stable.[1]

B) To Oxidation

In nearly all practical applications of paraffin waxes, they must be melted at one time or another. This is particularly true in the paper industry, which consumes the bulk of the production. For melting waxes,

[1] V. I. Shile, *Neftyanoe Khozyaistvo* **25**, 59 (1933).

Table 1:18

SOLUBILITY OF PARAFFIN WAX (M.P. 53°C.)

1 cc of	Dissolves at 20°C. mg
Ethyl Acetate	1.1
Acetone	1.2
Alcohol (96%)	1.9
Ether	83.4
Petroleum Ether	200.0
Ligroin	244.0
Chloroform	246.0
Benzene	285.0
Carbon Tetrachloride	317.0

The solubility of paraffin wax in Pliolite increases with increase in temperature to the melting point of the wax. At the melting point, they become miscible in all proportions. These mixtures (Pliowax) are tougher and give more flexible and lustrous films than the wax alone.

Table 1:19

SOLUBILITY OF PARAFFIN WAX (M.P. 124.5°F.) IN ORGANIC SOLVENTS

Solvent	Temperature °F.	Grams Wax per 100 cc Solvent
Cyclohexanol	87	0.8800
	68	0.2940
	58	0.1420
	47	0.0823
Isopropyl Alcohol	85	0.2400
	66	0.0600
	52	0.0240
	39	0.0094
n-Propyl Alcohol	80	0.2140
	60	0.0535
	46	0.0214
	32	0.0086
tert-Butyl Alcohol	80	0.2920
	58	0.0730
n-Butyl Alcohol	77	0.3600
	59	0.0900
	45	0.0360
Fusel Oil	80	0.4620
	58	0.1150
	42	0.0460

Table 1:19(continued)

Solvent	Temperature °F.	Grams Wax per 100 cc Solvent
tert-Amyl Alcohol	72	0.5520
	53	0.1380
	42	0.0552
Diacetone Alcohol	97	0.2560
	78	0.0640
	65	0.0256
Ethyl Alcohol	98	0.3580
	83	0.0895
	72	0.0358
Pentasol	83	0.8280
	61	0.2070
	50	0.0828
n-Propyl Formate	79	1.0480
	65	0.3160
	47	0.0790
Isopropyl Formate	75	0.4120
	57	0.1030
	45	0.0412
n-Butyl Formate	75	0.8980
	57	0.2250
	46	0.0898
	37	0.0475
Methyl Acetate	85	0.2780
	66	0.0695
	57	0.0278
	35	0.0086
Phenyl Acetate	96	0.3500
	75	0.0875
	62	0.0438
Ethyl Oxalate	112	0.6600
	93	0.1650
	80	0.0660
Chloroethyl Carbonate	70	0.5240
	50	0.1310
	39	0.0524
β-β-Dichloroethyl Ether	90	0.2800
	70	0.0700
	57	0.0280
Trichloroethylene	59	2.8600
	36	0.9160
	7	0.2290
Methyl *n*-Hexyl Ketone	65	1.0640
	48	0.2660
	38	0.1064
	30	0.0546
Isobultyl Methyl Ketone Hexone	83	2.0700
	72	0.8280
	46	0.2020

Table 1:19 (continued)

Solvent	Temperature °F.	Grams Wax per 100 cc Solvent
Cyclopentanone	74	0.5160
	57	0.2290
	45	0.0516
Methyl Cyclohexanone	83	3.0800
	54	0.3520
	35	0.0880
	22	0.0352
Dipropyl Ketone	61	0.8820
	43	0.2210
	33	0.0882
	23	0.0414
Cyclohexane	62	7.7600
	50	2.4480
Benzol (90%)	32	0.7440
	18	0.1860
Cresol	96	0.6080
	75	0.1520
	62	0.0608
Tetralin	56	2.5100
	43	1.1840
	26	0.3940

Table 1:20

SOLUBILITY OF REFINED PARAFFIN WAX (M.P. 130–132°F.) IN ORGANIC SOLVENTS

Solvent	Temperature °F.	Grams Wax per 100 cc Solvent
Butane	68	3.6700
	43	0.5480
	45	0.5960
	42	0.4880
	14	0.0660
	–4	0.0150
Propane	39–40	0.2380
	45	0.2600
	23	0.0548
	31	0.0824
	6	0.0189
	9	0.0189
	8.5	0.0189
	17.5	0.0330
	–2 to 2	0.0105
	–10	0.0056

Table 1:20 (continued)

Solvent	Temperature °F.	Grams Wax per 100 cc Solvent
Cylinder Stock Naphtha (P.B. 128–348°F.)	39	0.5000
	44	0.7500
	27	0.2500
	17	0.1250
	60	2.4900
	50	1.2500
	40	0.6200
	6	0.0610
	-10	0.0200
35% Acetone + 65% Benzol	62	0.507
	41	0.0995
	21	0.0205
	12	0.0100

Table 1:21

SOLUBILITY—GRAMS PARAFFIN WAX PER 100 cc SOLVENT

Temperature °C.	Pentane	Hexane	Heptane	Octane	Isodecane
0	—	2.77	1.37	0.99	—
5	—	3.69	2.18	1.69	0.94
10	5.11	4.81	3.55	2.90	1.44
15	6.94	6.07	5.06	4.24	2.74
20	9.53	8.31	7.18	5.93	4.98
25	17.16	16.23	14.36	11.66	9.17

Table 1:22

SOLUBILITY—GRAMS PARAFFIN WAX PER MOL SOLVENT

Temperature °C.	Pentane	Hexane	Heptane	Octane	Isodecane
0	—	3.61	2.01	1.60	—
5	—	4.81	3.22	2.73	1.84
10	5.83	6.28	5.23	4.67	2.84
15	7.92	7.91	7.48	6.84	5.40
20	10.87	10.83	10.57	9.58	9.80
25	19.48	21.33	21.06	18.81	18.03

many types of equipment are employed, but, in general, they consist of iron or steel assemblies heated with steam, direct flame, or electricity. As steam is the commonest source of heat and as even exhaust steam has a temperature of at least 212°F, the temperature of the molten wax is frequently about 200 or 210°F. At such elevated temperatures, the wax will oxidize with great rapidity, unless extreme care is exercised.

Exposure of paraffin wax to oxidation (air or chemical) or high-frequency discharges reduces its hardness and lowers its melting point.

Oxidation of paraffin wax depends on a number of variables encountered in usual practice. These are temperature, time, surface of exposure, and contact with metals and impurities. It is surprising how little attention is paid to these factors in both the design and the operation of commercial equipment. As previously stated, paraffin wax is closely related in chemical composition to low-viscosity paraffin oils. Few operators would expose light oils, such as these, particularly if used as lubricants, to the conditions that are often encountered by paraffin wax. Yet they expect that wax will not be harmed by careless treatment. This is probably due to the erroneous assumption that a solid is much more stable than a liquid.

1) *Temperature*:

Paraffin waxes will show a gradual increase in stability as the melting point rises. This is a reasonable observation because products of higher molecular weight, or higher viscosity, are present in these grades. However, the difference in stability is not sufficient to permit a high-melting wax (such as one of 135°F AMP) to be treated with less care than one that melts at a lower temperature (122°F). Crude scale waxes are less stable than refined waxes of the same melting point and, in general, they will deteriorate about twice as fast as these more refined products. Most measurements of oxidation are relative, but they are usually based on darkening color, lowered melting point, loss of tensile strength, and the formation of products, such as aldehydes, ketones, organic acids, and esters, most of which have a pronounced odor. Initially, oxidation probably causes a breakdown of the small oily fraction of the wax, but eventually it results in degradation of the solid constituents as well.

The effect of temperature and other variables also depends on the conditions of exposure. Under normal conditions, little oxidation occurs at 150°F, but when a temperature of 200°F is reached, serious oxidation begins, and above this temperature it proceeds at a rapid rate. If it is permissible for the job in hand, temperatures should be held below 200°F. Best results are obtained with thermostatic control, whatever the source of heat may be. Low-pressure steam should always be used for heating to avoid local overheating and consequent oxidation.

2) *Time of Exposure*:

Similarly, the time of exposure to conditions of oxidation must be held at the minimum. Naturally, the longer a molten wax is held at elevated temperatures, the more extensive will be the oxidation. Equipment that is not used 24 hours a day should be shut down during idle periods. This should be usual practice in all plants and should be followed for all shut-downs over week-ends. Losses of time in reheating solid wax will be more than compensated by savings in wax and the maintenance of uniform high quality. In this connection, it should also be mentioned that waxing equipment should be cleaned periodically to remove accumulations of oxidized wax. The presence of highly oxidized material in fresh wax, aside from contaminating the good wax, will catalyze further oxidation and accelerate further degradation. Where the quantity of wax used is large, cleaning may be less frequent. The simplest means of control is to resort to cleaning when the odor of the wax bath has become distinctly rancid.

3) *Area of Surface*:

The third variable affecting rate of oxidation is the exposed surface for the molten wax. As it is the quantity of oxygen reacted with the wax that is important, the minimum area of contact between air and wax is desirable. Samples stored in containers of about the same surface area showed that oxidation in the open is four times as rapid as that in a closed container. Equipment used for handling molten wax should always be designed to provide for the least possible exposed surface area. Another precaution that should be observed is to reduce to the minimum any wax spray or foam.

In making waxed paper, drippings from coating rolls usually cannot be controlled, but often lines deliver wax to heating pans above the surface which causes spattering. Extending the lines below the surface will correct this fault. The spraying of paraffin wax is always conducive to excessive oxidation as the surface area is tremendously increased. If other means of application are available for a given job, they should be chosen in preference to spraying.

4) *Contact with Metals*:

The presence of certain metals markedly affects the speed and extent of many chemical reactions. Oxidation of wax is no exception and the type of metal used in the equipment is, therefore, of considerable importance in preventing oxidation. In general, ferrous metals are satisfactory, and copper-bearing metals are not. There is apparently a difference among the various ferrous metals, because cast and wrought iron are less deleterious than stainless steel. Similarly, copper is much less desirable than brass and bronze. In order of desirability, the metals usually encountered may be listed as follows:

Cast and wrought iron	Nickel
Carbon steel	Brass
Stainless steel	Bronze
Aluminum	Copper

From this it is apparent that fittings made of copper and copper-bearing metals should be eliminated from waxing equipment wherever possible. (If possible, items for replacement should always be of iron or steel.)

Samples of refined paraffin, white crude scale wax, and microcrystalline waxes have been subjected to storage at several temperatures and in contact with metals. In addition, several of these waxes have been stored with no surface exposure. From the data obtained, the following conclusions can be drawn.

1. Oxidation of petroleum waxes causes deterioration: This is evidenced by darkening, increase in acid number, development of saponifiable bodies, loss of tensile strength, lowering of melting point, and loss of hardness.

2. High-melting-point waxes are more stable than those of lower melting point: This is particularly true of the microcrystalline waxes which are only slightly affected even at 250°F for 7 days.

3. Refined paraffin waxes are more stable than crude scale waxes of similar melting points because the more oxidizable fractions have been removed during refining.

4. Refined paraffin and crude scale waxes stored at 150°F exhibit little deterioration even after 21 days of exposure. Serious oxidation does not occur until the temperature of storage reaches 200°F. Above this temperature, oxidation is very rapid.

5. The surface of exposure has a pronounced effect on oxidation. In actual practice, it should be reduced to the minimum. This refers to agitation, aeration, spraying, and construction of equipment.

6. There is a period in the oxidation of waxes in which little change, other than darkening, occurs. Eventually, drastic oxidation commences with the results mentioned in 1. This is found when paraffin waxes are stored at 250°F, at which temperature only 3–5 days are required for serious oxidation to begin.

7. Drastic oxidation of microcrystalline waxes does not commence at 250°F for a considerably longer period than for paraffin waxes. Less care, therefore, is needed in the handling of these waxes.

Antioxidants for Paraffin Wax[1]

Baths of paraffin wax, such as those used commercially for the waterproofing, sizing, or impregnating of paper, tend to foul and cause discoloration or undesirable odors which are due to oxidation of the paraffin, particularly after long continued use at elevated temperatures. The addition of certain substances, even in minute quantities, to a bath of paraffin waxes or paraffin oils, prevents such oxidation and permits use of the bath for long periods of time without fouling.

Oxidation of a paraffin bath, operated at 235–240°F, can be prevented by the addition of extremely small percentages of natural and artificial

[1] Vore, U.S. Patents 2,325,085 and 2,325,167 (1943).

asphalts, pitches and their derivatives, and asphalt-like substances, such as gilsonite, manjak and other asphaltites, and petroleum residue asphalts. It is understood that the percentage of antioxidant required increases in accordance with the temperature of the bath, and at the higher temperatures the increase is much more rapid.

Table 1:23

ANTIOXIDANTS FOR PARAFFIN BATHS

Antioxidant	*Approximate % required to prevent oxidation of industrial paraffin bath at* 230–240°F	160–210°F
Egyptian Asphalt	0.020	0.010
Cuban Gilsonite	0.040	0.020
Petroleum-Residue Asphalt	0.006	0.003
Isthmian Asphalt	0.020	0.010
Manjak	0.020	0.010
Utah Gilsonite	0.020	0.010
Fatty-Acid Pitch	0.500	0.250
Cottonseed Pitch	0.100	0.050
Stearin Pitch	0.500	0.250
Crude Montan Wax	0.040	0.020
Crude Ozokerite	0.100	0.050
California Asphalt	0.200	0.100
Carnauba Wax	0.600	0.300
Candelilla Wax	0.600	0.200
Esparto Wax	0.800	0.400

In addition to the antioxidants listed in Table 1:23 it has been found that about 0.6% ceresin breaks down the crystalline structure of the paraffin and tends to make the wax harder and more resistant to water. A bath of paraffin wax melting at 133–138°F, containing about 0.6% ceresin and about 0.04% petroleum-residue asphalt, is particularly suitable.

The patent literature contains many references to other antioxidants, principally organic compounds.

Autoxidation of paraffin wax, at ordinary temperature, is slight and slow and the products formed diffuse into the air before detection is possible. Small quantities of hydroquinone, malachite green, or sulfur prevent the oxidation.

Table 1:24

SPECIFIC REFRACTION OF COMMERCIAL PARAFFIN WAXES IN THE LIQUID STATE

M.P. of Wax °F	°C	*n*	d	*t*°C	*Mol. Wt.*	*Approx. Formula*	r_D At *t*°	At 20°C	Calcd.
121	49.4	1.4349	0.7780	60	332	$C_{22.55}H_{49.1}$	0.3353	0.3343	0.3341
131	55.0	1.4355	0.7788	60	340	$C_{24.1}H_{50.2}$	0.3352	0.3342	0.3338
126	52.2	1.4352	0.7800	60	355	$C_{25.2}H_{52.4}$	0.3349	0.3339	0.3340
136	57.8	1.4370	0.7815	60	372	$C_{26.4}H_{54.8}$	0.3351	0.3341	0.3337
141	60.6	1.4294	0.7690	85	376	$C_{26.7}H_{55.4}$	0.3353	0.3337	0.3337

Table 1:25

SPECIFIC REFRACTION OF COMMERCIAL PARAFFIN WAXES IN THE SOLID STATE

Temperature		r_D *of Wax Melting at:*				
°F	°C	121°F (49.4°C)	126°F (52.2°C)	131°F (55.0°C)	136°F (57.8°C)	141°F (60.6°C)
30	−1.1	0.3434	0.3405	0.3413	0.3400	—
40	4.4	0.3429	0.3403	0.3410	0.3410	—
50	10.0	0.3430	0.3401	0.3412	0.3403	0.3415
60	15.6	0.3437	—	0.3410	0.3405	0.3414
70	21.1	—	—	0.3411	0.3405	0.3413
80	26.7	—	—	—	—	0.3414
100	37.8	—	—	—	—	0.3414

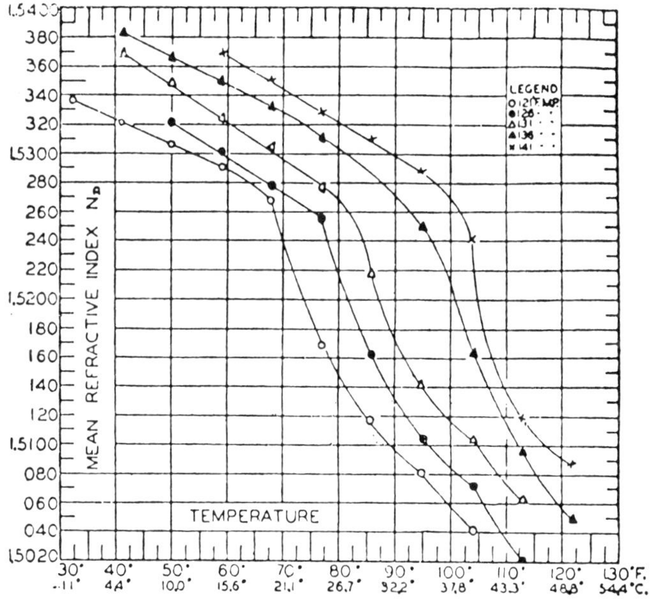

FIG. 1:44. MEAN REFRACTIVE INDEX OF COMMERCIAL PARAFFIN WAXES IN SOLID STATE.[1]

WAX – OIL MIXTURES

In view of the relationship between light-stability and oil content, it would perhaps be well to explain the term "oil content." A determination of the so-called "expressible oil and moisture" was given tentative recognition by the ASTM. This method measured the quantity of liquid (chiefly oil and low-melting wax) squeezed from a sample of wax under standard conditions at pressure 1000 psi. Because of poor reproducibility, this test has been abandoned and an improved method suggested instead.

[1] J. M. Page, *Ind. Eng. Chem.* **28**, 856 (1936).

The new test actually measures the quantity of oil present in the wax by recrystallizing the wax from a solvent under prescribed conditions and weighing the oil held in solution. The oil is defined in terms of pour point to make certain that excessive quantities of wax are not removed at the same time. Highly reproducible results can be obtained by this method, which shows that refined waxes consistently contain less than 0.5% oil. As determined by the old "expressible oil and moisture" method, values ranging from 0–1.0% were often reported for the same wax.

The solubility of paraffin wax in oil increases very markedly at temperatures approximately 10°C below the melting point of the wax. The low-melting waxes are more soluble in oil than the high-melting waxes. Mixtures of high- and low-melting waxes have solubilities intermediate between the solubilities of the components. The increase in solubility of the mixtures in oil at temperatures approximately 10°C below their melting point is as marked as for pure waxes.

***Table* 1:26**

RELATION OF OIL CONTENT TO MELTING POINT OF PARAFFIN WAX[1]

Type of Wax	*Pressed Oil, %*	*R.I. at* 25°C	*M.P.* °C
Parowax	0.16	1.4481	51.7
Refined Wax	0.28	1.4487	50.0
Refined Wax	0.25	1.4485	50.0
Refined Wax	0.17	1.4481	51.7
Refined Wax	0.16	1.4479	51.7
Refined Wax	0.09	1.4476	54.4
Refined Wax	0.08	1.4475	54.4
Scale Wax	2.58	1.4570	48.3
Scale Wax	2.56	1.4568	48.3

The melting point under pressure of 1–200 atm. can be calculated from the following formula:

$$t_p = t_o + 0.029776(P-1) - 0.0000523(P-1)$$

Effect on Plastic Properties of Waxes[2]

Oil has a considerable effect on the properties of petroleum wax, particularly the microcrystalline waxes. Thirty blends of technical white oil with three typical waxes were examined with respect to melting point, penetration, crystallinity index, plastic point, ductility, tensile strength, compressive strength, and flexibility.

Equipment and testing procedures were developed for most of the tests although several of the methods were only modifications of standard testing methods. Equipment and procedure for a test for flexibility have been developed in which a wax stick is bent through a 60° angle until it

[1] L. D. Myers and G. Stegeman, *Ind. Eng. Chem.*, **20**, 640 (1928).
[2] W. L. Nelson and L. D. Stewart, *Ind. Eng. Chem.*, **41**, 2231 (1949).

breaks. Tests for compressive strength were found to be more dependable than those for tensile strength. Oil appears to enter the space between the interlocking crystals of paraffin wax, altering the structure only by dissolving some of the wax; whereas in microcrystalline waxes, oil enters and dilutes the continuous phase gradually altering the properties of the wax, and possibly precipitating additional microcrystals. Only with extremely hard or high-melting microcrystalline waxes does oil improve the plasticity or flexibility.

Properties of Wax - Oil Blends

The melting point of waxes is reduced in a nearly linear manner by the addition of oil. (See Figure 1:45). This indicates that the oil merely reduces the relative amount of crystalline phase in proportion to the amount of oil added. The steeper slope of the brown-wax blends indicates that the brown waxes inherently contain more continuous phase than the other microcrystalline waxes. The position of the white-oil blends below corresponding black-oil blends may be explained by the lower viscosity of the white oil. The more rapid rate of reduction of melting point for the paraffin-oil blends is partly explained by the dissolving of paraffin wax in the oil. Also, the oil acts as a lubricant and does not adhere to the thermometer bulb as do the other wax-oil blends.

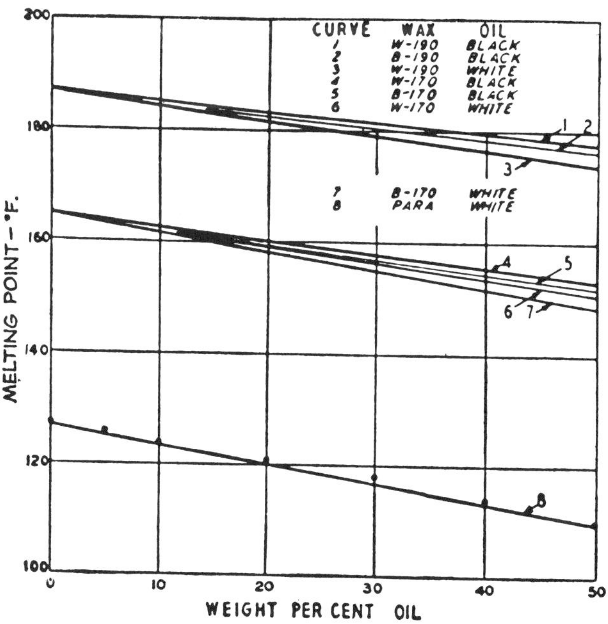

FIG. 1:45 MELTING POINTS OF WAX-OIL MIXTURES.

Table 1:27

MELTING POINT AND PENETRATION OF BLENDS OF BLACK OIL AND MICROCRYSTALLINE WAX

% Oil	*White 170 Blends* M.P.	Pen.	*Brown 170 Blends* M.P.	Pen.	*White 190 Blends* M.P.	Pen.	*Brown 190 Blends* M.P.	Pen.
None	165	18	165	18	187	4	187	4
5	164	22	164	22	187	7	187	7
10	163	28	163	28	186	10	186	10
20	161	47	161	47	184	17	184	19
30	159	74	158	76	182	27	182	30
40	156	123	155	142	180	40	180	48
50	153	250	152	240	178	56	177	73

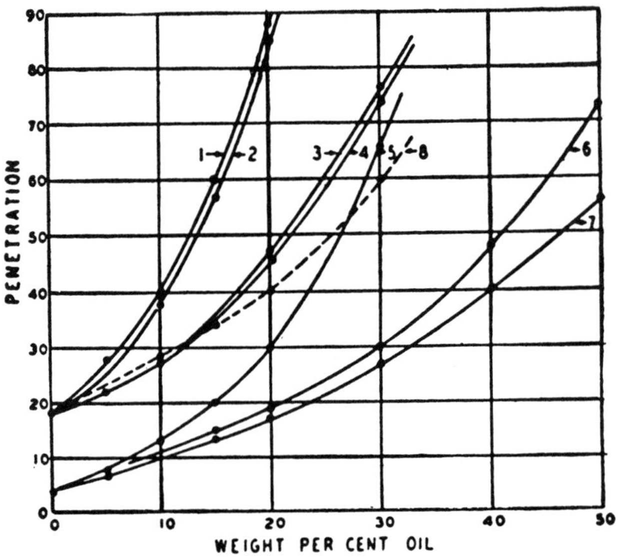

FIG. 1:46 PENETRATION CURVES OF WAX-OIL BLENDS.

Curve	*Wax*	*Oil*
1	B-170	White
2	W-170	White
3	B-170	Black
4	W-170	Black
5	W-190	White
6	B-190	Black
7	W-190	Black
8	Paraffin	White

The penetration curves shown in Figure 1:46 exhibit the same general arrangement or relationship to one another as the melting-point curves of Figure 1:45 except that the paraffin-wax curve has a flatter slope than most of the others. But for the dissolving that occurs when oil is added to paraffin wax, the structure of the wax remains substantially the same.

Table 1:28

Physical Properties of Wax—Oil Blends

Percentage of Oil in Blend	M.P. °F.	Penetration 1/100 cm.	Contraction %	Plastic Point	Tensile Strength lb/sq in.	Compressive Strength lb/sq in.	Ductility cm	Flexibility, Number of Bends
Paraffin Wax and Technical White Oil								
None	127	18	13.30	100	140	216	0	0
5	126	23	12.50	100	10	75	0	0
10	124	29	11.50	100	0	0	0	0
20	121	40	10.50	100	0	0	0	0
30	118	60	9.05	100	0	0	0	0
40	114	200	7.25	100	0	0	0	0
50	100	—	6.00	100	0	0	0	0
White 170 Wax and Technical White Oil								
None	165	18.0	11.00	70	224	232	1.0	86
5	164	27.5	10.70	50	128	132	1.3	112
10	163	38.5	10.10	35	80	80	1.3	60
20	161	86.0	9.17	–5	40	40	1.7	40
30	159	209.0	8.64	–34	24	24	1.8	30
40	155	300.0	7.55	–47	10	10	1.4	—
50	151	—	6.86	–52	—	—	—	—
Brown 170 Wax and Technical White Oil								
None	165	18	10.55	45	176	184	1.5	220
5	164	28	10.34	35	84	88	1.5	130
10	163	40	10.28	25	50	48	1.5	92
20	161	88	9.16	–20	16	16	1.5	64
30	158	215	8.18	–43	—	—	1.4	40
40	154	—	7.49	–58	—	—	1.0	—
50	149	—	6.79	–65	—	—	0.9	—
White 190 Wax and Technical White Oil								
None	187	4	11.40	118	300	720	0	0
5	186	8	11.40	110	226	416	0	0
10	185	12	11.04	100	180	282	0.3	40
20	183	30	10.52	84	96	120	0.7	92
30	180	66	10.24	30	48	56	1.0	80
40	177	161	8.76	–20	28	26	0.8	30
50	174	300	8.08	–43	16	16	0.6	16

Table 1:28 shows the results of all tests on white oil blends and Figures 3–7 show the same data plotted as curves.

The dotted contraction curves marked *A* in Figure 1:47 were computed from the total contractions (solid lines) of the blends by subtracting the contraction of the oil that had been added. The contraction due to cooling and solidification of wax, C_w per cent, in terms of total contraction, C_t per cent, contraction due to cooling of oil, C_o per cent, and the percentage of oil, *P*, is:

$$C_w = \frac{100\,C_t - PC_o}{100 - P}$$

C_t and C_o were measured experimentally and C_o is approximately 2% for 60°F.

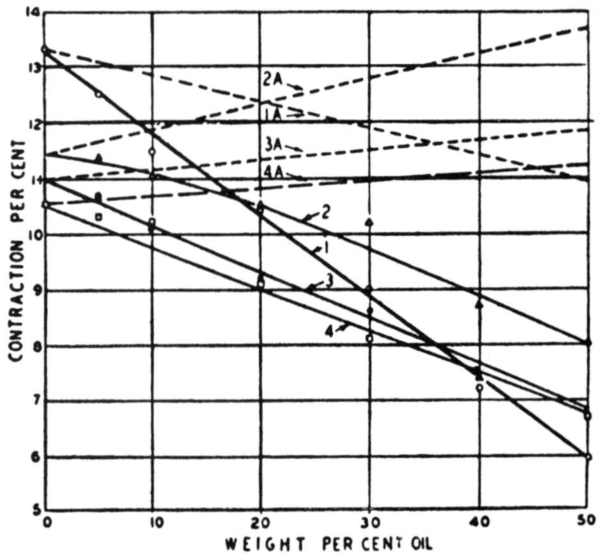

FIG. 1:47 PERCENTAGE CONTRACTION OF WAX-OIL MIXTURES.

Curve	*Wax*	*Oil*	*Curve*	*Oil Free*
1	Paraffin	White	1*A*	Paraffin
2	W-190	White	2*A*	W-190
3	W-170	White	3*A*	W-170
4	B-170	White	4*A*	B-170

The addition of oil in all cases reduces the total contraction of the wax blends (Fig. 1:47). The apparent contraction of the wax part of the microcrystalline wax blends (dotted lines) increases as oil is added but decreases for paraffin-oil blends. The dotted lines for the contraction of wax alone are smoothed data, presented only to indicate the general behavior of the wax portion of the blend. The behavior of the microcrystalline wax blends appears to indicate that additional crystalline substance separates as oil is added, because of more and more contraction due to crystallization. The reverse behavior of the dotted line for paraffin-wax indicates that oil dissolves some of the paraffin wax and thus there is less crystallization and less contraction.

The plastic point curves of Figure 1:41 are of interest primarily because the addition of oil did not alter the plastic point of the paraffin wax. The plastic points of the microcrystalline waxes are sharply reduced by the addition of oil and the three waxes generally arrange themselves in the order of their melting points. In microcrystalline waxes, the oil appears to enter and to alter the properties of the continuous phase, whereas in paraffin wax the original crystalline structure continues to dominate the behavior of the blend.

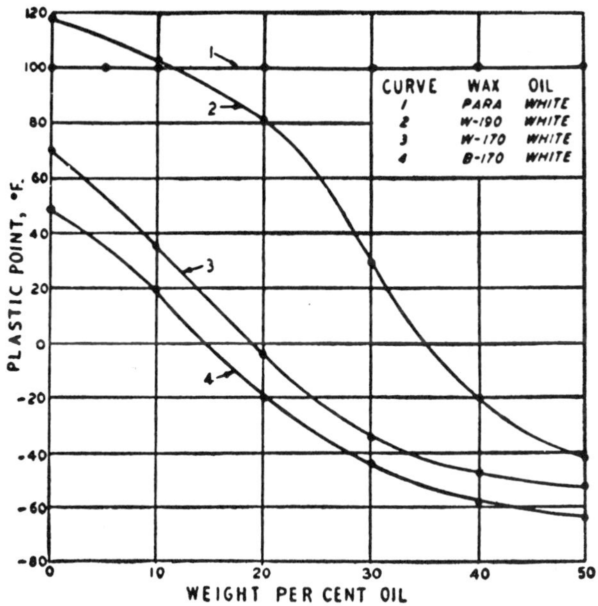

FIG. 1:48 PLASTIC POINT OF WAX-OIL MIXTURES.

The typically plastic White 170 and Brown 170 waxes have nearly the same tensile and compressive strengths, and the White 190 wax shows the same when enough oil has been added (30–50%) to soften its inherently plastic continuous phase. The tensile strength of paraffin-wax blends is always lower than the compressive strength (see Figure 1:49).

The ductility of White 170 and 190 waxes reaches its maximum with about 30% oil, that of Brown 170 wax is reduced by the addition of oil, and none of the paraffin-wax blends exhibits ductility. The continuous phase of waxes W-170 and W-190 is apparently so hard at 77°F that the oil dilutes it to a phase that is more plastic or ductile, whereas the continuous phase of wax B-170 is already sufficiently soft so that oil dilutes it to a phase that is too thin to exhibit maximum ductility (see Figure 1:52).

Flexibility at 77°F, as shown in Figure 1:53, is not imparted to paraffin wax by additions of oil. Oil increases the flexibility of white microcrystalline waxes to the maximum but if more is added, the adhesiveness of the continuous phase is inadequate for continued flexure. The results for flexure are similar to those for ductility except that less oil is required in attaining maximum flexibility. The flexure test is a more severe one and, therefore, it requires a somewhat stronger or harder matrix than that used for ductility.

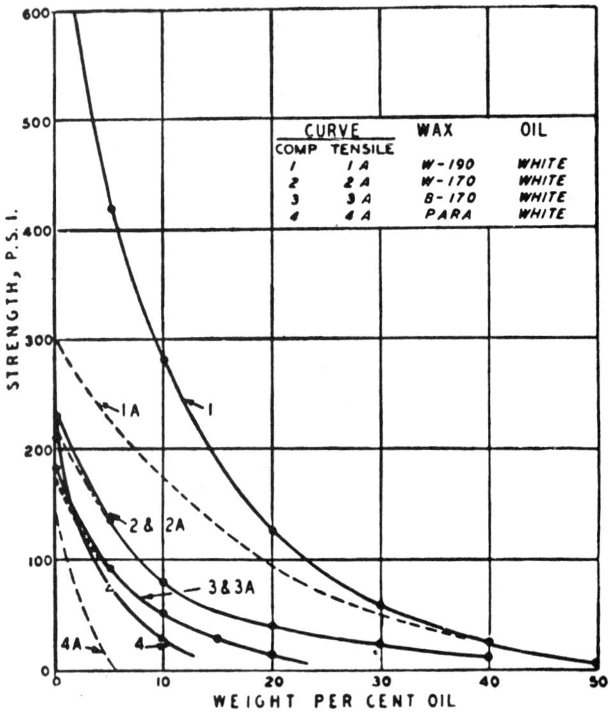

FIG. 1:49 TENSILE AND COMPRESSIVE STRENGTH OF WAX-OIL BLENDS.

Conclusions

Oil alters the melting point and percentage of contraction of waxes in nearly direct proportion to the amount of oil. Softness is increased by the addition of oil as evidenced by penetration, plastic-point, and compressive-strength tests. Paraffin wax is partly dissolved by oil, whereas the crystalline portion of some microcrystalline waxes may be increased by the addition of oil. Apparently, additional crystalline substance separates from microcrystalline waxes as the continuous phase of these particular waxes is diluted with oil. (See further discussion in Chapter 2).

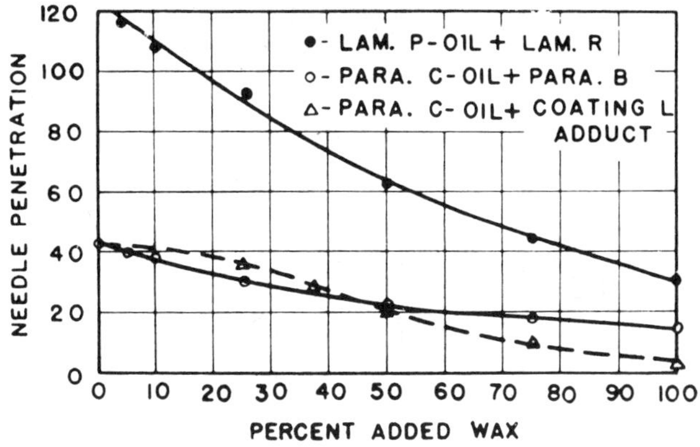

FIG. 1:50 Penetration of blends of similar waxes and oil

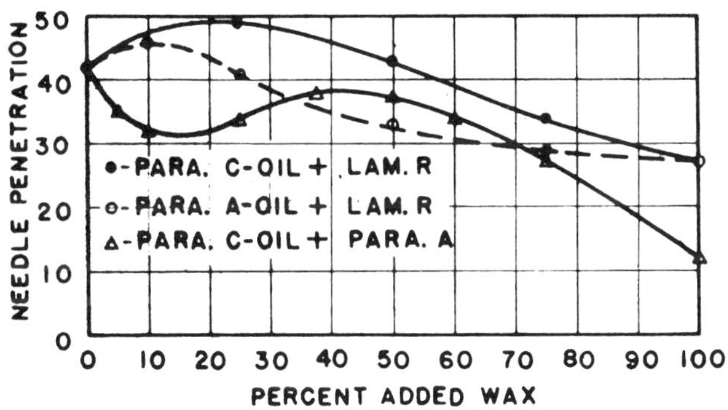

FIG. 1.51 Penetration of blends of dissimilar waxes and oil

Table 1:29

Bleeding Number of Microcrystalline Wax–Oil Blends

	Wax					
	Laminating P		Laminating R		Coating L	
Added oil, %	Cone[a] pen.	Bleed. no., mm.	Cone[a] pen.	Bleed. no., mm.	Cone[a] pen.	Bleed. no., mm.
0	18	0	15	0	4	0
20	48		38		15	2
30	74	3	55	4	42	2
40	120	8	86	9	60	3
50	156	13	138	11	76	8
60	194	2	164	18	95	10
70	Soft	33	210	24	151	21
80	Soft	54	Soft	4[illegible]	180	32

[a] ASTM (D 937-58).

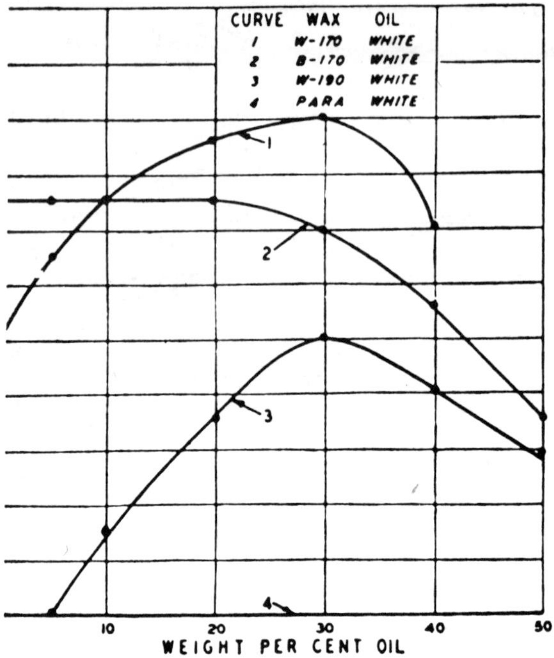

FIG. 1:52 DUCTILITY OF WAX-OIL BLENDS.

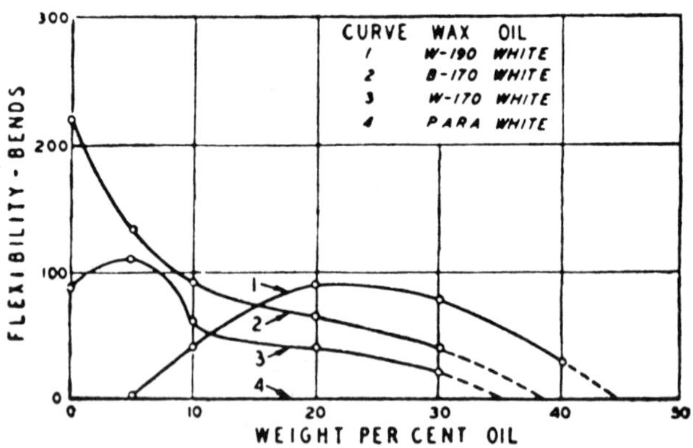

FIG. 1:53 FLEXIBILITY OF WAX-OIL MIXTURES.

Blends of microcrystalline wax and oil appear to exhibit a noncrystalline continuous phase throughout. Paraffin wax contains substantially no amorphous phase and oil merely dissolves some wax and pervades the space between the interlocking crystals. The noncrystalline phase dominates the behavior of the microcrystalline waxes and the oil appears to enter only the continuous phase. The exact nature of the continuous phase cannot be ascertained from the results of tests, but the behavior after contraction indicates that this phase contains wax as well as oil, and in an amount larger than can be explained by the plain solubility of wax in oil. Explanations for the large content of wax in this phase are the existence of gels, supersaturated solutions, or the great solubility of the lowest-melting wax that may be present in the original mixture. Also, crystallization inhibitors suppress crystal formation and thus, on solidification, more wax may be left in the continuous phase. The behavior of brown waxes as compared with white ones of the same melting point, in all tests but the one for ductility, appears to indicate a larger continuous phase. Colored petroleum substances are generally considered to act as inhibitors of crystallization. Oil imparts ductility and flexibility to hard microcrystalline waxes and the harder the wax the more oil is required to impart these properties.

Tests for compressive strength are more reproducible and sensitive than those for tensile strength in indicating the structure and the plasticity of waxes. The predominance of the continuous phase of typical plastic substances permits them to exhibit the same tensile as compressive strength. Nonplastic waxes tend to shatter during the compression test whereas plastic waxes flow.

The test for flexibility presented here is thought to be the best method of those investigated for indicating plasticity. It involves a combination of compressive strength and pliability. The results for compressive strength and plastic point indicate that the maximum flexibility of a given wax is attained when these two properties are adjusted so that the ratio of the compressive strength to the melting point less the plastic point is about equal to unity:

$$\frac{C.S.}{M.P. - P.P.} = 1 \text{ (approximately)}$$

Two conditions seem necessary for the existence of a plastic wax: 1. Small crystals that do not interlock, and 2. A continuous noncrystalline phase throughout which the crystalloids are dispersed. Given these two conditions, plasticity is governed by the ratio of crystalline to noncrystalline matter as affected by inhibitors, oils, petrolatums, etc.; the melting point and the melting range of the wax components; and the temperature at which the wax is used as compared to the melting point.

MISCELLANEOUS DATA

Contact Angles of Water Against Paraffin Wax[1]

Advancing Angle	*Std. Dev.*	*Receding Angle*	*Std. Dev.*
110.7°	1.5°	103.3°	2.6°

[1] J. L. Ihrig, *J. Chem. Educ.* **34**, 196 (1957).

Table 1:30

AIR CONTENT OF PARAFFIN WAX[1]

M.P. (°F)	*Volume % of Air*
121	2.7
126	1.8
131	2.2
136	1.9
141	2.3

Table 1:31

ANILINE POINTS OF PARAFFIN WAX

ASTM Melting Point		*Apparent Oil*	*Observed Mol. Wt.*	*Aniline Point*	*Calculated Mol. Wt.*	*Corresponding Hydrocarbon*
°F	°C	%		°C		
121	49.4	0.5	332	116.7	330	$C_{23.55}H_{49.1}$
126	52.2	0.5	340	117.8	342	$C_{24.1}H_{50.2}$
131	55.0	0.3	355	119.0	359	$C_{25.2}H_{52.4}$
136	57.8	0.3	372	121.0	370	$C_{26.4}H_{54.8}$
141	60.6	0.3	376	122.2	375	$C_{26.7}H_{55.4}$

The apparent Poisson constant of paraffin wax is 0.50.

Electrical Properties

The electrical conductivity of paraffin wax rises to the maximum at 25°C, then falls to the minimum at 35°C, and again rises sharply at 40°C. In the softening zone (40–49°C), paraffin behaves like amorphous polar substances. The conductivity of "B.P." paraffin wax is 0.00030 at 20°C.

Explosives

Paraffin wax is used for filling up spaces in explosives so that premature explosion does not occur; for converting explosives to a phlegm-like state to make them less susceptible to shock; for protecting metals against powder and explosives; for coating papers and cartons used to protect fireworks against humidity.

Fibrous Paraffin Wax

Paraffin wax, after drawing by hand or extrusion through a die, is shown by X-ray diffraction to have a distinct fibrous structure. The axis of the fibers may have either of two crystallographic directions, depending on the conditions of formation, but in both cases the long axes of the paraffin molecules are perpendicular to the axis of the fibers.[2]

Effects of Irradiation

If paraffin wax is irradiated with gamma rays in the range of $2.0 \times 10^6 - 100 \times 10^6$ roentgens, its antistatic properties are substantially increased.[3]

---

1 J. M. Page, *Ind. Eng. Chem.*, **28**, 7 (1936).
2 H. Rivsan, *Mem. Coll. Sci. Kyoto Imp. Univ.* **A19**, 279 (1936).
3 W. W. Elliott, U.S. Patent 3,014,854, Dec. 26, 1961.

The *n*-paraffin content of petroleum waxes is increased by irradiation.[1]

Exposure of paraffin wax to a beam of fast electrons (irradiation) causes increased electrical conductivity.[2]

Toxicity of Petroleum Wax

A four year study of all grades of commercial petroleum waxes using chemical and animal feeding tests shows the former to be safe for use in or in contact with foods. Up to 10% in the animal diets were used as well as applications to the skin from a solvent solution.[3]

Wetting

Paraffin wax is "wetted" with water containing 0.01% or more saponin.

The addition of fatty acids lowers the interfacial tension of paraffin wax against water.[4]

PHYSICAL PROPERTIES OF SOME COMMERCIAL PARAFFIN WAXES[5]

AMO Waxes

Type	*Melting Point* ASTM D-127	*Congealing Point* ASTM D-938	*Penetration* (CONE) @ 77°F	*Viscosity* @ 210°F	*Gr. API*
A-110	110–120	100–110	135 Max.	90–120	31–34
A-130	140–150	130–140	40 Max.	70–100	32–36
A-150	150–160	135–145	80 Max.	80–120	31–34

AMSCO Waxes

Melting Point						
	ASTM	122–124	130–132	135–137	138–140	149–155
	AMP	125–127	133–135	138–140	141–143	152–158
Color, Saybolt		30+	30+	30+	30+	21+
Odor		Nil	Nil	Nil	Nil	Nil
Taste		Nil	Nil	Nil	Nil	Nil
Tensile Strength (ASTM D-1320) p.s.i.		400	430	500	520	320
Breaking Strength @ 73°F, p.s.i.		515	545	500	450	375
Oil Content % Typical (ASTM D-721)		0.20	0.20	0.20	0.20	0.10
Blocking Point °F (ASTM D-1465)		96	100	110	115	118
UV Absorptivity (ASTM E-131) @ 290 μ, Max.		0.005	0.005	0.007	0.010	0.015

[1] Brit. Patent 821, 114, Sept. 30, 1959.
[2] V. L. Talroze, *Doklady Akad. Nauk.* S.S.S.R., **129**,858 (1959).
[3] Philippe Shubik, Chicago Medical School, 1962.
[4] Bhatnagar and Gardner, *J. Soc. Chem. Ind.*, **39**, 185 (1920).
[5] All commercial products mentioned throughout this book are identified in the Appendix.

Atlantic Waxes

	No.	Melting Point °F		Color Saybolt ASTM D-156	Penetration ASTM D-1321		Modulus of Rupture psi 73°F	Tensile Strength ASTM D-1320	Flash Point °F ASTM D-92	Oil % ASTM D-721	Taste and Odor	Blocking Temp. °F ASTM D-1465	Viscosity, cs ASTM D-445	
		Amp Range	American Melting Point		100°F	110°F							160°F	210°F
I. Fully Refined	131	124–127	126.5	+30	110	135	250	200	385	0.45	nine	95	4.90	3.20
	151	128–132	130.5	+30	95	125	245	200	390	0.35	nine	100	5.15	3.25
	171	133–136	134.5	+30	50	80	475	425	405	0.35	none	100	5.55	3.50
	173	133–136	134.5	+30	55	100	310	200	405	0.35	none	100	5.45	3.45
	191	138–140	139.5	+30	40	90	300	275	415	0.60	none	107	6.20	3.85
	1115	143–148	145.5	+30	13	22	400	325	425	0.15	none	115	6.75	4.20
	1116	153–158	154.0	+30	13	18	385	325	460	0.15	none	130	8.60	5.10
II. Scale Waxes	242	124–127	126.0	+24	180	215	—	—	390	1.25	slight	—	4.90	3.10
	252	128–130	129.0	+22	175	210	—	—	400	1.25	slight	—	5.10	3.25
	342	124–127	126.0	+24	180	215	—	—	390	1.25	slight	—	4.90	3.10
	351	128–130	129.0	+22	175	210	—	—	400	1.25	slight	—	5.10	3.25
	362	133–136	134.5	+22	105	170	—	—	410	1.25	slight	—	5.75	3.50
	371	136–140	137.5	+16	78	160	—	—	410	1.25	slight	—	6.25	3.75
	372	136–140	137.0	+24	85	165	—	—	410	1.25	slight	—	6.05	3.75
	392	143–145	144.5	+22	30	60	—	—	425	1.25	slight	—	6.80	4.10
	444	123–128	127.5	•	225	250	—	—	400	2.0	slight	—	5.20	3.20

• Color has ASTM D-1500 rating of L 1.5.

Canfield Waxes
(I)

	Board Coat A	*Board Coat B*	*Dairy Coat B*	*Paper Coat*	*Scale Wax*	*128–130*	*125–127*
ASTM Melting Point, °F (D-87)	142	141	127	147	122.5	126	122.0
ASTM Melting Point, °F (D-127)	—	—	—	—	—	—	—
Oil Content Wt. % (D-721)	0.4	0.4	0.4	0.5	2.0	0.4	0.4
Penetration (D-1321)							
77°F	—	—	10	21	—	15	14
90°F	—	—	—	—	—	—	—
100°F	—	—	75	43	—	239	279
105°F	—	—	—	—	—	—	—
110°C	—	—	92	92	—	275	320
Blocking Point	133	126	96	135	—	92	87
Pick Point	—	—	93	126	—	91	86
Tensile Strength	Good	Good	350	230	—	220	172
Odor	Good	Good	2.5	2.5	Good	2.5	2.5
Color	30	30	28	8	10	30	30
Flash, O. C., °F	—	—	425	485	400	400	400
Gloss Stability	Good	Good	—	Good	—	—	—

Canfield Waxes
(II)

	Dairy Coat C	*133–135*	*140–142*	*147–150*	*S-4*	*S-15*	*157–167*
ASTM Melting Point, °F (D-87)	137	131	138	147	134	120	—
ASTM Melting Point, °F (D-127)	—	—	—	—	—	—	159
Oil Content Wt. % (D-721)	0.50	0.10	0.01	0.3	4	15	0.5
Penetration (D-1321)							
77°F	16	12	14	8	—	—	11
90°F	—	—	—	—	—	—	—
100°F	60	45	24	16	—	—	21
105°F	64	—	—	—	—	—	—
110°F	—	179	107	22	—	—	38
Blocking Point	—	103	108	122	—	—	130
Pick Point	—	103	108	122	—	—	130
Tensile Strength	—	360	360	300	—	—	300
Odor	2.5	2.5	2.5	2.5	Good	Good	2.5
Color	17	30	30	30	Light	Light	30
Flash, O. C., °F	—	425	420	485	410	410	540
Gloss Stability	—	Good	Very Good	Good	—	—	—

Cardis Waxes

Name of wax	*M.P. ASTM D-127-30*	*Penetration* 100 g/77°/ *5 sec.*	*Color N.R.A.*	*Acid Number*	*Saponification Value*	*Type*
"Cardis" One Ⓟ	195–200	1–3	0–2½	12–16	45–55	Emulsifiable Petroleum Wax
"Cardis" 314 ®	184–189	4–6	1–3	13–16	45–55	Emulsifiable Petroleum Wax
"Cardis 319 ®	180–185	5–7	2–4	18–20	65–70	Emulsifiable Petroleum Wax
"Cardis 320 ®	180–185	5–7	1–3	28–30	75–80	Emulsifiable Petroleum Wax
"Cardis 262 ®	195–200	3–6	Brown	14–17	40–45	Specially Processed Petroleum Wax
Polymekon Ⓟ	195–205	0–3	Yellow	0–0	0–0	Specially Processed Petroleum Wax

Essowaxes[1]

Grade of Wax	*ASTM*	*2210*	*2530*	*3050*	*3250*	*3510*	*3550*	*3710*	*4030*	*4210*	*4610*	*5010*
Melting Point, EMP, °F	D 87	123	127	131	134	135	137	139	142	143	146	151
Odor		nil	nil	nil	nil	nil	nil	nil	nil	nil	nil	nil
Taste		nil	nil	nil	nil	nil	nil	nil	nil	nil	nil	nil
Color, Saybolt	D 156	+30	+30	+30	+30	+30	+30	+30	+30	+30	+30	+27
Oil Content, wt %	D 721	0.20	0.20	0.20	0.20	0.20	0.20	0.15	0.20	0.10	0.15	0.15
Refractive Index at 67 °C	D 1218	1.4315	1.4320	1.4330	1.4340	1.4340	1.4345	1.4350	1.4355	1.4330	1.4375	1.4385
Flash Point, C.O.C., °F	D 92	380	400	400	425	445	440	455	435	445	455	480
Viscosity, Kinematic, cs at 210 °F	D 445	3.0	3.2	3.8	4.1	3.8	4.2	3.9	4.3	4.3	4.7	5.6
at 180 °F		3.8	4.2	4.9	4.9	5.0	5.2	5.2	5.6	5.8	6.3	7.8
Pounds/Gallon at 150 °F	D 287	6.43	6.48	6.50	6.50	6.51	6.51	6.51	6.51	6.51	6.53	6.56
10 mm Vacuum Distillation, °F	D 1160											
5%		435	450	455	460	480	460	490	500	515	520	550
50%		465	475	500	510	505	520	515	535	530	550	590
95%		500	545	585	580	550	580	565	585	570	590	620
Width of Cut, 5%–95%, °F		65	95	130	120	70	120	75	85	55	70	70
Tensile Strength at 73 °F, psi	D 1320	190	230	340	380	310	370	310	320	260	250	340
Hardness, Abraham												
at 90 °F		26	30	34	35	41	35	41	35	35	35	35
at 100 °F		17	12	15	15	32	27	34	29	31	31	31
at 110 °F		14	11	13	13	14	14	23	21	26	25	26
Needle Penetration	D 1321											
at 77 °F		14	13	12	13	3	13	7	12	15	15	16
at 100 °F		186	83	35	47	20	40	20	34	20	38	33
at 110 °F		195	120	118	111	74	103	56	69	51	53	51
Gradient Blocking Point, °F	D 1465	115	95	98	102	108	105	111	110	116	125	125

[1] Distilled refined paraffin waxes which contains an oxidation inhibitor approved for foods.

Mobilwaxes

(Dairy Waxes)

Mobilwax Name	*1058*	*TW 38*	56	*D*	*55*
Color, Saybolt	+22	+22	+23	+30	+25
Melting Point, ASTM, °F	131	134	130	128	130
Viscosity, SSU @ 210 °F	42	42	40	39	39
Tensile Strength, (T.O.), psi	260	260	275	300	290
Elongation, in. @ 40 °F	.026	.026	.023	.015	.018
Cloud Point, °F	172	172	167	154	165

Shellwaxes

(Fully refined)

Shellwax Number	*Color Saybolt*	*R.I.* n_D	*M.P.* D-87°F	*Visc. SSU* @ 210°F	*Penetration* @ 77°F	*% Oil Content*	*Tens. Str.* D-1320	*Pick./Block* Temp. °F	*Seal. Str.* 73°F *Yield/Aver.*
100	30	1.4184	124.3	36.3	11	0.1	268	96/96	72/42
120	30	1.4207	131.2	38.3	14	0.1	304	98/98	82/36
125	30	1.4245	135.2	39.4	14	0.4	279	96/101	113/49
130	30	1.4212	132.8	39.3	11	0.2	364	98/103	99/41
200	30	1.4239	139.1	41.5	10	0.1	350	109/109	89/45
230	30	1.4244	140.2	41.1	10	0.3	290	111/114	128/>36
270	30	1.4249	139.6	41.8	9	0.1	220	116/126	94/>36
300	30	1.4305	159.6	51.4	12	1.6	186	123/128	19/19
700	13	1.4368	181.6	75.0	6	0.1	245	>135	7/7

Chapter 2

MICROCRYSTALLINE WAXES

The designations *microcrystalline* and *amorphous waxes* were long used synonymously, although the first is more accurate. These waxes differ from refined paraffin wax in the size and structure of crystals, and in that they are tougher, more flexible, and have a higher tensile strength and melting point. They are also more adhesive and less lustrous and greasy. They bind solvents, oils, etc., much better than paraffin wax does and thus prevent their sweating-out.

It should be noted in general that:

1. High penetration value and/or high refractive index indicate flexibility.
2. Waxes of high penetration value generally have more "tack."
3. Flexibility is not a function of melting point.
4. Oil content influences flexibility only to a limited degree.

DIFFERENCES BETWEEN MICROCRYSTALLINE AND PARAFFIN WAXES[1]

Both paraffin and microcrystalline waxes are separated from crude petroleum, but the process of manufacture and the resulting products are quite different.

As crude petroleum is subjected to distillation by heating in a still at atmospheric pressure, the following products are removed in the order of their increasing boiling points: light petroleum gases, gasoline, naphtha, kerosene, gas oil, paraffin wax distillate, light, neutral lubricating-oil fractions, and a residue in the still that will not distill overhead at atmospheric pressure without decomposition. Paraffin wax is separated from the paraffin-wax distillate by a relatively simple process, while the microcrystalline wax, which cannot be distilled without decomposition, is

[1] Bareco Wax Co., Division of Petrolite Corp., Tulsa, Okla.

separated by a complex series of solvent separations from the residue remaining in the still.

Paraffin wax, which has very little affinity for oil, is separated from the paraffin-wax distillate by cooling and filtering in a common plate-and-frame filter press where the wax is retained on canvas filter leaves and the distillate oil passes through the canvas and is removed. Oil remaining in the wax filter cake is removed by a sweating process which involves casting the wax in thin sheets and gradually raising the temperature to a point slightly below the melting point of the wax. In this process, the remaining oil runs out of the wax sheet leaving a paraffin wax which usually contains less than 0.5% oil.

Microcrystalline wax is present in the residual fraction from the still along with heavy residual lubricating oil and asphalt. Asphalt must be removed from the residue by conventional sulfuric acid treatment or one of the newer selective solvent processes before the wax can be separated. The microcrystalline wax is then removed as crude petrolatum from the residual lubricating oil by any one of several dewaxing processes which involve dilution with an organic solvent, chilling at 20–40°F., and separation of the crude petrolatum at high-speed centrifuges. Microcrystalline wax has a great affinity for oil and thus the crude petrolatum still contains 40–70% oil. Therefore, it is then mixed with another portion of organic solvent, usually a different solvent from the one used in the initial dewaxing step, and heated to dissolve the wax and oil. The blend is then cooled to precipitate the wax which is separated from the oil and solvent on a filter. The solvent used for deoiling microcrystalline wax must be polar, whereas for deoiling paraffin wax other types of solvent may also be used. Since the microcrystalline wax holds oil very strongly, the last solvent step is usually repeated to give a wax with lower oil content.

The oil content of the microcrystalline waxes varies with the grade of wax but is usually 2–12% as contrasted with 0.5% in paraffin wax.

Chemically, both paraffin and microcrystalline waxes consist of saturated hydrocarbon.

Paraffin and microcrystalline waxes are both long-chain compounds but microcrystalline waxes have much higher molecular weights. The common commercial paraffin waxes have molecular weights of 360–420, which means that the average molecule in these hydrocarbons contains 26–30 carbon atoms. The molecular weight of high-melting point paraffins may run as high as 600. The molecular weight of commercial microcrystalline waxes is 580–700, that is, the average molecule contains 41–50 carbon atoms. This molecular weight difference accounts for the fact that the paraffin waxes can be distilled at atmospheric pressure while the microcrystalline waxes decompose when distillation is attempted.

Extensive X-ray diffraction work has indicated differences in the chemical structure of the paraffin and microcrystalline wax molecules in addition to their molecular weight. Paraffin waxes have been found to consist mainly of straight chain molecules

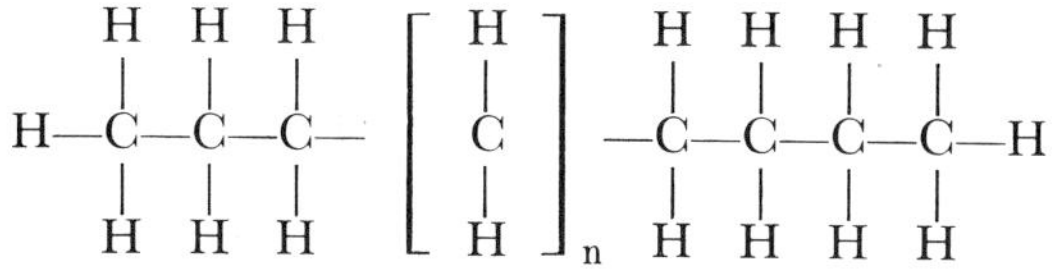

and a lesser amount of molecules containing branched chains.

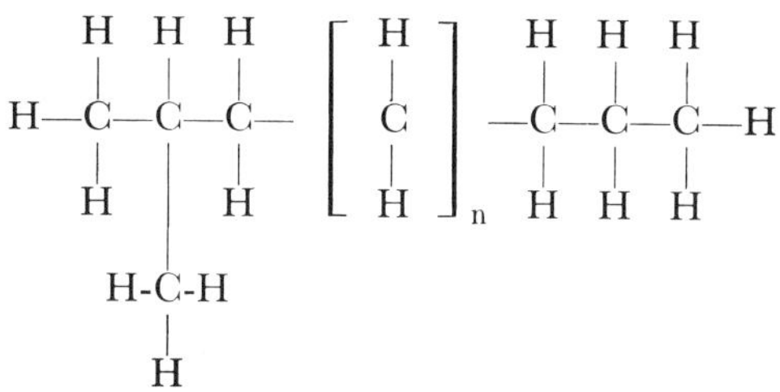

In some paraffin waxes, a very small quantity of aromatic hydrocarbons may also be present. The microcrystalline waxes contain some straight-chain molecules, but a large proportion of branched-chain molecules. The branched chains in the microcrystalline waxes are probably located at random along the carbon chain while in paraffin wax, they are near the end of the chain. There are indications that microcrystalline wax also contains a much higher concentration of ring-type compounds than paraffin wax.

Microscopic examination of paraffin and microcrystalline wax shows a striking difference. Paraffin wax has large, well-formed crystals both when crystallized from the melted wax and from solvents. Microcrystalline wax forms small irregular crystals from the melted wax, but no well-formed crystals of any size from solvents.

Table 2: 1

CRYSTAL TYPES IN MICROCRYSTALLINE WAXES

Boiling Point, °F at 10 mm	*Type of Crystals*		
	Plates	Malcrystalline	Needle
180	100	0	0
200	99	1	0
220	98	2	0
240	91	7	2
260	78	12	10
280	54	17	29
300	37	27	36

It can be seen that in the highest-boiling fraction, the malcrystalline and needle types are present in a greater total amount than the plate type. These waxes yield a much lower melting point for a given boiling point. For this reason, they are usually almost completely lost in paraffin refining.

These types represent Buchler and Graves' "soft wax" and they also constitute the bulk of microcrystalline or amorphous waxes obtained from higher-boiling distillate and residues.

FRACTIONAL CRYSTALLIZATION OF PETROLEUM WAXES[1]

A new method of fractionating waxes, called *zone precipitation*, separates crystalline from non-crystalline petroleum waxes through differences in solubility when a solution of the waxes is cooled.

The following tables indicate the possibilities of this method. The wax used was a microcrystalline wax (m.p. 175°F) from a North Louisiana crude.

Table 2:2

EFFECT OF SOLVENT:WAX RATIO ON PROPERTIES OF LIQUID IN MOLTEN ZONE

Solvent: sec-butyl acetate

Solvent/ Wax Wt. Ratio	*Temperature*		*Viscosity Solvent + Wax*	*ΔT after 4 Zone Passes*
	Point measured	°F	*Cp*	°F
None	Cloud point + 5°F	190	22.05	0
	Cloud point	185	24.87	
	Opaque point	177	27.93	
	Solid point	173	—	
1/1	—	190	2.62	15
	Cloud point + 5°F	165	3.96	
	Cloud point	160	5.48	
	Opaque point	155	7.92	
	Solid point	151	—	
3/1	—	190	0.84	45
	Cloud point + 5°F	159	1.14	
	Cloud point	154	1.16	
	Opaque point	150	1.27	
	Solid point	111	—	
4/1	—	190	0.70	—
	Cloud point + 5°F	158	0.94	
	Cloud point	153	1.08	
	Opaque point	148	1.50	
	Solid point	108	—	
6/1	—	190	0.55	45
	Cloud point + 5°F	156	0.74	
	Cloud point	151	0.77	
	Opaque point	146	1.44	
	Solid point	73	—	

The direction of motion of the dissolved waxes was characteristic. Components of low melting point and high solubility moved toward the hot zone; those of high melting point and low solubility moved in the opposite direction.

[1] Work of I. A. Eldib, *Ind. Eng. Chem.* Proc. Des. and Dev., **1**, 2–9 (1962).

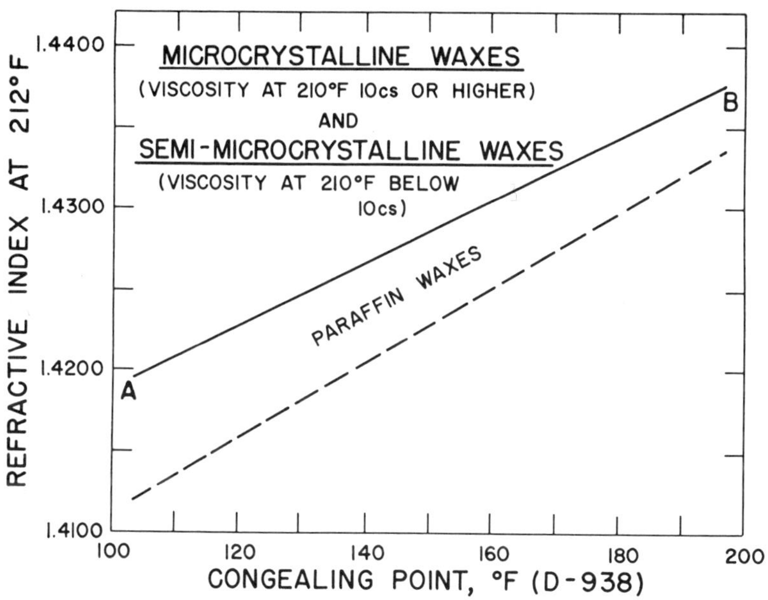

FIG. 2:1. Relationship between congealing point and refractive index, demonstrating classification by position of wax on the chart.

Table 2:3

EFFECT OF TYPE OF SOLVENT ON PROPERTIES OF LIQUID IN MOLTEN ZONE

Solvent/wax wt. ratio = 3/1

Solvent	*Dipole Moment of Solvent, E (Debye Units)*	*Temperature* Point measured	°F	*Viscosity of Solvent + —ax, Cp.*	*ΔT after 4 Zone Passes, °F*
Methyl ethyl ketone	2.77	Cloud point + 5°F	168	0.8772	16
		Cloud point	163	0.9718	
		Opaque point	157	1.3810	
		Solid point	132	—	
sec-Butyl acetate	1.75	Cloud point + 5°F	161	1.420	10.5
		Cloud point	154	1.5310	
		Opaque point	149	2.2220	
		Solid point	121	—	
Ethylene dichloride[1]	1.89 (*cis*)	Cloud point + 5°F	154	2.1090	10.5
	0.0 (*trans*)	Cloud point	149	2.3990	
		Opaque point	143	5.0560	
		Solid point	137	—	
Chloroform	1.02	Cloud point + 5°F	143	—	7
		Cloud point	138	4.0540	
		Opaque point	130	7.9910	
		Solid point	117	—	
Toluene	0.36	Cloud point + 5°F	133	1.4140	11.2
		Cloud point	128	1.9690	
		Opaque point	123	3.1150	
		Solid point	75	—	
Carbon tetrachloride	0.0	Cloud point + 5°F	137	4.2310	10
		Cloud point	132	6.0290	
		Opaque point	126	12.200	
		Solid point	110	—	

[1] Commercial solvent generally contains both *cis* and *trans* forms, predominantly *cis* form.

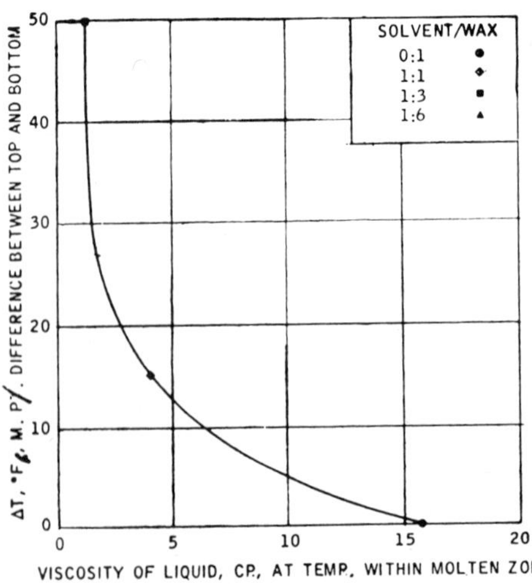

FIG. 2:2. IMPROVEMENT IN FRACTIONATION WITH DECREASING VISCOSITY OF LIQUID.

Table 2:4

EFFECT OF NUMBER OF ZONE PASSES ON FRACTIONATION

Ingot length = 15 in.
Zone length = 1.5
Zone travel rate = 1.5 in./hr

Solvent/wax wt. ratio = 3/1
Solvent = *sec*-butyl acetate
Dry wax feed m.p. = 175°F

Product Analysis	*% of Ingot*[1]	*M.P., °F*	*Product Analysis*	*% of Ingot*[1]	*M.P., °F*
ZONE PASSES[2] = 1 × 2			ZONE PASSES[2] = 2 × 1		
Fraction 1 (top)	28.4	175.5	Fraction 1 (top)	18.4	176.0
2	16.7	176.5	2	18.4	175.5
3	19.6	176.0	3	20.2	174.0
4	15.7	175.5	4	19.3	173.5
5	19.6	169.0	5	23.7	170.0
ΔT[3,4]		7.5	ΔT[4]		7.5
ZONE PASSES[2] = 1 × 4			ZONE PASSES[2] = 2 × 4		
Fraction 1 (top)	16.2	175.5	Fraction 1 (top)	24.8	180.5
2	9.5	174.0	2	20.4	175.0
3	27.6	173.0	3	18.6	174.0
4	30.5	171.0	4	19.5	168.5
5	16.2	165.0	5	16.7	155.5
ΔT[4]		10.5	ΔT[4]		26.0
ZONE PASSES[2] = 1 × 10			ZONE PASSES[2,3] = 2 × 10		
Fraction 1 (top)	25.3	181.5	Fraction 1 (top)	21.4	186.5
2	21.2	177.0	2	21.4	177.0
3	11.1	175.0	3	21.4	168.0
4	20.2	168.5	4	19.7	155.5
5	22.2	161.5	5	16.1	145.0
ΔT[4]		22.7	ΔT[4]		45.0

Table 2:5

EFFECT OF SOLVENT:WAX RATIO ON FRACTIONATION

Ingot length = 15 in.
Zone length = 1.5 in.
Zone travel rate = 1.5 in./hr.

Solvent = *sec*-butyl acetate
Dry wax feed m.p. = 175°F

Product Analysis	*% of Ingot*[1]	*M.P., ° F.*	*Product Analysis*	*% of Ingot*[1]	*M.P., °F.*
ZONE PASSES = 2 × 18[2] SOLVENT: WAX = 1/1			ZONE PASSES × 2 × 10[2] SOLVENT: WAX = 6/1		
Fraction 1 (top)	8.6	184.0	Fraction 1 (top)	29.7	184.5
2	10.4	181.0	2	35.9	168.0
3	13.4	177.5	3	17.2	159.5
4	12.6	176.5	4	17.2	141.0
5	13.7	175.0	ΔT[4]		45.0
6	13.5	172.5			
7	14.4	170.5			
8	13.4	168.5			
ΔT[4]		15.0			

[1] Dry basis.
[2] Zones × passes.
[3] Hydrogenated Mikrovan 1750 used in this run.
[4] Difference between melting point of two end cuts (top and bottom), each 15% of total ingot treated by ne precipitation.

Figure 2:2 shows that fractionation, as measured by ΔT, improved as liquid viscosity in the molten zone is decreased.

Fractionation also improved with an increasing number of zone passes, and when the ratio of solvent to wax was increased from 1:1 to 3:1; but further increase to 6:1 did not produce further improvement. The increase to 3:1, however, caused a decrease in viscosity in the molten zone. Results in zone precipitation are better with poor solvents.

Table 2:6

DIFFUSIVITY OF MICROWAX IN WAX:*sec*-BUTYL ACETATE SOLUTIONS

Concn. on Capillary Wt. %	*Solvent: Wax Ratio in Bulk Soln.*	*Diffusivity, Sq. Cm./Sec. × 10²*	*Temp. of bulk Soln.,[1] °F*	*Visc. Liq. in Capillary[2] Cp.*
60	1/1	1.95	165	5.50
25	4/1	5.18	159	1.14
20	~6/1	5.83	158	1.00

Table 2:7

EFFECT OF TYPE OF SOLVENT ON FRACTIONATION

Ingot length = 15 in.
Zone length = 1.5 in.
Zone travel rate = 1.5 in./hr
Solvent: wax wt. ratio = 3/1
Dry wax feed m.p. = 175°F.
No. of zone passes[3] = 4

Product Analysis	*% of Ingot[4]*	*M.P., °F.*	*Product Analysis*	*% of Ingot[4]*	*M.P., °F.*
METHYL ETHYL KETONE			*sec*-BUTYL ACETATE		
Fraction 1 (top)	14.4	183.5	Fraction 1 (top)	16.2	175.5
2	21.7	174.5	2	9.5	174.0
3	21.7	175.0	3	27.6	173.0
4	24.7	173.0	4	30.5	171.0
5	17.5	168.0	5	16.2	165.0
ΔT[5]	—	13.9	ΔT[5]	—	10.5
ETHYLENE DICHLORIDE			CHLOROFORM		
Fraction 1 (top)	20.4	176.5	Fraction 1 (top)	17.4	177.5
2	21.2	175.0	2	15.6	175.0
3	21.2	175.0	3	20.0	174.5
4	18.6	173.0	4	24.4	173.0
5	18.6	167.5	5	22.6	171.5
ΔT[5]	—	10.5	ΔT[5]	—	7.5
TOLUENE			CARBON TETRACHLORIDE		
Fraction 1 (top)	29.6	176.5	Fraction 1 (top)	20.5	179.0
2	14.9	175.0	2	23.9	180.5
3	15.7	173.0	3	15.2	175.5
4	17.6	171.0	4	19.2	170.5
5	22.2	166.5	5	21.2	169.5
ΔT[5]	—	11.2	ΔT[5]	—	9.5

[1] Cloud point + 5°F.
[2] Viscosity of solvent:wax solution at cloud point + 5°F.
[3] 1 zone × 4 passes.
[4] Dry basis.
[5] Difference between melting point of two end cuts (top and bottom), each 15% of total ingot treated zone precipitation.

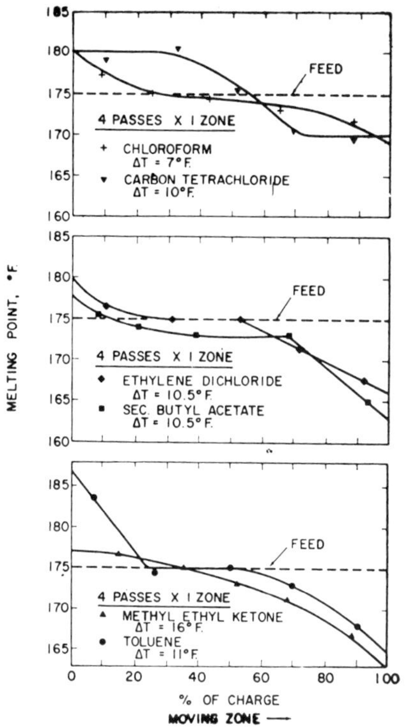

FIG. 2:3. PROFILES OF MELTING POINTS WITH VARIOUS SOLVENTS.

PROPERTIES OF MICROCRYSTALLINE WAXES

Ielting Point

The melting point of commercial paraffin waxes is, in general, much ower than that of microcrystalline waxes. Paraffin waxes usually melt at 26–134°F, and microcrystalline waxes at 145–195°F. The melting point, owever, is not sufficient for distinguishing them because paraffin waxes elting up to 170°F are available, and microcrystalline waxes have been repared with a melting point as low as 130°F.

Paraffin waxes are brittle but microcrystalline waxes are tough, resisting acture. Most grades of microcrystalline wax are plastic and tend to flow ıder compression, whereas all grades of paraffin wax will shatter under mpression.

Oil added to most grades of microcrystalline wax will increase its asticity. Addition of oil to a paraffin wax will yield not a plastic product, ıt one with greatly reduced strength which tends to crumble when ndled.

Paraffin wax is transparent; microcrystalline wax is opalescent. Various grades may be white, brown or black in color.

Films of paraffin and microcrystalline waxes of the same color and thickness transmit different amounts of light and thus films of paraffin wax appear more transparent than those of microcrystalline-wax films.

Paraffin wax forms a stronger gel in turpentine than does microcrystalline wax in the same ratio of melting point and solvent.

If a glass rod is dipped into a solution containing a precipitate of microcrystalline wax in mineral oil and is withdrawn, the solution will drain away, leaving only a thin layer of oil on the rod.[1] If the solution contains a precipitate of paraffin wax, crystals will remain on the rod, because the surface tension of the crystals of paraffin wax is less than that of the oil, whereas that of the microcrystalline-wax crystals is greater than that of the oil.

Paraffin wax contracts to a greater extent when passing from the liquid state to the solid state than do microcrystalline waxes.[2] A test to differentiate between paraffin and microcrystalline waxes has been based on this behavior.

The Saybolt Universal viscosity at 210°F for microcrystalline waxes is usually 70–100 seconds; that of paraffin wax, 40–50 seconds.

Table 2:8

PHYSICAL PROPERTIES OF SOME MICROCRYSTALLINE WAXES

Properties	*Brands*				
	Quaker State Microwax	Barnsdall Special Wax	Syncera Wax	Socony No. 2300	Be Square Special
Melting Point °F	145–146	160–165	155–160	155	160–185
Specific Gravity at 60°F	0.90–0.92	0.92–0.94	0.912	—	0.92–0.94
Specific Gravity at 210°F	0.78–0.80	0.80–0.82	—	—	—
Viscosity at 210°F (Saybolt)	75–85	75–100	51	65	75–100
Flash Point °F	500 min.	500 min.	425	495	500
Fire Point °F	575 min.	575 min.	—	550	575
Saponification Number	0.05–0.10	<2	0	—	<2
Acid Number	0.1 max.	0.1–0.2	0	—	0.1–0.2
Iodine Number					1.5
Color	white to amber	white to amber	yellow	yellow to brown	white to black
Penetration Number (ASTM) (Petrolatum method –77°F; 100 g wt.)	—	5–10	—	20–25	5–20

[1] Jones and Blachly, *Ind. Eng. Chem.* **21**, 318 (1929).
[2] Kinsel and Phillips, *Ind. Eng. Chem.* **17**, 152 (1945).

The refractive index at 212° for paraffin waxes is usually 1.430–1.433, and for microcrystalline waxes, 1.435–1445.

The tensile strength of paraffin wax of very low oil content, (0.1% or less) is high, but the addition of even 0.5–1% oil will greatly reduce the tensile strength (up to 50%). The reduction of tensile strength of microcrystalline wax with the same addition of oil will be about 5%. This phenomenon is due to the fact that oil weakens the structure of paraffin wax, but it is so readily absorbed by microcrystalline waxes that a more plastic product results which imparts plastic flow under tensile stress.

The extensibility is much greater with microcrystalline wax than with paraffin wax. Addition of oil will increase the extensibility of most microcrystalline waxes, but does not affect that of paraffin wax.

Cast in small rods, most microcrystalline waxes will bend without breaking at temperatures considerably below their melting point. Paraffin waxes cast in the same way usually break at or near their melting point.

Compatibility

Microcrystalline waxes are compatible in all proportions with other mineral waxes, most vegetable waxes, and resins. This property has made them valuable as agents for increasing melting point and hardness, and for decreasing blocking, tack, sweating, and other undesirable qualities in wax blends.

Laminating Properties of Microcrystalline Waxes

Adhesion Test

The adhesion test measures the force required to tear apart a test specimen prepared by laminating a square of unwaxed cellophane on either side of a square of waxed cellophane. An unlaminated edge of each of the outer cellophane squares is placed in a clamp on a machine imilar to that used for measuring the strength of paper. One clamp is lowly drawn away from the other by a motor-driven screw until the quares have been separated. A lever attached to the stationary clamp noves along a scale and indicates the relative amount of force required to ear the laminated squares apart. The higher the adhesive strength of the ax, the higher is the test value obtained.

leed Test

In this test, a sample of the wax is melted, and five droplets (6–8 mm diameter) are placed on a filter paper (Whatman No. 1). When the roplets have solidified, the paper is placed in an oven at 130°F. for 24 ours. The average value (in millimeters) obtained by subtracting the ameter of each droplet from the diameter of the oil ring that surrounds is recorded as the bleeding number. The less the tendency of the wax bleed, the lower is the number obtained.

Table 2:9

PHYSICAL PROPERTIES OF MICROCRYSTALLINE (AMORPHOUS) WAXES

Name or Grade	ASTM Penetration at 77°F.	Refractive Index at 80°C.	ASTM Melting Point Drop Method	Oil Content	Flexibility: Room Temperature	Flexibility: at 34°F.	Flexibility: at 20°F.	Taste	Odor 200°F.	*Remarks*
Benowax C	21	1.4508	140.2	4.01–3.99	good	good	good	bad	burnt	very sticky
Micro Wax, Dark Amber	22	1.4491	145.4	1.26–1.56	good	good	good	poor	oily	—
Micro Wax, White	23	1.4451	146.2	0.66–0.93	good	good	good	none	none	—
Standard Oil Syncera, 145/50 Orange	19	1.4341	146.4	—	good	brittle	—	good–very little	slight	hard
Standard Oil Syncera, 145/50 Yellow	17	1.4330	146.4	0.88–1.06	good	brittle	—	good–very little	slight	hard
Standard Oil Syncera, 155/60 Yellow	19	1.4382	155.8	—	brittle	—	—	good	slightly oily	hard
Aristowax 160/65	14	1.4362	160.4	2.53–2.55	brittle	—	—	none	slight propane	—
Deep Rock, 160 Lemon Pale	33	1.4480	161.2	11.19–10.87	good	fair	brittle	poor	kerosene	mealy texture
Wadhams 2305	31	1.4461	162.7	5.40–5.27	good	good	good	poor	slightly oily	sticky
Standard Oil Superla, 165/70 Yellow	12	1.4446	165.8	—	brittle	—	—	good	slight	—
Wadhams Petrosene A	23	1.4448	167.8	8.61–8.01	good	brittle	—	good	slightly oily	—

The adhesion and bleed properties of Multiwax ML 445* as compared to those of other microcrystalline waxes on the market are presented in Table 2:10

Table 2:10

ADHESION AND BLEEDING OF MICROCRYSTALLINE WAXES

Wax	ASTM M.P	ASTM Needle Penetration	Adhesion at 77°F.	Bleed Test
Multiwax ML 445	163	30	33	1
Wax A	175	12	1	0
Wax B	156	21	2	4
Wax C	155	24	6	5
Wax D	167	27	5	6
Wax E	164	34	43	4
Wax F	158	34	32	4
Wax G	157	36	35	5

The bleed values might indicate that low needle-penetration has been overstressed as a criterion of the suitability of a wax for laminating. It has been held that waxes with relatively high penetration have a greater tendency to strike through the laminating paper. These values do not support this theory.

It may be noted from the preceding table that some waxes have high adhesion at the expense of their bleed characteristics (Wax E, for example). Others have good bleed properties, but low adhesion (Wax A, for example). Obviously, the most suitable laminating wax would be one that has a combination of relatively high adhesion and relatively low bleed characteristics. Such a combination which depends on the proper selection of raw materials and the manufacturing process is represented in Multiwax ML 445, a wax which is used in increasing quantities as a paper laminant.

Miscibility with Additives

In many applications of microcrystalline waxes, it is essential to know their miscibility with various resins and other additives. The solubilities of a number of products in Petrosene C (microcrystalline wax) were determined on a semiquantitative basis. Table 2:11 presents data indicating the solubility-temperature relationship for these mixtures. In all cases, 10% of additive was employed so that no attempt was made to determine the maximum solubility. It is expected that the solubility of any resin in paraffin wax will be greater than in Petrosene C and in other microcrystalline waxes it will be of the same order as in Petrosene C. Table 2:11 is divided into two sections, one listing the additives soluble up to 10% in Petrosene C at all temperatures above its melting point and the second listing those which are insoluble or only partially soluble.

Solubility

The solubility of microcrystalline waxes in solvents is a function of the solvent used, the temperature and the amount of wax in the solvent.

When mineral oil is added to microcrystalline waxes the bleeding or staining tendency of the resultant blends increases at a slowly accelerating rate with the amount of oil added. However, when oil is added to conventional paraffin waxes, the bleeding tendency increases rapidly at first and then more slowly. At the same wax–oil blend penetration, the paraffin waxes show a much higher bleeding value.

The ductility of blends of mineral oil and paraffin waxes is usually lower than that of blends of oil and microcrystalline waxes. The former yield zero values, but some microcrystalline waxes also yield zero values.

When waxes are softened with oil to a constant penetration value, the paraffin waxes have a lower flexibility at low temperature than do microcrystalline waxes. Conventional paraffin waxes yield zero flexibility.

When a microcrystalline wax is added to a blend of paraffin wax and oil, the resultant penetration is higher than the weighted arithmetic average of the two components. Curves of penetration versus per cent added microcrystalline wax show a maximum, which is not the case when microcrystalline wax is added to a microcrystalline wax–oil blend or when paraffin wax is added to a paraffin wax–oil blend.

Table 2:11

MISCIBILITY OF VARIOUS ADDITIVES WITH PETROSENE C (MICROCRYSTALLINE WAX)

Percentages Used: Petrosene C—90%; Additive—10%.

Additives Completely Miscible above and at Melting Point

Product	Product
1. Cumar Resin P-25	9. Chromium Oleate
2. Amberol—Grade ST-137	10. Diglycol Stearate
3. Opalwax 10	11. Lead Linoleate
4. No. 1 Singapore Damar Gum	12. Copper Stearate
5. Batavia Damar A/D	13. Manganese Naphthenate
6. Santo-Resin	14. Manganese Resinate
7. Burgundy Pitch	15. Calcium Stearate
8. Calcium Stearate	16. Zinc Stearate

Additives Insoluble or Partially Miscible

Product	Limit of Miscibility, °F.
1. Cumar Resin V 1½	above 250
2. Cumar Resin P-70	above 240
3. Cumar Resin MH 1½	above 230
4. Amberol M-16	at least above 300
5. Amberol M-82	at least above 300
6. Resin XR9517	above 270
7. Resin BR8900	above 230
8. Residuum Pitch	at least above 300
9. Acacia Gum	at least above 300
10. Gum Arabic	at least above 300
11. Gum Kauri	at least above 300
12. Cobalt Linoleate	at least above 300
13. Nuba No. 2	at least above 300
14. Nubalene Resin	at least above 300
15. Nuba No. 3	at least above 300
16. Resin No. 465	at least above 300
17. Nuba No. 1	at least above 300
18. Cobalt Resinate	at least above 300
19. Aluminum Oleate	at least above 300
20. Aluminum Naphthenate	at least above 300
21. Calcium Oleate	at least above 300

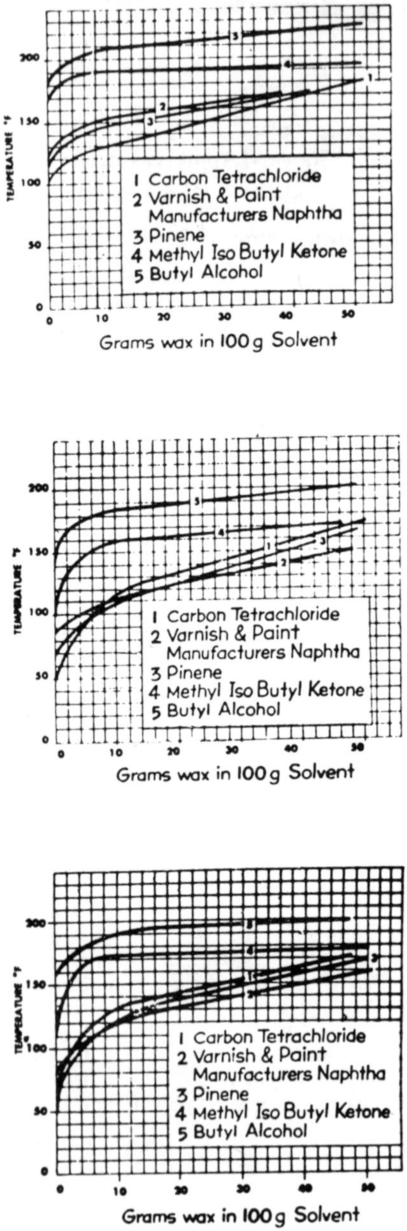

FIG. 2:4. SOLUBILITY OF BARECO WAXES IN COMMERCIAL SOLVENTS.

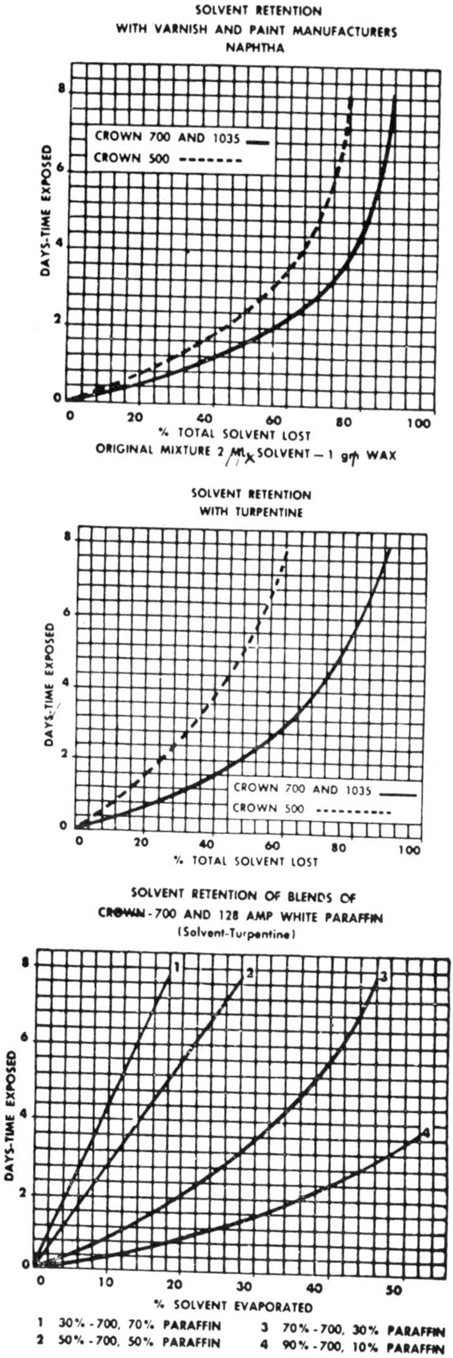

FIG. 2:5. RETENTION OF SOLVENTS BY BARECO WAXES.

The cloud point is a simple and accurate means of expressing the solubility of waxes in solvents. Above the temperature of the cloud point the wax is completely soluble, whereas at the cloud point and below the cloudiness indicates at least partial insolubility.

To determine cloud points, the wax was weighed in the solid state and placed in a 250 ml flask. 100 ml of the solvent was then measured into the flask and the flask set up under a reflux condenser to eliminate any possible loss of solvent. Heat was then applied with a burner until the wax was all in solution. The solution was then allowed to cool gradually with constant, gentle swirling and the temperature was noted when the first cloud appeared in the solution. The temperature was taken on a thermometer inserted through the stopper in the flask and extending into the solution.

After the first temperature of cloud was noted the solution was reheated, again allowed to cool, and a second temperature was noted. This operation was repeated until three check readings were obtained. The higher the melting point of the wax, the lower its solubility in any given solvent.

Specific Gravity at Various Temperatures

When microcrystalline waxes solidify, they contract and their specific gravity changes, this has a bearing on their performance in many applications. In the impregnation of electrical or other equipment, for example,

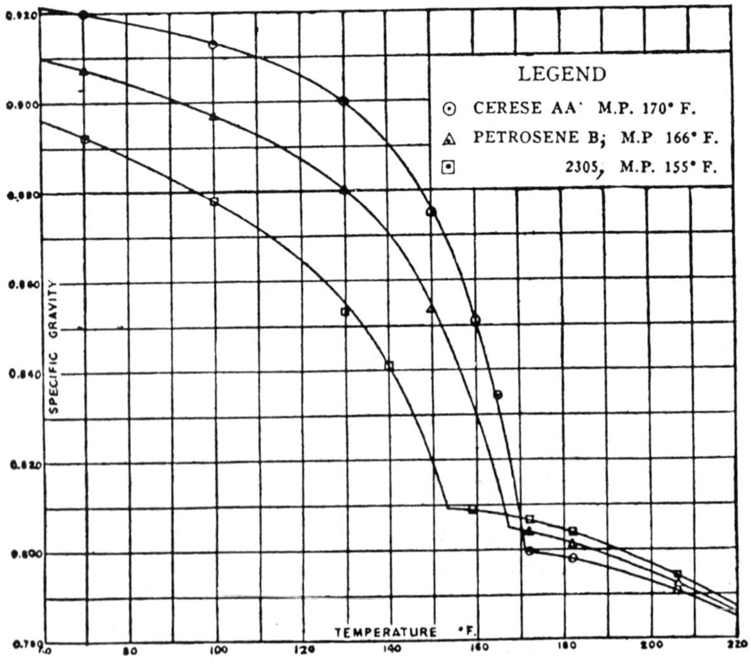

FIG. 2:6. SPECIFIC GRAVITIES OF MICROCRYSTALLINE WAXES.

where voids are filled with molten wax, it is frequently essential to determine how much contraction will occur on solidification so that adjustments may be made to avoid loss in waterproofing value. In addition, it may be necessary to determine what volume a given quantity of solid wax will occupy when melted and heated to a definite temperature.

In order to obtain this information, representative samples of microcrystalline waxes were held at definite temperatures and their specific gravities measured. The data so obtained are given in Figure 2:6. Cerese Wax AA, Petrosene B, and Product 2305 were tested. These are the intermediate colored waxes in each of the three series. Since the other waxes differ from these only in color, their specific gravities may be considered comparable. Due to slight variations from batch to batch, other samples of any wax will probably show slightly different values from the figures noted. Based on the available data, however, these variations are not expected to exceed 0.005, except for samples of widely different melting points, in which case, the greatest divergence will be at temperatures near the melting point.

It is interesting to note that, in the solid state, Cerese Wax AA has the highest specific gravity, whereas in the molten state, it has the lowest. Petrosene B is intermediate between Petrolite C 2305 and Cerese Wax AA.

Thermal Expansion and Melting Point of Waxes

Pure crystalline substances are very often identified by their melting point. Waxes, being complex mixtures, do not exhibit sharp melting points, but there is a gradual transition from the liquid to the solid state. This gradual softening results in an abnormally high apparent coefficient of

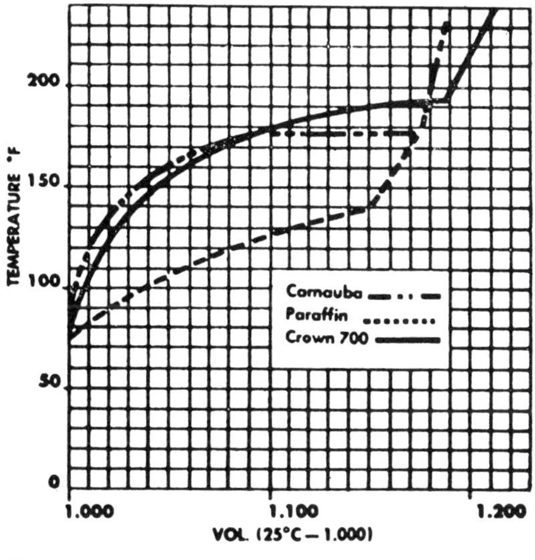

FIG. 2:7 THERMAL EXPANSION AND MELTING POINT OF WAXES.

expansion while the wax is still in the solid state. This characteristic of waxes has a bearing on their performance in many applications. The curves in Figure 2:7 compare this effect for a crystalline paraffin, carnauba wax, and a commercial product, Petrolite 700. The breaking point in the curves may be interpreted as the liquefication temperature in each case.

Viscosity

The viscosity of microcrystalline waxes, when melted, is greatly lowered by small additions of ordinary paraffin wax.

Modification of Paraffin Wax with Microwax

Very many uses for paraffin wax can be improved by the addition of microcrystalline wax. All paraffin grades have serious shortcomings due to their high crystallinity and to the low intercrystalline forces bonding one crystal to another. Microcrystalline wax provides a means of modifying the functional properties of paraffin waxes by controlling:

1. The crystal size
2. The amount of amorphous intercrystalline material
3. The molecular weight
4. The 2nd order transition point.

Friction

Table 1 illustrates typical coefficient of friction data of several micro–paraffin wax blends. These data are typical of paraffin waxes having similar melting points. Generally the higher melting grades of micro-wax reduce the friction of those paraffins which have moderately high friction (of the order of 0.36 or higher), and moderately increase the friction of paraffin waxes having low friction, such as the 130/32 FR paraffin in the table.

Gloss

Microwaxes play a very important role in the gloss of coatings. Microwaxes function by inhibiting the formation of large paraffin crystals and by increasing the amount of amorphous material.

Gloss is a surface phenomenon associated with texture. The reflection of light from an optically smooth, mirror-like surface is specular, i.e. all light rays are reflected at an angle precisely equal to the incident angle. As a surface becomes less smooth, the total amount of light reflected may remain the same, but the reflected light is more diffused and appears less glossy. On gloss measuring instruments, roughening the surface also reduces gloss by scattering a portion of light in directions other than the detector.

The loss of gloss of paraffin waxes appears to be caused by surface roughening which occurs either under long term aging at 73°F., or accelerated conditions. The surface roughness is a result of stresses frozen into the surface by progressive cooling, starting with the outside layer and gradually working into the center. The combination of smaller crystals, and increased amorphous material permits release of the surface stress by slippage along crystal boundaries. At higher levels (approximately 40% micro), formation of the hexagonal crystal structure between the first and second order transition points is eliminated.

Gloss in EVA–WAX Systems

Another form of gloss decay occurs with EVA–paraffin systems. Most users are familiar with the extreme decay which occurs with EVA–paraffin blends. This occurs most dramatically with paraffin–EVA blends containing less than 30% EVA and with normal fully refined paraffin wax. It is less apt to occur with the so-called intermediate paraffin waxes. This type of gloss decay results in a powdery surface which has extremely poor gloss and involves an entirely different mechanism than the loss of gloss of paraffin waxes. The loss of gloss of EVA blends is due to a rejection from the crystal lattice of a specific normal paraffin. The mechanism by which microwax prevents this type of gloss decay is unknown but may be due either to the formation of more stable solid solutions of EVA–paraffin wax–microwax or due to the amorphous microcrystalline wax fractions acting as solvents for the rejected normal paraffin fractions.

Typical of this loss of gloss and the stabilizing effect of microwaxes is shown below.

EVA Resin	*FRP 135/37°F*	*Starwax 100*	*Resin (alpha methyl styrene)*	*Gloss (Initial/Aged)*
20%	80%			82/4
20%	65%		15%	82/3
20%	45%	20%	15%	79/4
20%	30%	35%	15%	82/99
20%	15%	50%	15%	83/83

It should be emphasized that the above type of gloss decay is a peculiar phenomenon. Although it is dependent largely upon the chemical composition of the paraffin wax, its occurrence is erratic. When it will occur and under what conditions it will occur are not predictable because of the lack of analytical data as to the precise chemical composition of commercial paraffin waxes. The following points are important in this regard.

1. The decay is more serious with lower melting paraffin wax than with higher melting paraffin wax.
2. Decay does not occur with microcrystalline waxes.
3. The addition of micro wax to paraffin wax prevents the decay from occurring. The amount of micro wax required is dependent upon the particular commercial paraffin wax involved.

Seal Strength of EVA–Wax Systems

The decay of seal strength of EVA/paraffin blends is another phenomenon which may be prevented by the use of microwaxes. The loss of seal strength or adhesive strength upon aging appears to be caused by the same mechanism which results in gloss decay, i.e. the rejection of specific normal paraffin molecules from the co-crystalline lattice composed of paraffin and copolymer molecules. However, because of the complications involved in measuring seal strength, the measurement of seal strength decay is quite difficult. The sealing strength measurement is made up of many factors including substrates and their tensile and flexural characteristics, and the adhesive compound and its cohesive and adhesive characteristics. Because of these factors, seal strength decay may be masked by either the base stock or by the conditions of measurement. Consequently many conflicting statements may be found in the trade literature because of these complicating factors.

Regardless of the complicating factors involved in measuring this property it is important that we keep in mind the physical changes which do occur with EVA/paraffin blends and what these mean in an adhesive system. The rejection of normal paraffin hydrocarbons from the crystal lattice comprised of copolymer-paraffin wax results in serious changes in tensile adhesion characteristics. These changes may result in all of the following:

1. Reduced density of the adhesive mass due to loss of the rejected normal paraffins.
2. Loss of specific adhesion by the migration of the rejected normal paraffinic hydrocarbons to the adhesive-substrate interface. This can be particularly serious with film substrates.
3. Loss of tensile strength by rejection of the normal-paraffinic hydrocarbons and concentrations of these low molecular weight materials at the crystal boundaries. This will be most serious with rigid substrates such as folding cartons or corrugated containers.

Typical changes in the systems are shown in the following table.

EVA Resin (MI 20, VA. 28%)	*Resin (alpha methylstyrene)*	*FRP 135/37°F*	*Starwax 100*	*% Loss Seal Strength*	*Seal Strength Rating on folding carton stock Initial/aged*
a. 20		80		70	Excellent/poor
b. 20	15	65		13	Excellent/fair
c. 20	15	45	20	8	Excellent/excellent
d. 25	15	60		20	Excellent/poor
e. 25	15	40	20	12	Excellent/excellent

Blocking Point

The hard high melting microwaxes are commonly used to increase the blocking and picking points of low melting paraffin waxes. Usually, the reported values are determined by means of ASTM D-1465. It should be pointed out that this procedure is intended to provide relative data only, under the condition prescribed by this method, and does not always necessarily correlate with actual practice. This method was designed to provide relative data on straight-run paraffin waxes unmodified by the addition of microwaxes. The fault does not lie with the method, but is due to the definition of the term "blocking point." Many industrial problems associated with the blocking of a wax coated product occur without any discernible surface disruption. An example of this is often found with so-called "blocked" folding cartons which feed poorly from a magazine. When these are peeled apart they will not exhibit any surface disruption which can be called either "blocking" or "picking." In such a case there are small, but very isolated, spots where one carton is bonded to another. These can cause serious problems, particularly with the bottom combing type feeder, but will feed well when fed via suction feeds. Very similar problems occur with both waxed cups and waxed papers. Our experience indicates that high melting microwaxes, particularly "Be Square" 195, are extremely effective in eliminating the practical blocking problems described above and in raising the Blocking Point as measured by D-1465.

Table 2:12

	Coefficient of Friction			
% "Be Square" 195	130/132 FRP	135/137 FRP	140/142 FRP	150/155 FRP
5	0.28	0.34	0.33	0.45
10	0.33	0.35	0.35	0.43
25	0.29	0.32	0.34	0.35
50	0.23	0.25	0.25	0.24
75	0.21	0.22	0.22	0.21
0	0.29	0.39	0.34	

Effect of Microwax on the Cloud Point of Paraffin Wax (132/135 Amp)

Fig. 2:8

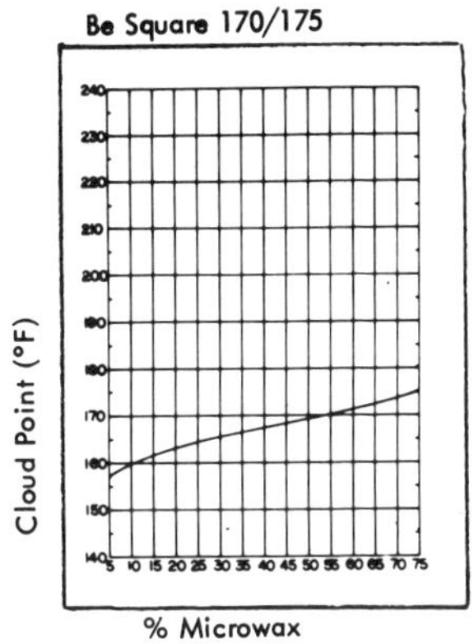

Fig. 2:9

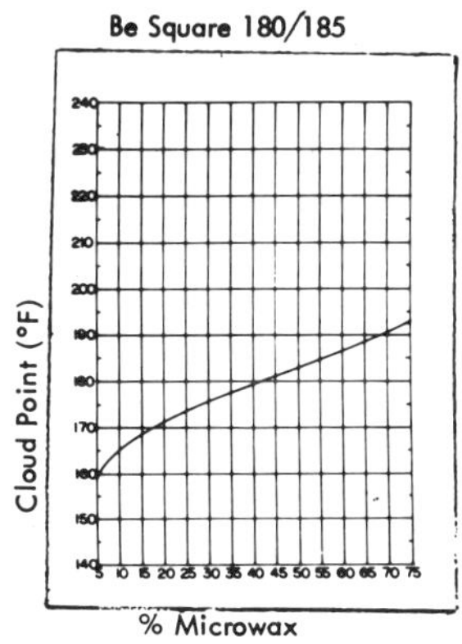

Fig. 2:10

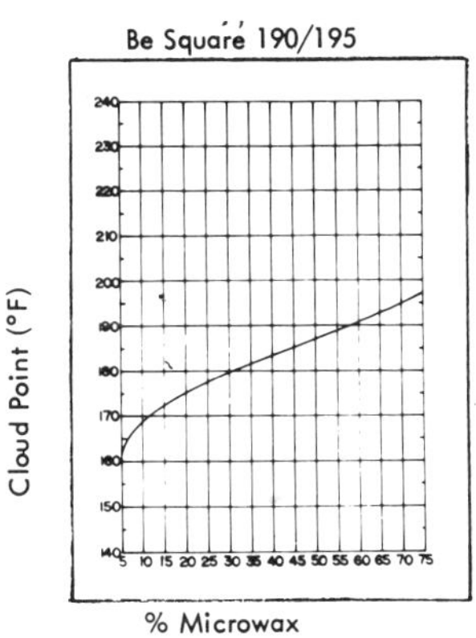

EFFECT OF MICROWAX (25% BY WT. MICROWAX) ON THE PENETRATION HARDNESS (ASTM D-1321) OF PARAFFIN WAX (132/135 AMP).

FIG. 2:11

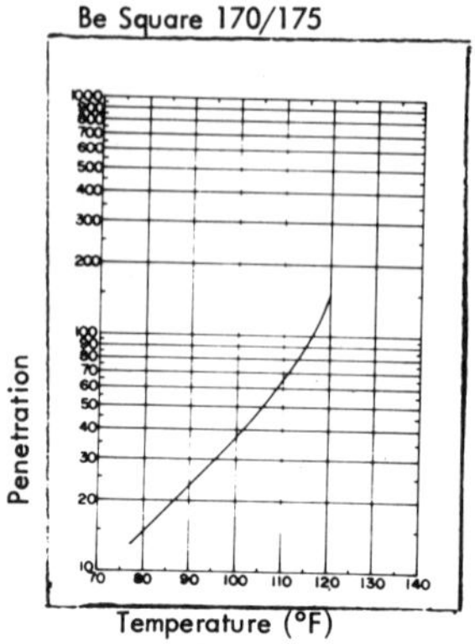

FIG. 2:12

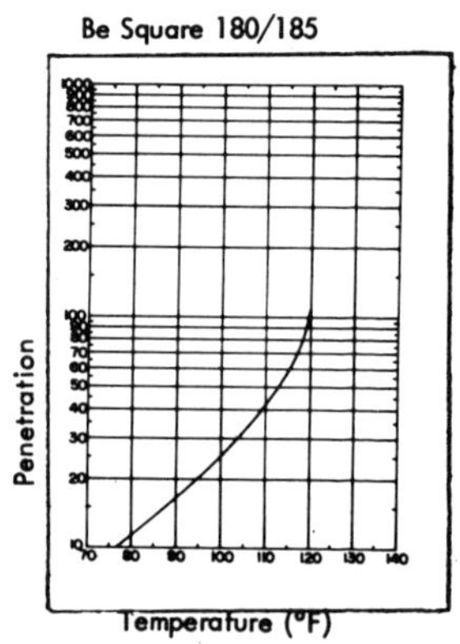

FIG. 2:13

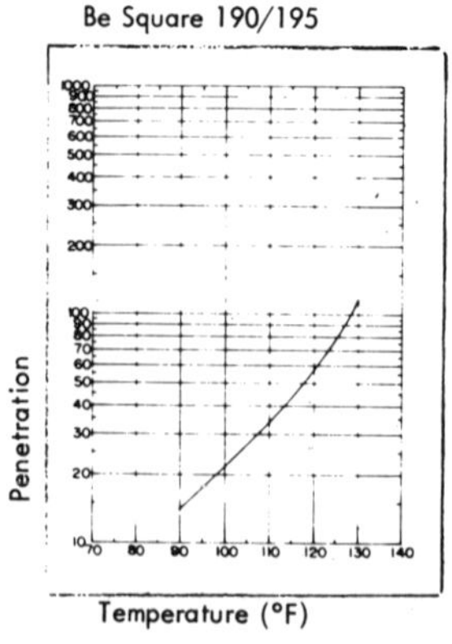

EFFECT OF MICROWAX ON THE MELTING POINT (ASTM D-127) OF PARAFFIN (132/135 AMP)

FIG. 2:14

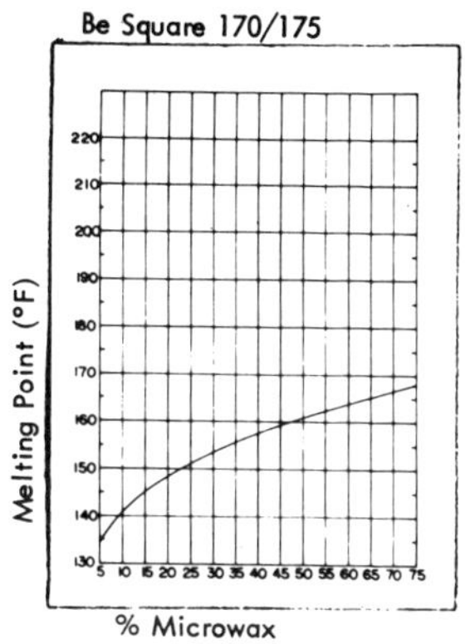

FIG. 2:15

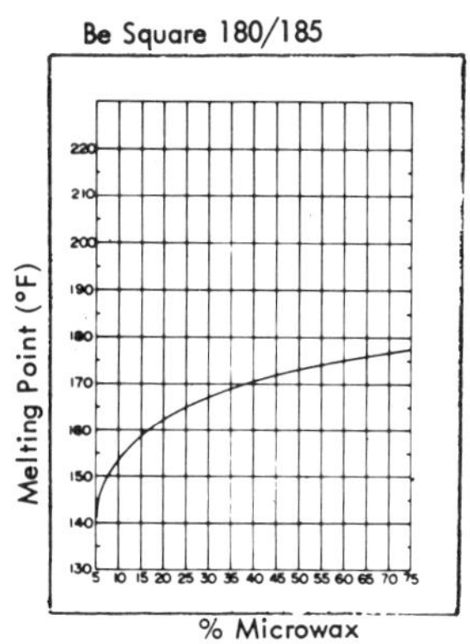

FIG. 2:16

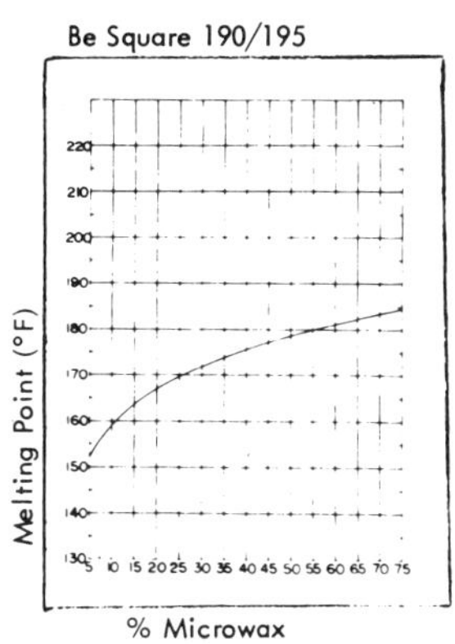

Table 2:13

% "Be Square" 195	Blocking Point 130/132 FRP	Blocking Point 135/137 FRP
0	102	100
5	105	102
10	107	107
25	114	118
50	117	126
75	126	130

Table 2:14

% Be Square 170/175A	5	10	25	50	75	5	10	25	50	75
F.R. Paraffin 135/137	95	90	75	50	25					
F.R. Paraffin 143/145						95	90	75	50	25
Congealing Point (°F)	136	137	141	153	160	142	142	142	152.5	160
Melting Point (°F)	142.5	146.5	151.5	164	172	142.5	146	152	165.5	171.5
Flash (°F)	420	425	425	445	470	440	440	445	460	480
Viscosity (SUS) 210	40.9	42.2	45	50.5	62	41.2	42.4	46.8	52.5	61.8
Penetration @ 77	13	13	15	16	17.5	12.5	12.5	12.5	13.5	15
@ 90	23.5	23.5	23	24.5	25.5	16.5	16	16	18	18.5
@100	45	43	37	39	39	25	24.5	24	27.5	28.5
@110	82	80	77	74	74	52	48	46	49	55
@120	208	190	240	158	139	125	120	120	102	103
@130	--	--	--	--	--	--	--	310	288	209
@140	--	--	--	--	--	--	--	--	--	--
Cloud (°F)	158	160	166	170	176	158	160	166	170	176

% Be Square 170/175A	5	10	25	50	75	5	10	25	50	75
F.R. Paraffin 150/155	95	90	75	50	25					
F.R. Paraffin 132/135						95	90	75	50	25
Congealing Point (°F)	150	150.5	152	155	161	133	133.5	135	150	159
Melting Point (°F)	151	151.5	154.5	166	172	134	138.5	149	164	168
Flash (°F)	460	460	470	485	500	420	420	425	445	490
Viscosity 210 (SUS)	45.2	45.4	48.7	53.8	63.3	39	39.8	43.2	49.5	59.4
Penetration @ 77	9	9.5	11.5	14	15.5	13	13	13	14	15
@ 90	11.5	11.5	13.5	18	20	22	21	21	21	21
@100	15	15.5	17.5	25	29	37	37	37	35	35
@110	21	22	25	35	46	76	67	68	65	64
@120	39	40	39	55	87	147	157	149	136	140
@130	96	97	98	117	139	--	--	--	--	--
@140	--	--	--	--	--	--	--	--	--	--
Cloud (°F)	158	160	166	170	176	157	159	165	169	175

Table 2:15

% Be Square 180/185A	5	10	25	50	75	5	10	25	50	75
F.R. Paraffin 135/137	95	90	75	50	25					
F.R. Paraffin 143/154						95	90	75	50	25
Congealing Point (°F)	137	144	153	165	169	142.5	143.5	150	163	171
Melting Point (°F)	151	155	165	175	182	146	154	166	175	182
Flash (°F)	420	425	425	445	470	440	440	445	460	475
Viscosity 210 (SUS)	40.	40.8	44.3	51.3	60.8	41.5	42.4	45.8	51.3	60.6
Penetration @ 77	13	12.5	12	11.5	11.5	11.5	11.5	12	11.	10.
@ 90	20	19.5	16	15.	14.5	15	15	15	13.5	13.5
@100	40	36	26	23	22	23.5	23.5	22	20	20
@110	75	64	55	42.5	40.5	48	46	34.5	33	33
@120	202	166	147	108	82	108	98	72	63	59
@130	--	--	--	--	--	--	--	207	163	124
@140	--	--	--	--	--	--	--	--	--	--
Cloud (°F)	160	166	174	184	194	160	166	174	184	194

% Be Square 180/185A	5	10	25	50	75	5	10	25	50	75
F.R. Paraffin 150/155	95	90	75	50	25					
F.R. Paraffin 132/135						95	90	75	50	25
Congealing Point (°F)	151	152	155	165	170.5	134	137	149.5	162	170
Melting Point (°F)	153	159	166.5	175	182	141	154	164.5	172.5	178
Flash (°F)	460	460	470	485	500	420	420	430	445	470
Viscosity 210 (SUS)	45.1	45.5	48	53	63	39.2	39.8	43.3	48.8	58
Penetration @ 77	9.5	9.5	10	10.5	10.5	12	11	10	9.5	9
@ 90	12.	12	12	13	13.5	20	18.5	15	14.	14
@100	16	16	15.5	17.5	18.5	37	30	25	24	24
@110	20	20	22	25	28	66	53	47	40	38
@120	38	35	33	39	44	121	118	110	90	72
@130	82	73	--	--	--	--	--	--	--	--
@140]	178	180	181	154	143	--	--	--	--	--
Cloud (°F)	160	166	174	184	194	159	165	173	183	193

Table 2:16

% Starwax 100	5	10	25	50	75	5	10	25	50	75
F.R. Paraffin 135/137	95	90	75	50	25					
F.R. Paraffin 143/145						95	90	75	50	25
Congealing Point (°F)	137	141	152.5	157	164	142.5	143	140	158	165
Melting Point (°F)	147	151	159.5	169.5	177	143	151	160	170	179
Flash (°F)	420	425	430	450	475	440	440	445	460	480
Viscosity 210 (SUS)	40.8	41.8	45	50.6	61.2	41.5	42.4	46	51.5	61.4
Penetration @ 77	13	13	13	13.5	14.5	12	12	12.5	13	13.5
@ 90	23.5	23	20	20	20	16	16	16.5	17	17
@100	41	39	33	33	34	25	24	24	24.5	25
@110	72	64	77	77	69	51	46	43.5	43.5	47
@120	200	193	220	148	120	118	112	99	81	81
@130	--	--	--	--	--	--	--	251	229	170
@140	--	--	--	--	--	--	--	--	--	--
Cloud (°F)	160	165	170	176	180	160	165	170	176	180

% Starwax 100	5	10	25	50	75	5	10	25	50	75
F.R. Paraffin 150/155	95	90	75	50	25					
F.R. Paraffin 132/135						95	90	75	50	25
Congealing Point (°F)	150	150.5	153	157	165	133.5	135	141	156	164
Melting Point (°F)	151	152	158.7	169.5	177	140	150	159.5	169.5	176.5
Flash (°F)	460	460	470	485	505	420	420	440	460	480
Viscosity 210 (SUS)	45.2	45.4	48.7	54.	65.5	39.2	40.	43.6	49.3	60.4
Penetration @ 77	9.5	9.5	10	12	13	12.5	12.5	12.5	12.5	13
@ 90	12.5	12.5	12.5	15	16.5	21	21	20	18.5	18
@100	16	16	16.5	20	24	35	35	35	30	28
@110	21	21	25	31	38	68	64	65	51	48
@120	40	40	38	48	61	148	150	132	105	95
@130	91	92	--	--	--	--	--	--	--	--
@140	178	190	220	198	169	--	--	--	--	--
Cloud (°F)	160	165	170	176	180	159	164	170	175	179

Table 2:17

% Be Square 190/195A	5	10	25	50	75	5	10	25	50	75
F.R. Paraffin 135/137	95	90	75	50	25					
F.R. Paraffin 143/145						95	90	75	50	25
Congealing Point (°F)	138.5	148	158	168	175	143	147	157	167	175
Melting Point (°F)	154	160	171.5	179	184.5	156	162	171	180	185
Flash (°F)	420	425	445	460	480	440	440	455	465	480
Viscosity 210 (SUS)	41.2	42.	46	53.8	61.	41.5	42.5	46.3	52	61.5
Penetration @ 77	12.5	11.5	10	8	6.5	11.5	11.5	10.5	9.5	8.
@ 90	17.5	16.5	14	11	9.	16.	16	13	11	9.5
@100	30	28	22	16	12.5	25.	25	17.5	14	12.5
@110	64	58	40	28	21	47	47	27	21	19
@120	170	140	98	59	42	98	80	50	38	30
@130	--	--	75	206	89	228	210	150	88	59
@140	--	--	--	--	--	--	--	--	--	--
Cloud (°F)	162	170	178	188	198	162	170	178	188	198

% Be Square 190/195A	5	10	25	50	75	5	10	25	50	75
F.R. Paraffin 150/155	95	90	75	50	25					
F.R. Paraffin 132/135						95	90	75	50	25
Congealing Point (°F)	151	156	158	170	178	135	141	156	167	174
Melting Point (°F)	154	163	171	180	186	152.5	159.5	169	179	184.5
Flash (°F)	460	460	470	485	500	420	420	445	460	480
Viscosity 210 (SUS)	45	45.4	48.6	54.5	63.7	39.2	40.	43.5	49.4	60.
Penetration @ 77	9.5	9.5	8.5	7.5	6.5	11.5	11.	9.5	8.5	7.5
@ 90	12	12	9.5	9.	7.8	18	17.5	14	12	10.5
@100	15.5	15	12	11.5	10.5	31	27	21	18	16
@110	20	19	16.5	16	15	60	48	34	32	24
@120	36	33	25	23	21.5	101	92	62	58	40
@130	80	68	42	38	34	--	--	--	102	67
@140	158	148	109	93	68	--	--	--	--	--
Cloud (°F)	162	170	178	188	198	161	169	177	187	197

Blending Microcrystalline Wax with Resins

The wax is melted in a heated mixing vessel and brought up to a temperature of about 250°F. At this point the antioxidant stabilizer should be added. With power stirring, the resins are added. If there are two or more resins, they should be added in descending order of melt index, always allowing the first to be thoroughly dissolved before adding the second. After the EVA resins are thoroughly dissolved and mixed, the modifiers can be added, usually in descending order of thermal stability. Slip agents are usually added last.

Table 2:18

MULTIWAX 180-M (MICROCRYSTALLINE)
PROPERTIES

Melting Point, (ASTM D-127), °F	185
Needle Penetration, (ASTM D-1321), @77 °F	16
Saybolt Viscosity, SUS, (ASTM D-88), @210 °F	85
Color, (ASTM D-1500)	2 Max.
Flash Point, COC, (ASTM D-92), °F	525 Min.
Specific Gravity, @220 °F	0.79
Pounds Per Gallon, @220 °F	6.3
Odor and Taste	None
FDA Status	Complies with CFR 121.1156 & CFR 121.2586

ELECTRICAL PROPERTIES

Dielectric strength	@30 °C	460 volts/mil
Dielectric constant		
1,000,000 cycles/sec.	@30 °C	2.05
	@100 °C	2.07
Dissipation factor		
100,000 cycles/sec.	@30 °C	3.5×10^{-4}
	@100 °C	0.6×10^{-4}
DC volume resistivity		
500 volts, Ohm-cm	@30 °C	$>2.7\times10^{17}$
	@100 °C	3.6×10^{13}

Table 2:19

MULTIWAX 195-M
PROPERTIES

Melting Point, (ASTM D-127), °F	195
Needle Penetration, (ASTM D-1321), @77 °F	7
Saybolt Viscosity, SUS, (ASTM D-88), @210 °F	86
Color, (ASTM D-1500)	1 Max.
Flash Point, COC, (ASTM D-92), °F	550 Min.
Specific Gravity, @240 °F	0.78
Pounds per Gallon, @240 °F	6.2
Odor and Taste	None
FDA Status	Complies with CFR 121.1156 & CFR 121.2586

Table 2:20

MULTIWAX 200-M
PROPERTIES

Melting Point, (ASTM D-127), °F	200
Needle Penetration, (ASTM D-1321), @77 °F	5
Saybolt Viscosity, SUS, (ASTM D-88), @210 °F	85
Color, (ASTM D-1500)	1 Max.
Flash Point, COC, (ASTM D-92), °F	550 Min.
Specific Gravity, @240 °F	0.78
Pounds per Gallon, @240 °F	6.2
Odor and Taste	None
FDA Status	Complies with CFR 121.1156 & CFR 121.2586

Table 2:21

MULTIWAX ML-445
PROPERTIES

Melting Point, (ASTM D-127), °F	173
Needle Penetration, (ASTM D-1321), @77 °F	30
Saybolt Viscosity, SUS, (ASTM D-88), @210 °F	85
Color, (ASTM D-1500)	2 Max.
Flash Point, COC, (ASTM D-92), °F	525 Min.
Specific Gravity, @220 °F	0.80
Pounds per Gallon, @220 °F	6.4
Odor and Taste	None
FDA Status	Complies with CFR 121.1156 & CFR 121.2586

Table 2:22

MULTIWAX W-835
PROPERTIES

Melting Point, (ASTM D-127), °F	170
Needle Penetration, (ASTM D-1321), @77 °F	70
Saybolt Viscosity, SUS, (ASTM D-88), @210 °F	85
Saybolt Color, (ASTM D-156)	+16 Min.
Flash Point, COC, (ASTM D-92), °F	500 Min.
Specific Gravity, @200 °F	0.81
Pounds per Gallon, @200 °F	6.5
Odor and Taste	None
FDA Status	Complies with CFR 121.1156 & CFR 121.2586

Table 2:23

MULTIWAX X-145A
PROPERTIES

Melting Point, (ASTM D-127), °F	150
Needle Penetration, (ASTM D-1321), @77 °F	40
Saybolt Viscosity, SUS, (ASTM D-88), @210 °F	83
Color, (ASTM D-1500)	2 Max.
Flash Point, COC, (ASTM D-92), °F	500 Min.
Specific Gravity, @200 °F	0.81
Pounds per Gallon, @200 °F	6.5
Odor and Taste	None
FDA Status	Complies with CFR 121.1156 & CFR 121.2586

OXIDIZED MICROCRYSTALLINE WAXES

Oxidized microcrystalline waxes are prepared from the residue on the bottom of the tanks in petroleum refining. This is extracted with methyl ethyl ketone, chilled, filtered, and refined. It is then oxidized in air in the presence of a catalyst, often cobalt or manganese soaps.

Estawax is an oxidized microcrystalline wax with properties as follows:

	Estawax 20	*Estawax 25*
Color, NPA	3 Max.	3 Max.
Melting Point, °F (ASTM E28–42T)	210–220	210–220
Penetration (100/5/77)	2 Max.	2 Max.
Viscosity, centipoises at 300°F	450–550	200–300

Uses: Phonograph-record component; paste and liquid polishes.

Cardis One is an oxidized microcrystalline wax with a hardness of 1–2 (ASTM penetration).

It is easily emulsifiable and yields films of the highest gloss and wear resistance.

Its emulsions dry to highly scuff-resistant films.

It gives translucent and opaque emulsions of the highest quality. Other properties are:

Penetration (ASTM)	1–2
Color	4–5
Melting Point	195–200°F
Acid Number	13–15
Saponification Number	55–60

Mycrox-Wax 114 is an oxidized microcrystalline wax, used as a substitute for beeswax which it resembles in plasticity and emulsifiability. Its properties are as follows:

Setting Point	69°C
Penetration I.P.T.	16–18
Acid Value	20–21
Sap. Value	66–65
Iodine Value	4–6
Color	Almost White
Foreign Matter	0

Shanco W-1102 Wax

SHANCO W-1102 is a pale oxidized microcrystalline wax designed for use in emulsion type polishes.

Melting Point—Air Drop Method	181–184°F
Acid Number	20–22
Color—Gardner	8
Saponification Number	40–45
Penetration—100 grams, 77°F, 5 sec.	5–7

Shanco W-1105 Wax

SHANCO W-1105 is an oxidized and emulsifiable microcrystalline wax.

Melting Point—Air Drop Method	181–183°F
Acid Number	20–22
Color—Gardner	12
Saponification Number	40–45
Penetration—100 grams, 77°F, 5 sec.	4–5

Shanco W-1124 Wax

SHANCO W-1124 is a light colored, high melting, oxidized microcrystalline wax.

Melting Point—Air Drop Method	182–184°F
Acid Number	14–16
Color—Gardner	7
Saponification Number	27–32
Penetration—100 grams, 77°F, 5 sec.	4–6

Emulsifiable Microcrystalline Waxes

Emulsifiable waxes result from the oxidation of high-molecular-weight microcrystalline waxes. This oxidation forms esters and fatty acids which impart saponifiability. The emulsifiable waxes are widely used in the manufacture of polishes. They are also used to reduce substantially the amount of vegetable waxes. In many cases they have displaced vegetable wax entirely.

PETROLATUM WAX

Petrolatum wax (mineral jelly) consists of a natural mixture of microcrystalline wax plus 10% of mineral oil. Freed from most of its liquid and semi-liquid hydrocarbons, petrolatum yields an amorphous, white, translucent, plastic wax. It is commercially available under many names. Its plastic and noncrystalline properties make it very useful. It also has good oil-holding capacity and prevents leakage or sweating out of oils. Thus, it can replace ceresin and ozokerite, where their low melting point is not undesirable.

Table 2:24

PROPERTIES OF COMMERCIAL PETROLATUM

Brand of Petrolatum	*White Fybrene*	*Olive Fybrene*
Specific Gravity at 60°F	0.896–0.899	0.896–0.899
API Gravity	26.0–26.5	26.0–26.5
ASTM Melting Point, °F	128–133	130–135
ASTM Consistency (Method D-217)	50–85	30–50
ASTM Consistency (Method D-5)	150–300	75–200
Flash Point, °F	460–470	460–470
Fire Point, °F	520	520
Saybolt Vis. at 210°F	50–55	50–55
Color	White	Olive Green
Ash, %	0.03	0.03
Saponification Number	0.3	0.3

Raising the Melting Point of Petrolatum

The maximum increase in melting point of petrolatum is usually obtained by adding 3–5%, of a variety of waxes. As 5% white beeswax increases the melting point 2°, 10% increases it 5°, and 20% increases it 9°, obviously, the optimum amount is somewhere between 10 and 20% beeswax.

An addition of 5% yellow beeswax increases the melting point 2°, 10% increases it 10° and 20% increases it 17°. That its, yellow beeswax is better than white beeswax for increasing the melting point of petrolatum.

USES OF MICROCRYSTALLINE WAXES

Microcrystalline waxes are used for the laminating of paper, cloth, etc.; for waterproofing of paper, boxboard, textiles, wood, etc.; potting

Table 2:25

MELTING POINTS OF MIXTURES: WHITE PETROLATUM AND BEESWAX

Petrolatum, %	*Beeswax, %*	*M.P., °F, with* White	Yellow
97	3	110	112
95	5	112	114
90	10	115	122
80	20	119	129
100	0	124	142

compounds for condensers; electric wire insulation; binder for pipe coverings; finishes for leather; polishes for floors, furniture, skis, leather; rust prevention; compounding of rubber; pattern-making (sheets and fillets); cosmetics, printing inks, lubricants, records.

They are steadily increasing in value and application in the packaging industry.

Some specialized uses are:

Drum Linings

Microcrystalline waxes are widely used for lining drums. The application of the wax is quite simple. The drum is heated to a temperature that will give the thickness desired. Heated, molten wax is poured into the bung, the opening is closed, and the drum is rolled in such a way as to bring the molten wax into contact with the entire inner surface. The excess is then poured out or removed by other means. Any desired thickness of film can be evenly distributed over the inner surface, with considerable reinforcement at the corners. The quantity required will vary from 0.5 to 3 or 4 lb, depending on the desired thickness. Adhesion will be excellent with the proper wax, if the interior of the drum is absolutely dry.

Vat Linings

Wood or steel vats can be lined with wax in a number of ways. The surface to be coated must be absolutely dry and preferably warm. The wax may then be applied by trowelling (hot, plastic), flushing, mopping, or spraying. In the case of mopping or trowelling, it is good precaution to apply a preliminary thin spray coat and then the second coat while the surface is still warm enough to insure good adhesion.

These are approximate figures, but they will serve as guides and multipliers. They are for vertical tanks, with only the bottoms and vertical sides waxed.

Several testing methods are required for exploring commercial applications and many of the methods must yet be standardized. The properties of five additional commercial waxes, without the addition of oil, are shown in Table 2:6. They may be of assistance to commercial users of wax and may help in the formulating of specifications and testing methods.

Table 2:26

WAX QUANTITIES FOR VAT LININGS

Diameter and Height, ft	*Volume gal*	*Coatings Thickness, in.*	*Quantity of Wax, lb*
5 × 5	785	0.25	119
5 × 10	1,469	0.25	214
10 × 5	2,937	0.25	285
10 × 10	5,875	0.25	475
10 × 20	11,750	0.25	855
20 × 10	23,501	0.25	1,140
20 × 20	47,002	0.25	1,900

Table 2:27

PROPERTIES OF COMMERCIAL MICROCRYSTALLINE WAXES BY SPECIAL TESTING METHODS

Type of Wax	*A*	*B*	*C*	*D*	*E*
Color	Amber	Brown	Yellow	White	Brown
Melting Point, °F	156.0	156.0	162.0	163.0	158.0
Penetration, 0.01 cm	40.0	37.0	31.0	25.0	29.0
Contraction, %	10.3	10.7	11.0	11.4	10.0
Plastic Point, °F	13.0	20.0	48.0	55.0	15.0
Tensile Strength, lb/sq in.	78.0	80.0	100.0	116.0	140.0
Compressive Strength, lb/sq in.	78.0	84.0	102.0	120.0	140.0
Ductility, cm	1.6	1.6	1.6	1.6	1.8
Flexibility, number of bends	164.0	168.0	160.0	150.0	200.0

MISCELLANEOUS DATA

Blocking, hardness, and tensile strength of microcrystalline waxes are adversely affected by increased content of branched-chain paraffins. Sealing strength and flexibility are improved by increased content of cyclic and branched paraffins.[1]

Small amounts of Paraflow inhibit excessive growth of crystals and change the shape from the needle type to small equiaxial grains of uniform size.[2]

Microcrystalline waxes mixed with 5–10 percent polypropylene at 140°C give products of tensile strength and better flexibility and adhesion.[3]

Sealing strength is improved and brittle point is lowered in microcrystalline wax by addition of 2–8.5% of hydrogenated phenanthrenes and chrysenes.[4]

PROPERTIES OF SOME COMMERCIAL MICROCRYSTALLINE WAXES

AMSCO Waxes

Microcrystalline

Properties

Melting Point, (ASTM D-127-49)	155	164	170	175
Color, NPA (ASTM D-1500-57)	2	2	2	2
Viscosity, SUS @ 210°F	63.2	98.0	93.7	90.4
Penetration, Std. Needle @ 77°F (ASTM D-1321-57T)	21	41	21	13
Stain Test (KVP) °F	135	122	Above 140	Above 140
Oil Content % (ASTM D-721-56T)	.80	1.36	0.24	0.12
Bareco Bend Test °F	+32	+2	+43	+55

[1] Tappi, **41**, 257–67 (1958).
[2] G. H. B. Davis, *Nat. Petroleum News* **24**, 32 (1932).
[3] Ger. Patent 1,050,052, Feb. 5, 1959.
[4] A. Schaerer, U.S. 2,780,556, Feb. 5, 1957.

Aristowaxes

Type of Wax	*130/134*	*143/150*	*143/150 NC*	*165*	*MB*	*Canco 134*
American Melting Point (ASTM D-87 + 3°F)	132	146	148	—	—	—
Congealing Point, ASTM D-938, °F	129	143	145	163	152	131
Oil Content, ASTM D-721, % by weight	0.2	0.15	0.04	0.2	0.2	0.3
Saybolt Color, ASTM D-156	+28	+28	+28	+28	+18	+23
Kinematic Viscosity, ASTM D-445, Centistokes at 210°F	3.7	5.4	4.8	7.3	6.3	3.9
Saybolt Universal Viscosity, ASTM D-446, SSU at 150°F	48.5	63.2	56.3	—	—	50.0
SSU at 175°F	43.0	52.2	48.4	62.0	57.5	44.0
SSU at 210°F	38.5	44.0	42.0	50.0	47.0	39.1
SSU at 250°F	35.4	39.1	37.9	43.0	40.8	35.9
Needle Penetration, ASTM D-1321, mm × 10 at 77°F	12	12	10	12	13	13
mm × 10 at 90°F	25	15	14	14	20	24
mm × 10 at 100°F	75	29	25	18	35	59
mm × 10 at 110°F	—	50	55	26	58	—
Brinell Hardness Number, UTM 197, Kg/cm^2 at 60°F	80	75	95	80	65	75
Kg/cm^2 at 100°F	soft	14	30	29	9	soft
Cylinder Penetration, UTM 203, in./min × 10^4 at 100°F	soft	10	0.5	—	30	—
Blocking Temperature, ASTM D-1465, °F	97	116	119	135	116	101
Modulus of Rupture, ASTM 1958 App VI, psi at 32°F	450	440	440	350	340	460
psi at 73°F	410	350	400	325	290	420
Tensile Strength, ASTM D-1320, psi at 73°F	380	340	360	400	280	350

Sealing Strength, ASTM 1958						
App V, g/in.	10	5	5	28	20	—
with 4% Polyethylene, g/in.	13	5	19	tear	tear	—
Thermal Shock Resistance Test, UTM 256	—	—	—	—	—	pass
Specific Gravity, 212°F/39.2°F	0.7525	0.7660	0.7627	0.7737	0.7695	0.7573
Refractive Index at 212°F	1.4214	1.4275	1.4255	1.4305	1.4305	1.4245
COC Flash Point, ASTM D-92, °F	430	490	480	515	465	435
Food Packaging Laws	pass	pass	pass	pass	pass	pass

Bareco Waxes

Bareco microcrystalline waxes are featured by high melting points, excellent heat-sealing characteristics, plasticity over a wide temperature range, hardness, high gloss, and a wide range of colors.

Bareco microcrystalline waxes are classified into four groups: (1) hard, (2) plastic, (3) emulsifiable and (4) synthetic.

General Properties

Type of Wax	*Laminating*	*Coating*	*Hardening*
Melting Point, °F ASTM D-127-49	130–155°	140–165°	165–210°
Needle Penetration at 77°F ASTM D-1321-54T	20–35	18–28	below 5
Color, NPA	1–3	1–3	1–3
Minimum Flex Temperature, °F	0–35°	0–35°	above 75°
Oil Content, per cent	1–5	0.5–2	0.0–1
Urea Reactible, per cent	20–35	20–40	50–90
Odor	Slight/None	Slight/None	Slight/None

Bareco Hard Waxes

The Bareco hard microcrystalline wax group includes the hardest and highest-melting-point petroleum-derived waxes available.

Specifications

Product	*Melting Pt. Minimum °F*	*Penetration 77°F Maximum*	*Color N.P.A. Maximum*	*Acid No.*	*Sap. No.*
Petrolite C-1035	195	3	1½	Nil	Nil
Petrolite C-200-7	190	4	†	Nil	Nil
Petrolite C-700	190	4	1½	Nil	Nil
Petrolite C-200	190	8	†	Nil	Nil
Petrolite C-500	190	8	2½	Nil	Nil
Be Square 190/195	190	7	1½*	Nil	Nil
Be Square 180/185	180	11	1½*	Nil	Nil
Starwax 100	180	16	2*	Nil	Nil

Bareco hard waxes are used in polishes, printing ink, phonograph records, carbon paper, paper coating, sizings, protective coatings.

Bareco Plastic Waxes

Bareco plastic microcrystalline waxes are characterized by ductility, high sealing strength, freedom from odor and taste, low temperature flexibility and strong resistance to moisture-vapor transmission.

Specifications

Product	*Melting Pt. Minimum °F*	*Penetration 77°F Maximum*	*Color N.P.A. Maximum*	*Acid No.*	*Sap. No.*
Be Square 170/175	170	20	1½*	Nil	Nil
Ceraweld	155	25	2½*	Nil	Nil
Ceratak	155	30	2½	Nil	Nil
Victory	155	30	2½*	Nil	Nil
Ultraflex	143	30	2*	Nil	Nil

* Also available in other colors.
† Brown

"Bareco" Polywax E-2020 Emulsion

"Bareco" Polywax E-2020 must be emulsified with pressure equipment since its melting point is very much above the boiling point of water at atmospheric pressure.

	Cationic LX-368	*Nonionic LX-380*	*Anionic LX-370*
"Bareco" Polywax E-2020	100	100	100
"Ethoquad 18/12	20		
Triton N-101		20	
KOH (85%)		2.5	
Oleic Acid			8
Diethylethanolamine			8
Water (Initial)	50	50	40
Water (To be added)	230	234	204
% Polywax Solids	25%	25%	25%

1. The 'Polywax,' emulsifiers and enough water to produce approximately 70% solids are charged to the pressure vessel.
2. With the vents open, heat the vessel to approximately 210°F to expel air.
3. On reaching 210°F, close the vessel and continue heating until a temperature of 260°F to 270°F is reached.
4. Begin agitation and stir while adding the remaining water which has been pressurized and heated at 240°F. During this step the addition of the water should be very slow until the "water in oil" emulsion inverts to an "oil in water" emulsion. This inversion occurs close to the 40% solids level and is characterized by

a thick gel. After this gel has "broken," the remainder of water may be added as rapidly as desired.

5. Allow the emulsion to cool to 212°F at which point the reaction vessel is vented to prevent a vacuum from forming and possible foaming. The emulsion at this stage is 30% solids and can now be further cooled to less than 120°F for storage; however, it is preferred to cool to less than 100°F before storing.

Bareco Emulsifiable Waxes

Bareco emulsifiable waxes result from the oxidation of high-molecular-weight microcrystalline waxes. This oxidation forms esters and fatty acids which impart saponifiability.

Specifications

Product	*Melting Pt. Minimum °F*	*Penetration 77°F Maximum*	*Color N.P.A. Maximum*	*Acid No.*	*Sap. No.*
Petrolite C-15	180	6	5	15/17	45/55
Petrolite C-23	180	6	5	20/25	55/65
Petrolite C-36	180	8 max.	6	30/35	75/85
Petrolite PE-100	195	3	6	15/20	45/55
Petronauba C	180	7	3	22/28	50/60
Petronauba D	185	5	6	20/28	50/60
Petronauba F	180	5	3	15/25	50/60
Petronauba H	195	3	4	15/25	50/60

The Bareco emulsifiable waxes are used in floor, furniture and leather polishes, leather dressings, self-polishing compounds.

Bareco Synthetic Waxes

Bareco microcrystalline waxes are combined or reacted with various resins or other chemical materials to incorporate characteristics not attainable in the waxes alone.

Specifications

Product	*Melting Pt. Minimum °F*	*Penetration 77° Maximum*	*Color N.P.A. Maximum*	*Acid No.*	*Sap. No.*
Petrolite C-6500	200	3	4	10/13	30/40
Petrolite C-7500	205	3	2	10/13	25/30
Petrolite R-50	190	2	4½	40/50	65/80
Petrolite P-20	210	2	3	Nil	Nil
Petrolite P-25	210	2	3	Nil	Nil
Petrolite WB2	180	5	†	23/30	70/85
Petrolite WB5	185	3	†	19/27	50/65
Petrolite PC13	185	8	5	Nil	Nil
Petrolite PC44	160	4	5	Nil	Nil

† Brown.

Many of the Bareco synthetic waxes are tougher, more flexible and higher melting than carnauba. Certain of the Bareco synthetics are extremely hard and have excellent molding characteristics without being brittle.

Bareco synthetic waxes are used in polishes, carbon paper, casting wax.

ELECTRICAL PROPERTIES OF BARECO WAXES

	Petrolite C-1035			*Petrolite* C-500		
			DIELECTRIC CONSTANT			
°C	60	10^3	10^6	60	10^3	10^6
80	2.46	2.30	2.47	2.38	2.49	2.48
60	2.37	2.39	2.44	2.42	2.53	2.50
40	2.31	2.25	2.44	2.56	2.74	2.74
20	2.16	2.09	2.45	2.62	2.79	2.79
			% POWER FACTOR			
80	2.20	2.05	0.11	0.11	0.04	0.12
60	2.10	1.98	0.08	0.04	0.03	0.08
40	1.69	1.64	0.08	0.22	0.04	0.20
20	1.79	1.51	0.33	0.06	0.03	0.17

POTENTIAL BREAKDOWN (V/MIL)

°C	20	40	60	80
Petrolite C-1035	2050	710	550	675
Petrolite C-500	1500	780	625	420

VOLUME RESISTIVITY

Petrolite C-1035	3.8×10^{15} ohms-cm 22°C (55% humidity)
Petrolite C-500	6.4×10^{15} ohms-cm 22°C (55% humidity)

Be Square Waxes

The following table gives the solubility of Be Square Waxes in terms of cloud point for various weights of each wax in 100 ml of solvent.

Solvent	*Cloud Point, °F* Be Square 170–175 1 g	5 g	10 g	Be Square 190–195 1 g	5 g	10 g
n-Butly Alcohol	173	196	198	190	203	205
sec-Butyl Alcohol	136	150	155	154	167	171
n-Amyl Alcohol	158	174	178	176	192	194
Isopropyl Acetate	143	154	155	164	175	176
Chloroform	94	108	114	103	126	138
Ethylene Dichloride	128	142	144	151	158	160
Carbon Tetra-chloride	90	104	108	102	120	125
Benzene	99	114	118	114	130	135
Toluene	94	109	116	110	128	134
Xylene	98	112	120	112	132	136
Nitrobenzene	147	165	170	166	180	184
Turpentine	92	120	124	112	134	140
V.M.P.Naphtha	105	114	122	120	131	138
Stoddard Solvent	98	115	128	116	134	145
Kerosene	100	122	132	122	140	143
Chlorox	174	5 g will not dissolve at b.p.		184	5 g will not dissolve at b.p.	
Aniline	260	,, ,, ,, ,, ,,		264	,, ,, ,, ,, ,,	
Ethyl Alcohol	1 g will not dissolve at b.p.			1 g will not dissolve at b.p.		
Isopropyl Alcohol	,, ,, ,, ,, ,,			,, ,, ,, ,, ,,		
Furfural	,, ,, ,, ,, ,,			,, ,, ,, ,, ,,		
Phenol	,, ,, ,, ,, ,,			,, ,, ,, ,, ,,		
Cellosolve	,, ,, ,, ,, ,,			,, ,, ,, ,, ,,		

Mobilwaxes

(Microcrystalline)

I

Name of Mobilwax	*Cerese*	*100*	*105*	*2300*	*2305*
Description	Coating, Electrical Wax	Laminating Waxes			
Color, ASTM	Lt 2	4½ Dil	Lt 3	4½ Dil	Lt 2
Melting Point, °F	180	169	169	170	170
Viscosity, SSU @ 210°F	80	95	95	80	80
Needle Penetration					
@ 77°F	16	27	27	27	27
@ 108°F	40	85	85	105	105
@ 115°F	50	125	125	160	160
Oil Content, %	—	4.5	4.5	3.5	3.5
Flash Point, °F	540	580	580	540	540

II

Name of Mobilwax	*22*[1]	*24*[1]	*105BR*	*1305BR*	*R*	*RB*	*RF*
Description	Special Laminating Waxes				Micro-Paraffin Blend		
Color, ASTM	White	White	3	2½	1½	2	1½
Melting Point, °F	170	170	169	170	150	145	155
Viscosity, SSU @ 210°F	180	500	160	160	53	45	54
Needle Penetration							
@ 77°F	27	27	27	27	—	30	21
@ 108°F	—	—	85	105	—	—	—
@ 115°F	—	—	125	160	—	—	—
Ash Content, %	10	20	—	—	—	—	—
Flash Point, °F	540	540	580	540	—	—	—

[1] Mobilads.

Multiwaxes

Grade	*M.P.* (min) (ASTM D127-49)	*Penetration* (ASTM D1331-54T)	*Color*	*Viscosity* Saybolt @ 210°F
MOST WIDELY USED GRADES				
W835	155°F	60/80	White	80/90
W445	165°F	25/35	White	80/90
ML445	165°F	25/35	Pale Yellow	80/90
BH445	165°F	25/35	Light Amber	80/90
D445	160°F	45/60	Dark Olive	90 min.
D835	155°F	60/80	Dark Olive	90 min.
OTHER GRADES				
X145	140°F	35/45	Pale Yellow	60/70
X145A	140°F	35/45	Pale Yellow	70/85
180M	180°F	10/20	Pale Yellow	80/90
180R	180°F	10/20	Dark Brown	80/90
195M	190°F	2/7	Pale Yellow	80/90
195R	190°F	2/7	Dark Brown	80/90

Pearl Waxes

Color	From Amber 5 NPA to Yellow 3 NPA
Melting Point	180°F minimum
Penetration at 77°F 50 g 5 sec	30 minutes
Odor	None

Shell Waxes

Shellmax *No.*	*Color* ASTM D-1500	*Refractive* Index 100 ⁿD	*M.P.* D-127°F	*Viscosity* SSU@ 210°F	*Penetration* @ 77°F	*Oil Content* %W
500	L2.0	1.4435	141.3	80.0	22	0.9
400	1.5	1.4413	176.4	77.8	13	0.6
800	L6.5	1.4496	166.6	89.8	96	11.1

SOVAWAXES

These waxes are used for coating or laminating, their application being restricted only by the special hardness or melting-point requirements of isolated uses. They are recommended for all applications requiring inert, chemical-resistant, durable, adhesive, flexible coatings or impregnants.

Brand	*Sovawax 100*	*Sovawax 105*	*Sovawax 110*
Melting Point, °F ASTM, minimum	160	160	160
Needle Penetration at 77°F	20/30	20/30	20/30
Saybolt Viscosity at 210°F	85/95	85/95	85/95
Flash Point, °F (COC)	500	500	500
Odor	OK	OK	OK
Color, ASTM	3–4 dil.	2–3	2 max.
Flexibility at 0°F	OK	OK	OK

Superflex Waxes

Name	*M.P.* *ASTM*	*Penetration* @ 77°F	@ 100°F	*Visc.* @ 210°F	*Flexing* (°F)
S100	145–150	25 Max.	75 Max.	80–100	0/+10
L500	150–160	25 Max.	85 Max.	80–100	+10/+25
M85	160–170	25 Max.	95 Max.	80–100	+30/+50

Table 2:28

WARWICK WAXES

NAME OF WAX	A.S.T.M. D-127-30	100g/77°/ 5 sec.	COLOR N.P.A.	ACID NUMBER	SAPONIFICATION VALUE	TYPE
Polymekon®*	195-205	2-4	Yellow	0-0	0-0	Petroleum Wax
Fortex®	190-200	3-4	1 Max.	0-0	0-0	Hard & Brittle
Mekon® White	190-200	3-5	1 Max.	0-0	0-0	Hard & Brittle
Mekon® Amber	190-200	3-5	3 Max.	0-0	0-0	Hard & Brittle
Warco® 180 White	180-190	4-7	White	0-0	0-0	Hard & Brittle
Cardis® Polymer #8*	195-205	1-3	0-2.5	9-14	30-40	Polymer
Cardis® Polymer #10	210-215	1-2.5	0.5-1	10-15	25-30	Polymer
Cardis® 314*	180-190	4-7	1-3	13-18	45-55	Petroleum Wax
Cardis® 319*	180-185	5-7	2-4	15-20	65-75	Petroleum Wax
Cardis® 320*	180-185	5-8	1-3	25-30	70-80	Petroleum Wax
Cardis® 262*	195-200	3-6	Brown	15-20	40-45	Petroleum Wax
Cardis® One*	195-200	2-4	1-3	12-16	40-45	Petroleum Wax
Cardipol® N-50	260 Max.	1-4	1.5 Max.	0-0	0-0	Polymer
Cardipol® 0-10	240-250	1-2	0.5-1.0+	10-12	30-40	Emulsifiable Polymer
Cardipol® 0-20	225-235	2.0 Max.	1.5 Max.	17-20	50-60	Emulsifiable Polymer
Warcosine® White	155-165	30 Max.	White	0-0	0-0	Plastic
Warco® 150-A Yellow	160-170	25 Max.	2 Max.	0-0	0-0	Plastic
Warco® 150-A Brown	160-170	25 Max.	Brown	0-0	0-0	Plastic

Ceresine Wax

The name Ceresine Wax, is rather loosely applied when used commercially. The true Ceresines were Mineral waxes derived from the Ozokerites by refining and bleaching.

The Ceresines of domestic origin are basically an upgrading of the various types of paraffins, through the controlled admixing of materials that are either compatible or miscible with the original paraffin. Since the uses of Ceresines are dictated by their physical characteristics and not be their chemical specifications, the user through experimentation finally determines the specific type necessary for his end use.

Owing to the variety of paraffins available and the choice of several materials which may be admixed, there is naturally a considerable variation in the character and physical properties of Ceresines. Depending upon the additives, domestic Ceresines may be hard or soft and those with identical melting points can have entirely different penetrations, softening points, plasticity ranges and properties of solvent absorption and gel.

The following table shows the variations in penetrations of a few of our domestic Ceresines based on the wax and also on a 50/50 gel of wax and solvent. The penetrations are expressed in 1/10 mm. and are determined by ASTM procedure.

Table 2:29

No.	Color	Capillary M.P.	Penetration at 5/77/100 Wax	50/50 Wax : Solvent
252	White	128/135°F	12	137
265	"	133/143°F	8.5	106
1154/4	"	138/148°F	11.5	136
375	"	139/149°F	11	105
101	"	154/164°F	7	70
608	"	157/167°F	10	140
165	"	158/168°F	11	101
1502	"	158/168°F	11	113
1556	"	177/187°F	7.5	97
20	Lemon	132/142°F	9	125
15	Orange	132/142°F	8.5	112
1154/2	Amber	128/138°F	11.5	102
669	"	145/155°F	25	.....
1343	"	158/168°F	11	163
1135/7R	"	153/163°F	14	130
1188/7	"	155/165°F	10	118
901B	Lt. Tan	128/135°F	13	130

Chemical and Physical Properties:

Melting Point Range	128–185°F. (53.3–87.8°C.)
Specific Gravity at 25° C.	0.880–0.935
Acid Number	0
Saponification Number	2 (max.)
Unsaponifiables, percent	100
Refractive Index	1.425–1.435
Penetration—various	See table above

Compatibility:

Vegetable, Animal, and Mineral Waxes
Large variety of natural and synthetic resins.
Fatty Acids Glycerides Hydrocarbons

Chapter 3

OTHER MINERAL WAXES

In addition to the waxes derived from petroleum, discussed in the preceding chapters, are other mineral waxes, derived from coal in various stages of development, i.e., lignite, bitumens, peat.

Lignite is a low-grade coal, between peat and sub-bituminous coal. It is found in various regions, but the best deposits are in the west of the U.S.A., and in Germany. It is the source of Montan Wax.

MONTAN WAX

Sources

The principal sources of crude montan wax are located at, or near, Halle Zeitz, and the Oberroblingen area, Eastern Germany (Riebeck Works) and Ione, Amador County, California, U.S.A. The Hoechst Company, located in Western Germany, has developed a new source of lignite for the production montan wax. Czechoslovakia also claims a minor production of crude montan wax.

The A. Riebeck'sche Montanwerke A.G. is responsible for the major production of crude montan wax in Europe. It is believed that their production plants are located at Amsdorf and Wansleben, East Germany. The Amsdorf plant produced about twice the quantity of the Wansleben plant and it is believed that both plants were engaged in simple refining.

The Hoechst Company, at Gersthofen, near Augsburg, Western Germany, operates a modern extraction plant for crude montan wax, for their own exclusive consumption. Reportedly, their production of crude montan wax is not sold outside. Hoechst uses crude and deresined grades of montan wax as basic raw materials for the production of many oxidized grades which are synthetically reformed. Hoechst depends upon the Riebeck Works as its major supplier of montan wax.

The American Lignite Products Co., Inc., produces all the montan wax in the U.S.A., near Ione, California. They are the major sellers of montan waxes in the United States. A wide range of different grades is available.

The Czechoslovakian source of montan wax is considered to be of minor significance as compared to the German and American sources.

Shortly after World War II, there were two relatively small Czechoslovakian firms, both producing inferior grades of wax, very high in resinous matter and therefore not considered true montans.

Production of Montan Wax[1]

Montan wax is obtained by solvent-extraction of lignite or brown coal. Not all lignites have a high enough wax content to make the extraction economically feasible. The prefix *montan* refers to mountain or mining, and the unmodified term *montan wax* is usually reserved to the wax extracted from lignite deposits of Central Germany by the Riebeck Montan Works. Waxes obtained from other sources of lignite are either designated by source, such as American Montan Wax, or are sometimes referred to as montana waxes, although the latter term is also sometimes used for waxes obtained from peats or similar material.

Because of its origin, montan wax is classified as a mineral wax; chemically, however, it is more closely related to the hard vegetable waxes than to petroleum waxes or ozocerite waxes.

To obtain the wax, lignite is crushed and screened to remove fine powder, and dried to a moisture content of 10 to 18%. The crude wax is then extracted with a solvent or mixture of solvents, and recovered by distilling the solvent. In Eastern Germany, the wax is generally obtained by using the azeotropic solvent of about 85% benzene and 15% ethyl alcohol, a procedure also used by the U.S. Bureau of Mines.

American Montan Wax is usually extracted with a solution that is approximately half and half aromatic and aliphatic hydrocarbons. Although the yields obtained with the latter mixture are somewhat lower than those obtained with the benzene-alcohol solvent, for most applications the crude wax obtained is of a superior quality.

Composition

Ultimate analysis of the montan wax showed: water, 0.08–0.12%; ash, 0.2–0.3%; hydrogen, 12%; carbon 79%; sulfur, 0.3%.

The montan wax may be crudely fractionated by solvents into so-called resins, waxes, and asphalts. Exhaustive extraction of crude montan wax with refluxing isopropyl alcohol leaves approximately 7–12% undissolved material. This insoluble material is designated as *asphalt*. On prolonged heating, the black insoluble material polymerizes to a non-fusible black solid which contains most of the ash-producing components and sulfur compounds of the crude wax.

The yield of asphaltic material depends to some extent on the solvent used. When comparing crude waxes, the solvents used for various components should be specified.

---

[1] Amer. Lignite Products Co., Ione, California.

Table 3:1 Montan Wax Properties

	Pure Waxes	*Resins*	*Asphalts* +)
Acid No.	31	27	10
Ester No.	67	68	116
Sap. No.	98	95	126
Iodine No.	14	135	80
M.pt °C	88.6	++)	++)
Ash %	0.05	0.06	2.85
Viscosity (90° C) cp	52	++)	++)
Sulfur %	1.1	2.5	3.1
Carbon %	79.2	78.6	77.1
Hydrogen %	13.0	10.9	11.6
Color	light brown	red brown	black

+) Contains about 10% waxes

	German Montan Wax		*American Montan Wax*	
Chain Length	*Acids*	*Alcohols*	*Acids*	*Alcohols*
C-Atoms	%	%	%	%
22	1.8	—	—	—
22	2.2	0.3	7.3	18.5
23	1.6	—	—	—
24	6.2	4.9	13.3	12.3
25	3.1	0.5	—	—
26	13.7	13.3	10.8	34.8
27	4.7	1.0	—	—
28	21.3	22.9	42.7	34.4
29	5.7	2.3	—	—
30	23.5	28.8	25.9	—
31	3.4	1.8	—	—
32	9.1	17.7	—	—
32	3.7	6.5	—	—

+) According to investigations of W. Presting and Th. Kreuter as well as H. R. Fleck.

Table 3:2 SPECIFICATIONS OF MONTAN WAXES OF DIFFERENT ORIGIN.

Origin	*Czech*	*German*	*USA*	*Russian*
Technical Designation	*Bohemia*	*Romonta*	*Alpco 16*	—
Density	1.03	1.03	1.03	1.03
Penetration 0.1 mm	1	1	1	1
Melting point °C	82	86	87	86
Solidification point °C	76	78	78	80
Viscosity (90° C) cp	350	70	200	60
Acid number	35	32	48	26
Ester number	62	60	64	62
Saponification number	97	92	112	88
Ash %	0.4	0.2	0.5	0.2
Resin %	31	13	11	14
Asphalt %	8	5	12	2

Table 3:3

EFFECT OF THE ADDITION OF GERMAN (ROMONTA) MONTAN WAX TO PARAFFIN

Composition Montan Wax	*Paraffin*	*M. Pt.*	*Solidification Point*	*Penetration (25°C, 100 g, 5")*
%	%	°C	°C	*0.1 mm.*
0	100	54.5	53.5	17.0
5	95	74	57	13.0
10	90	78	59	11.0
20	80	81	61.5	8.0
30	70	81	63	6.5
40	60	81.5	64	5.0
50	50	81.5	65	4.0
60	40	82.5	66	3.5
70	30	83	68	2.5
80	20	84	71	2.0
90	10	85	74.5	2.0
100	0	87	77	1.5

Table 3:4

EFFECT OF TYPES OF SOLVENTS ON YIELD AND COMPOSITION OF YIELD OF LIGNITE

					Other Statistics				
Type of Solvent	*Solvent*	*Yield* %*	*Ether Soluble* (% *Resin*)	*Paraffin Insoluble* (% *Asphalt*)	*Ash* %	*Acid* #	*Ester* #	*Saponification* #	*Solidification* °C
Alcohols	Methanol	2.8	41	29	3.0	30	28	58	78
	Ethanol	6.0	27	23	1.5	46	34	80	74
	Propanol	13.5	20	14	0.3	40	50	82	73.5
	Iso-Amyl Alcohol	15.1	18	14	0.3	32	50	82	74
Esters	Ethylacetate	11.7	23	7	0.2	34	45	79	74.5
	Butylacetate	14.3	16	7	0.1	40	45	85	76
Ketone	Methyl-Ethyl-Ketone	12.7	21	21	0.2	47	56	103	73.5
Chlorinated Hydrocarbon	Carbon Tetra-Chloride	10.3	13	0	0.2	29	35	64	76.5
	Trichloroethylene	12.7	13	1	0.3	33	45	78	73.5
	Dichloroethylene	11.7	19.5	1	0.1	32	45	77	77

Aliphatic Hydrocarbon	Petroleum Benzene (80–95°C)	9.9	11	0	0.1	28	42	70	78
	Petroleum Benzene (90–105°C)	9.9	13	0	0.1	29	45	74	77
	Petroleum Benzene (107–135°C)	10.9	14	0	0.1	29	45	74	78
	Lignite Benzene (70–110°C)	11.1	14	0	0.1	32	44	76	77
	Lignite Benzene (103–125°C)	11.9	14	0	0.3	29	36	65	77
Cyclical Hydrocarbon	Cyclohexane	9.9	13	0	0.3	29	53	82	76
Aromatic Hydrocarbon	Benzol	12.0	14	0.4	0.2	31	42	73	75.5
	Toluol	12.6	14.5	0.4	0.3	32	50	82	77
	Xylol	12.7	15.5	0.1	0.3	29	47	76	77
Mixtures of Alcohols and Aromatic Hydrocarbons	Benzol (50%) Ethyl Alcohol (50%)	16.4	21	25	0.8	38	47	85	74
	Benzol (90%) Ethyl Alcohol (10%)	14.9	16	12	0.5	36	43	79	74

*Montan Wax

Table 3:5

EXTRACTION OF WAX FROM CALIFORNIA LIGNITE[1]

Expt. No.	Solvent	Yield %	Asphalt %	Resin %	Acid No.	Sap. No.	Ester No.	Melting Point °F	Congealing Point °F
1.	Methanol	4.0	0.9	28.3	95.2	169.4	74.2	169	156
2.	Ethanol	6.5	0.22	20.5	99.4	169.4	70.0	174	156
3.	Isopropanol	9.6	0.08	15.0	85.4	156.8	71.4	180	167
4.	*n*-Propanol	11.2	0.0	22.2	53.2	149.8	96.6	185	174
5.	Methyl ethyl ketone	11.4	1.73	11.2	75.6	155.4	79.8	188	175
6.	Cellosolve acetate	9.9	0.96	15.3	64.4	177.8	113.4	186	175
7.	Isopropyl acetate	13.0	0.95	18.5	64.4	137.2	72.8	180	167
8.	Methyl isobutyl carbinol	13.7	14.23	22.9	40.6	142.8	101.2	188	170
9.	Ethyl acetate	11.5	0.23	12.5	70.0	159.6	89.6	183	169
10.	Rubber solvent	5.75	0.95	15.4	49.0	113.4	64.4	188.5	175
11.	Hexane	4.4	0.64	18.5	49.0	97.8	48.8	182.5	175
12.	Toluene	9.9	6.75	12.2	56.0	134.4	78.4	188	178
13.	Aromatic Hydrocarbon	9.4	6.21	13.0	51.8	184.8	133.0	188	177
14.	Aliphatic Hydrocarbons	8.25	4.07	13.0	51.8	126.0	74.2	187	174
15.	Chlorothene	10.1	4.66	12.4	57.4	148.4	91.0	188	178
16.	Carbon tetrachloride	9.4	8.37	13.4	50.4	140.0	89.6	189	178
17.	Arom: Aliph Hydrocarbon	8.5	7.75	12.9	50.4	116.2	65.8	189.5	177

[1] American Lignite Products Co.

Resins are those components in the crude wax that are soluble in a specified volume of some cold solvent. Various solvents have been used to deresinify crude waxes. Both the percentage of the crude wax that is resin and the nature of the resins obtained depend on the solvent used. Among the solvents used are ethyl ether, acetone, methyl alcohol, and a variety of aliphatic hydrocarbons and esters.

The resin content of crude montan wax, using mixed hexanes as the deresinifying solvent, varies from 14–20%. This resinous material is an opaque brown material which is much softer than the crude wax. Fractionation of this material with methanol indicates that it contains approximately 50–60% of wax acids and esters and 4–10% ketones. Resins obtained from a methanol solvent form a sticky, transparent, orange-red viscous liquid-solid.

The deresinified, deasphaltenized wax, called a "dasder" wax accounts for approximately 68–75% of the crude product. The dasder wax which is appreciably lighter in color than the crude wax, is composed principally of free wax acids and wax esters, accompanied by smaller amounts of alcohols, ketones, and hydrocarbons.

Separation of ALPCO Wax Type 16 into various types of components shows the wax to have the following composition:

Wax Acids (Free and Esterified)	40–50%	(Mol. wt. 400–500)
Wax Alcohols (Esterified)	10–30%	
Ketones	0–5%	
Hydrocarbons (Resins and Asphaltenes)	7–12%	

Typical approximate compositions for crude montan waxes from Eastern Germany show:

(I)[1]

Free Wax Acids	17%
Wax Esters	53%
Free Wax Alcohols	1–2%
Ketones	3–6%
Resin	20–23%
Asphaltenes	3%

Except for a slightly higher asphalt content, ALPCO Type 16 compares favorably with such an analysis.

(II)

Wax Acid Esters	57–53.5%
Wax Acids (Free)	17
Alcohols (Free)	1–2
Ketones	3–6
Resins	20–23
Bitumens	3

[1] Warth, *Waxes*, 2nd Ed., p. 368, Reinhold Publishing Co.

Properties of Montan Wax[1]

American Montan Wax is a brownish black solid which shows a conchoidal fracture. The surface of the wax is so hard it can hardly be scratched with a fingernail.

Montan wax has a density (d^{15}) of 1.02 g/cc. It melts at 84°C and re-solidifies around 81°C. It is completely soluble in hot benzene, toluene, carbon tetrachloride, methyl ethyl ketone, and dioxane. It is incompletely soluble in hot alcohol, esters, aliphatic hydrocarbons, and hot water. Freezing point depression of naphthalene shows an average molecular weight of 825, ± 100 for the crude montan wax.

According to various reports this wax is soluble in benzene, carbon tetrachloride, chloroform, dichloroethylene, isopropyl ether, naphtha, tetralin, toluene, trichloroethylene, turpentine, xylene.

Montan wax solutions set more slowly than carnauba wax solutions, and, therefore, should be filled into containers at a lower temperature. Pure montan wax is a good solvent (when melted) for basic dyes and it shows good wetting and flow in oil solutions. It is the hardest of the natural nonvegetable waxes and is fairly resistant to oxidation.

The bleached or highly refined wax loses some of the listed characteristics, becoming softer and less emulsifiable.

Montan wax is compatible with vegetable waxes, hydrocarbons (except Bohemian), and resins.

The principal grades of Alpco Montan Waxes and their basic typical properties are:

Table 3:6 PROPERTIES OF ALPCO WAXES

Type No.	*Color*	*Melt. Point °F.*	*Acid Value*	*Sap. No.*
Alpco 10	Black-Brown	190	43	112
Alpco 16	Black-Brown	188	48	112
Alpco 20	Dark Amber	186	50	105
Alpco 400	Amber	181	54	100
Alpco 431	Light Amber	183	46	112
Alpco 434	Yellow-Amber	185	41	120
Alpco PX-100	Black-Brown	187	42	113
Alpco AX-200	Brown	187	48	141

Durmont Waxes

DURMONT 500 offers special features of solvent-retention, oil-binding, homogenizing, and mirror-like solidification. DURMONT 500 is free from any tack. It has excellent anti-blocking characteristics, a feature necessary for any coating of paper.

When DURMONT 500 is melted and allowed to re-solidify, the surface is mirror-like. No vegetable waxes, animal waxes, or strictly synthetic waxes have this property.

[1] Unless otherwise specified, the properties described for American montan wax are for those of ALPCO Type 16.

Table 3:7

SPECIFICATIONS FOR ALPCO WAX, TYPE 16

Property	*Maximum Minimum Tolerance*
Melting Point, ASTM, D-127-49	84°C ± 2°
Acid Value, M/8	51.0 ± 5
Ash, M/14	0.30% max. %
Insolubles, M/16, %	0.30% max. %
Penetration Value, ASTM, D-1321-54T	2.0 max.
Viscosity, Saybolt Seconds @ 99°C., ASTM-D-88-53	550 in. max.
Specific Gravity, ASTM, D-1217-54	1.02 – 1.03
Saponification Value, M/9	112.0 ± 5
Ester Value, M/10	(By diff.)
Asphaltic Matter, M/11	12.0% max.
Resinus Matter, M/12	20.0% max.
Benzol Solubility, M/17	99.7% min.
Compatibility, Paraffin, Aristowax 133/4, M/19	Good
Methylethylketone Solubility, M/18	99.7% min.
Oil Retention, 50/50, Wax/Oil, ASTM D-1321-54T	40.0 max.

Table 3:8

PROPERTIES OF SOME PURIFIED FRACTIONS FROM ALPCO TYPE 16 WAX

Fraction	*Acid No.*	*Hydroxyl No.*	*Av. Mol. Wt.*
Type 16 Wax	40–50	30–40	800
Iso-propanol dasder	50–60	20–30	900
Methanol dasder	80–95	40–50	550
Hexane resins	40–50	20–30	650
Methanol resins	40–50		400
Iso-propanol asphalts	—	30–40	—

Table 3:9

PROPERTIES OF DURMONT WAXES (LIGHT REFINED)

Typical Characteristics	*Durmont 500*	*Durmont 500 SX*
Melting point (Ubbelohde), °C	88–91	88–91
Solidification point at rotating thermometer, °C	68–72	70–75
Acid Value	5–7	3–5
Saponification Value	25–35	36–40
Penetration (100 g/5 sec/77°F	5–8	5–8
Color	Tan	Light Tan

Uses of Montan Wax

In Europe, montan wax may be regarded as the most important wax used in industry. Crude montan wax is used as an inexpensive substitute for carnauba wax when color is unimportant. In many other applications where color is not a stringent factor, crude montan wax can be used as an extender for carnauba and other expensive waxes. Refined grades of montan wax, such as the Hoechst and Alpco types, can replace carnauba in virtually all applications, if properly used.

Both crude montan wax and refined grades are used in the following industries: Carbon paper inks, car wax, polishes for floors, furniture, and shoes, explosives, investment castings, wire coatings, cable compounds, dielectric products, plastics, cosmetics, modifiers for waxes and resins, laminating, impregnations, surface coatings, anti-corrosives, waterproofing, wax blending, building materials, asphalt paints, lubricants and greases, dispersants, paper coatings, rubber, rust-prevention, mold-release agents, wax specialists, leather treatment, and casting waxes (records).

Modified Montan Waxes

Besides the removal of resins and asphaltenes from the crude waxes montan waxes may be further refined by a variety of methods to produce special waxes for specific applications.

For many applications, the dark color of montan wax prevents the use of it even though its other physical properties may be satisfactory. Although removal of resins and asphaltenes improves the color, it is still too dark for many applications. Adsorbents, such as decolorizing carbon, are effective in removing the color from montan wax, but the quantity of adsorbent necessary makes the process industrially uneconomical.

Although distillation with superheated steam has been used in Europe to produce a very light-colored wax from lignite, the physical properties of the wax thus produced are not so desirable for most applications as the waxes produced by solvent-extraction.

The montan wax from Central Germany (Riebeck or Romanto) is chemically modified to produce nearly colorless, synthetic waxes. Refined montan wax is oxidized to convert most of the constituents to acids. These nearly colorless acids are then used directly or are converted to ester-type waxes with a variety of alcohols, especially glycols. The type of wax produced depends on the application for which it is intended.

Sulfuric and nitric acids, or alkali, at elevated temperatures have also been used to bleach montan wax. As yet, no economical method has been devised to bleach it completely, without significantly changing its chemical properties.

LIGNITE WAX

The wax from British lignite is a satisfactory substitute for montan wax. The crude wax consists of a mixture of ester wax with a resinous and asphaltic matter. It is dark brown, hard, brittle, and odorless. The properties of the crude wax are:

Melting Point	69–84°C
Acid Number	29–34
Saponification Number	70–77
Ester Number	36–45

The resin content was 40%. In exceptional cases the crude wax contained 50–95% resin.

Table 3:10

EFFECT OF TEMPERATURE ON YIELD AND PROPERTIES OF LIGNITE WAX[1]

	At 1 At. 80°C	*At 35 At.* 250°C	*Commercial Montan Wax*
Yield of Wax, % Dry Lignite	4.2	6.9	—
Melting Point	76–82°C	58–68°C	74–82°C
Acid Number	34	25	33
Saponification Number	70	68	76
Ester Number	36	38	43
Ash, %	0.6	0.8	0.3

The wax used was Deon lignite, extracted by benzene.

The use of a higher extraction temperature increased the yield of wax, but changed its properties. All the waxes were dark brown, but while the montan wax and the lignite wax extracted at 80°C were hard and brittle, the wax extracted at 250°C was relatively soft and was readily indented by finger pressure.

The color of the purified wax is ivory yellow. The color varies with the amount of oxidant added. If less oxidant is used, a brown wax is obtained while with more oxidant the wax has a green color. The best color is obtained by clarifying the wax with Fuller's earth.

OZOCERITE WAX[2]

Occurrence

True ozocerite wax is a bituminous product occurring in miocene formations near petroleum deposits in Poland, Austria, Russia, Ukraine, and also in Utah and Texas. In recent years however, these true ozocerites have not been freely available in the United States.

Today the ozocerites of commerce are blends of true waxes which produce a dry type wax of high solvent absorption and gel values similar to the formerly imported ozocerites. They have long fibers, unlike the structure of the paraffins and microcrystalline waxes.

Manufacture of Ozocerite

Ozocerite is separated from earthy matter by melting in boiling water and then drawing off the upper wax layer. The crude wax is purified by treatment with sulfuric acid and bleaching earth to obtain lighter-colored products.

The commercial grades of ozocerite are:

1. Unbleached (Black)
2. Single bleached (Yellow)
3. Double bleached (White)

---

[1] C. M. Cawley, and J. G. King, *J. Soc. Chem. Ind.*, p. 237, Sept. 1945.

[2] This spelling of the word which is preferred to the European spelling, *ozokerite*, is used throughout this book.

American ozocerite has been produced in qualities equal to those of the imported varieties. Midcontinent (Oklahoma) stocks (long fiber) of 135–150°F melting point are used as starting material; Pennsylvania and Gulf stocks are far less suitable.

By centrifuging, a very plastic grade, melting up to 165°F, is obtained. By vacuum distillation, harder grades, melting up to 220°F, are produced. These are bleached to obtain the so-called American, white ozocerite. This differs from the plastic grade in being free from oil and color.

Composition

The chemical constitution of ozocerite is not definitely known. It appears to be a mixture of solid saturated and unsaturated hydrocarbons of high molecular weight, with some liquid hydrocarbons and oxygenated bodies.

Properties of Ozocerite

Ozocerite is characterized by its excellent absorbing power for oil, grease, and certain solvents. In this respect, it is superior to paraffin wax which tends to crystallize from mixture and allows fluid components, present in the composition, to sweat out. It is waterproof, alkali- and acid-resistant; only fuming sulfuric and concentrated nitric acids attack it. Low temperatures do not make it so brittle as paraffin wax. It is not so slippery as paraffin wax and actually has a slight "tack." It becomes harder on aging. It is less resistant to actinic rays than paraffin wax. Ozocerite sets more rapidly and shrinks more than beeswax. To offset these properties Venice turpentine, or another plasticizer, is often added.

Table 3:11

PROPERTIES OF OZOCERITE

Melting Point	149–176°F
Specific Gravity	0.85–0.95
Acid Number	0
Saponification Number	0
Iodine Number	7.8–9.2
Sulfur	0.2%
Ash	0.16%
Color	white, yellow, green, black
Refractive Index (60°C)	1.440
Dielectric Constant	2.37–2.55
Effective A.C. Conductivity	10–24
Volume Resistivity	7–40
Penetration at 77°F (100 g/5 sec)	14
Coefficient of Expansion 0.001 per 1°C	between 1° & 50°C.

Ozocerite is soluble in benzene turpentine, kerosene, toluene, mineral spirits, carbon disulfide, isopropyl ether, trichloroethylene. Insoluble in ethyl and methyl alcohols.

Table 3:12

SOLUBILITY OF OZOCERITE[1]

Solvent	*Wax 100 g Solvent*
Carbon Bisulfide	12.99
Petroleum Ether, Boiling up to 75°C	11.73
Turpentine, B.P. 158–166°C	6.06
Xylene, B.P. 135–143°C	3.95
Toluene, B.P. 108–110°C	3.83
Chloroform	2.42
Benzene	1.99
Ethyl Ether	1.95

The following table shows the penetration of a few ozocerites based on the wax and also on a 50/50 gel of wax and solvent. The penetrations are expressed in 1/10 mm. and are determined by ASTM procedure.

Table 3:13

PENETRATION OF OZOCERITES

Wax No.	*Color*	*Capillary M.P.*	*Penetration at 5/77/100* *Wax*	*50/50 Wax: Solvent*
1543W	White	156–160°F	12.5	102
1544W	White	170–175°F	11.5	92
1545W	White	178–182°F	10	80
64W	White	145–148°F	13	112
71W	White	160–163°F	10	105
77W	White	170–173°F	8	142
1543Y	Yellow	156–160°F	10	111
1544Y	Yellow	170–175°F	9.5	96
1545Y	Yellow	178–182°F	9.5	81
64Y	Yellow	145–148°F	13	130
71Y	Yellow	160–163°F	10	105
77Y	Yellow	170–173°F	8	147
American	White	145–180°F	12–17	—
American	Yellow	185–220	3–5	—

Compatibility

Ozocerite is compatible with all vegetable waxes (in certain proportions); with a large variety of natural and synthetic resins; all animal waxes; glycerides and fatty acids.

Uses

Ozocerite is widely used in cosmetics (creams, pomades); paints and varnishes; polishes for leather, automobile, floors; and printing inks.

[1] Pawlewski and Filemonewicz, *J. Chem. Soc.*, Abstr. 82, 1889.

Table 3:14

DURON PURE OZOKERITE 71/74

Color	Pale Yellow
Congealing Point (ASTM D-938)	70–71°C
Melting Point (ASTM D-127)	73–74°C
Penetration (ASTM D-1321)	16–20/100 g/25°C/5 sec.
Specific Gravity /90°C	0.7872
Refraction Index /90°C	1.4388
Flash Point COC	273°C
Fire Point COC	310°C
Acid Value	Nil
Saponification Value	Nil
Ester Value	Nil
Hydroxyl Value	Nil
Iodine Value	3

Solubilities (Cloud Point of 10 parts wax in 90 parts solvent):

Mineral Spirits	44°C
Spirits of Gum Turpentine	43°C
Dekalin	40°C
Dichlor Ethane	61°C
Butyl Acetate	60°C

ADULTERATIONS

Ozocerite is often adulterated with paraffin wax to reduce its cost. It is also sophisticated with carnauba wax, resins, and asphaltum to increase its melting point or hardness.

Table 3:15

OZOCERITE—PARAFFIN BLENDS

Yellow Ozocerite M.P., 75°C } mixed with { Paraffin Wax M.P., 50°C

Proportions	*New M.P., °C*
4 Ozocerite 1 Paraffin Wax	73.5
4 Ozocerite 2 Paraffin Wax	71.7
4 Ozocerite 3 Paraffin Wax	70.5
4 Ozocerite 4 Paraffin Wax	69.7

White Ozocerite M.P., 75.7°C } mixed with { Paraffin Wax M.P., 58.3°C

Proportions	*New M.P., °C*
4 Ozocerite 1 Paraffin Wax	74.4
4 Ozocerite 2 Paraffin Wax	73.2
4 Ozocerite 3 Paraffin Wax	72.5
4 Ozocerite 4 Paraffin Wax	72.0

CERESIN

Ceresin originally was a purified ozocerite. It was mixed with paraffin wax to reduce its cost. Today commercial ceresin contains 50–80% paraffin wax. Some European ceresins also contain rosin and coloring matter to approach the properties of beeswax.

Commercial ceresin varies in color from yellow to white and melts at 130–170°F, depending on its purity. Its hardness, plasticity, etc., also vary with purity. The grades containing large percentages of ozocerite naturally approach ozocerite in properties.

Table 3:16

PROPERTIES OF CERESIN
(High Grade)

Melting Point	130–175°F
Specific Gravity	0.88–0.92
Acid Number	0
Saponification Number	0–1
Iodine Number	7–10
Color	white, yellow
Refractive Index	1.4416–1.4465
Dielectric Constant	2.15–2.33
Effective A.C. Conductivity	1–12
Volume Resistivity	400 > 900

Soluble in the same solvents as paraffin wax.

UTAH WAX

Utah wax is an amorphous mineral (ozocerite-type) wax from Utah available in a number of grades that vary in hardness, melting point, and color.

Properties

Table 3:17

GENERAL PROPERTIES OF UTAH WAX
(Crude)

Melting Point	73°C	Acid Number	0
Cloud Point	76°C	Saponification Number	0
Iodine Number	<1	Sulfonation Residue	100%

Dielectric Strength: in excess of 30,000 volts across a 1/10-in. General Electric standard gap. Retains a high value when melted. *Dielectric Constant:* (or permittivity) 2.03. *Power Factor:* 0.03. *Resistivity:* 36,000,000 megohms across a 1 cm cube. *Coefficient of Expansion* (Volume): 0.001 per degree Centigrade between 10–50°C.

Table 3:18

HARDNESS OF UTAH WAX

(Data obtained by Penetrometer with Standard Needle, 100 g Weight, 15 Seconds Time)

Temperature	*Penetration*
°C	mm
20	0.43
30	0.64
40	1.27
50	2.2
60	3.8

A comparison of these figures with similar measurements of other waxes shows that Utah wax is softened less by any given increase in temperature.

Solubility: Slightly soluble in most organic solvents when cold. Solubility increases rapidly on heating, so that Utah wax becomes very soluble in practically all commercial organic solvents, with the exception of methyl and ethyl alcohol, glycol, and glycerine. Its best solvent is carbon disulfide.

Solutions made by heating retain a fluid character when cooled, if the concentration of wax is not too high. Otherwise they set into pastes or gels of the nonsweating variety.

Uses

Utah waxes are successfully utilized in dance-floor wax, floor and furniture polishes, linoleum polish, shoe polish, leather dressings, rubber goods, candles, wax figures, carbon paper, wax bottles, waxed paper, electrical insulation, waterproofing fabrics, impregnation of wood, electrotypers' molding compound, pattern-makers' wax, waxing artificial flowers, wax crayons, salves, and face creams.

PEAT WAX

The properties of the waxes obtained from the English and Scottish peats are similar in all respects. They are dark brown (almost black), and hard and brittle, but more readily fractured than lignite wax or montan wax.

Peat wax has a higher oxygen and a lower carbon content than lignite wax and montan wax. This property is associated with the higher acid and ester numbers of peat wax.

Table 3:19

WAXES FROM ENGLISH AND SCOTTISH PEATS

No.	*Yield of Crude Wax from Benzene, % by wt. of Dry Peat*	*Melting Point of Crude Wax °C*	*Properties of Crude Wax* Acid No.	Saponification No.	Ester No.	*Resin Content of Crude Wax, %*
ENGLISH PEATS						
1	6.0	62–68	49	—	—	—
2	9.4	63–68	51	—	—	—
3	6.8	60–66	37	—	—	—
4	5.8	63–68	52	—	—	—
5	5.4	62–67	54	—	—	—
6	5.5	63–68	54	—	—	—
7	6.3	62–67	52	—	—	—
8	6.8	63–68	55	—	—	—
9	6.3	62–66	55	—	—	—
Bulked Samples 1–9	—	63–68	50	119	69	—
SCOTTISH PEATS						
10	4.8	67–74	51	119	68	—
11	7.6	62–72	45	112	67	25
12	4.3	59–64	44	106	62	—
13	3.2	63–75	51	103	52	—
14	3.2	68–73	43	107	64	29
15	11.8	64–72	44	122	78	20
16	7.6	64–72	45	113	68	20
17	2.7	68–77	55	120	65	—
18	7.8	60–71	47	106	59	20
19	8.6	64–74	47	116	69	—
20	5.2	63–75	59	122	63	26
Bulked Samples 10–20	—	64–73	48	113	65	23

Table 3:20

COMPARISON OF PROPERTIES OF MINERAL WAXES

	German Montan	*Devon Lignite*	*English Peat*	*Scottish Peat*
Melting Point	74–82°C	73–83°C	63–68°C	64–73°C
Acid Number	33	30	50	48
Saponification Number	76	75	119	113
Ester Number	43	45	69	65
Resin Content, %	14	40	—	23
Specific Gravity at 20°C	1.05	1.04	1.03	—

Table 3:21

Properties of Crude Peat Wax (Irish)[1]

(Extracted by Benzene)

M.P. °C	75
Molecular Weight	686
Acid Value	56
Ester Value	60
Hydroxyl Value	86
Carbonyl Value	45
Resin %	21
Wax %	64
Asphalt %	15

Table 3:22

Yields and Properties of Crude Waxes Extracted from Peat with Different Solvents

Scottish Peat No.	*Solvent*	*Yield of Crude Wax % Dry Peat*	*Melting Point*	*Acid No.*	*Saponification No.*
14	Light Petroleum, B.P.				
	100–120°C	2.3	54–70	—	—
	Benzene	3.2	68–73	—	
	Benzene: Alcohol Azeotrope				—
	(65/35 by vol.)	6.6	>86 S.O.*	—	—
15	Light Petroleum, B.P.				
	100–120°C	10.4	55–68	—	—
	Benzene	11.8	64–72	44	122
	Benzene: Alcohol Azeotrope				
	(65/35 by vol.)	18.6	>70 S.O.	62	155
17	Light Petroleum, B.P.				
	100–120°C	2.4	70–78	—	—
	Benzene	2.7	68–77	55	120
	Benzene: Alcohol Azeotrope				
	(65/35 by vol.)	4.9	64–90	61	132
19	Light Petroleum, B.P.				
	100–120°C	6.8	60–71	—	—
	Benzene	8.6	64–74	47·	116
	Benzene: Alcohol Azeotrope				
	(65/35 by vol.)	12.8	65–90	—	—
20	Light Petroleum, B.P.				
	100–120°C	4.1	54–74	—	—
	Benzene	9.0	>70 S.O.	55	152
	Benzene: Alcohol Azeotrope	5.2	63–75	59	122
	(65/35 by vol.)				

* S.O. = Softened only

[1] A. J. Howard, *J. Amer. Oil Chem.*, p. 252, May, 1962.

Chapter 4

VEGETABLE WAXES

Waxes of various qualities are found as coatings on leaves, stems, berries, and grasses that grow wild in many countries. The largest amounts of wax are produced by plants in tropical and arid locations where this coating is provided by nature to reduce the evaporation of moisture.

True vegetable waxes are soluble in most fat solvents, such as chloroform, ether, or toluol, but are less soluble than fats. Since they contain no glycerides, they do not form acrolein on heating. They do not usually become rancid and are less easily hydrolyzed than fats. They can, however, be decomposed by prolonged heating with alcoholic potash. Because of the high molecular weight of their fatty acid components, their saponification numbers are lower than those of fats. They are also characterized by very high amounts of unsaponifiable constituents which consist of sterols and higher saturated alcohols, usually occurring as esters.[1]

BAYBERRY WAX

Source

Bayberry wax (also called *candleberry*, *myrtle wax*) is obtained from the berries of the shrub *Myrica Cerifera* or *Carolinensis*. It should be distinguished sharply from *bay*, which is the European laurel, used as a source of oil.

Preparation

Bayberry wax is obtained by boiling the berries in water until the wax melts and floats to the surface. It is then skimmed off and purified by being remelted and filtered.

Properties

Bayberry wax is a greenish-white fatty substance, more unctuous and brittle than beeswax. It burns with a pleasant odor.

Chemically, bayberry wax is a fat consisting of stearin, palmitin, myristin, and olein (glycerides). It saponifies readily with alkaline solutions. It is soluble in alcohol and turpentine (only slightly).

---

[1] C. J. Marsel, Department of Chemical Engineering, New York University.

Table 4 : 1

PROPERTIES OF BAYBERRY WAX

Melting Point	102–120°F.
Specific Gravity	0.875–0.980
Saponification Number	206–217
Iodine Number	1.9–3.9
Color	green-white
Odor	faint
Taste	bitter

Bayberry wax is compatible with all vegetable, animal, and mineral waxes, fatty acids, and glycerides, hydrocarbons, and a large variety of natural and synthetic resins.

Uses

Bayberry wax is used for buffing compounds, candles, cosmetics, crayons, laundry wax, leather polish, soaps and textile sizing.

CANDELILLA WAX

Candelilla, a vegetable wax, is found in the form of scales entirely covering a reed-like plant (*Euphorbia Antisiphilitica*, the *Euphorbia Cerifera* and the *Pedilanthus Pavonis*) which grows wild and in great abundance on the rocky slopes and plains of northwestern Mexico and southern Texas. Production is year-round with somewhat less wax obtained in the wet season, July to September. The plant grows on the average to a height of one to three feet and only the stalks of the adult plant are uprooted. The wax is extracted by immersing the plant in boiling water containing a small percentage of sulphuric acid. The wax is skimmed off, strained, and gently cooked to remove the excess moisture. Then it is shipped as prime quality candelilla Wax in lump form.

Much of this crude wax is remelted and processed through filter presses and bleached by means of filtering clays and chemicals. The resulting wax is known as Refined Candelilla Wax.

Table 4 : 2

COMPOSITION OF REFINED CANDELILLA WAX

Wax Acid Alcohol Esters	28–29%
Alcohol, Stearols, Neutral Resins	12–14
Hydrocarbons	50–51
Acids (Free)	7–9
Moisture	0.5–1
Inorganic Matter	0.7

Properties

The commercial, refined candelilla wax is hard, brittle, and lustrous and has an aromatic odor. Different grades and shipments exhibit varying degrees of stickiness (on pressure, in solution, and in compositions).

Table 4 : 3

PROPERTIES OF CANDELILLA WAX

Quality of Wax	*Standard*	*Double Refined*
Melting Point	149–156°F	158–162°F
Specific Gravity, 15°C	0.982–0.993	—
Acid Number	11–19	19.6
Saponification Number	46–66	58.6
Iodine Number	15–36	37.8
Ester Number	—	39.0
Ash	—	0.05%
Moisture	small percentage	0.73%
Unsaponifiable Matter	65–67%	—
Color	gray–brown	—
Refractive Index	1.4555	—
Odor	aromatic	—
Dielectric Constant	2.50–2.63	—
Effective Conductivity	19	—
Volume Resistivity	120	—

Candelilla wax is not quite so hard as carnauba wax and does not emulsify or saponify so readily. The commercial wax often contains a small amount of water which should be driven off by heating before use in non-aqueous compositions. It is very slow in setting in melted compositions and often does not reach maximum hardness for a few days. Its stickiness varies with its degree of refinement, different lots showing a variation. Mineral waxes and Acrawax C reduce its stickiness, but increase the aging and setting time. It is a good solvent for basic dyes. Additions of oleic or similar acids slow down crystallization and rapidly increase softness. It does not blend well with ouricury wax.

Candelilla is soluble in: Acetone, carbon tetrachloride, chloroform, turpentine, petroleum (forms a gel on cooling hot solutions in these solvents); alcohol-benzene (2:3 mixture).

Uses

Candelilla wax is widely used in leather dressing, furniture and shoe polishes, candles, cements, varnishes, sealing wax, electrical insulating compositions, phonograph records, paper sizing, celluloid, rubber; in water- and insectproofing containers; in dentistry, in paint removers, in rubber compositions, as a stiffener for soft wax, and as a substitute for carnauba wax and beeswax.

Compatibility

Candelilla is compatible with all vegetable and animal waxes, a large variety of natural and synthetic resins, fatty acids, glycerides, and hydrocarbons (in certain proportions).

Table 4:4

PROPERTIES OF STANDARD-GRADE MEXICAN CANDELILLA WAX

General		
Melting Point (Drop)		66.0–71.0°C
Specific Gravity at 15°/4°C		0.9820–0.9856
Refractive Index—71°C		1.4555
Saponification Number		47–64
Ester Number		40–47
Acid Number		12.3–20.6
Nonsaponifiable Matter		47–62
Hydrocarbons (Principally Hentriacontane and Tritriacontane)		40–60%
Resins		10–20%
Moisture and Volatile Matter		0.3–1%
Iodine Number		19–44
Color in Toluene Solution	Red	0.9–1.4
	Yellow	2.0–4.0
Myricilic Alcohol, Dihydroxymyricinoleic Acid, Lactones		30–40%
Reichert-Meissl Number		0.5–7.60
Electric Rigidity (Volts per mm of Section at 50 Cycles, Steel Spheres of 25 mm diameter)		23,500–45,400
Solubility		
Alcohol		insoluble
Alcohol (Hot)		soluble
Acetone (Hot)		slightly soluble
Benzene (Hot)		soluble
Petroleum Ether (P.E. 30–75°C) 20°C		insoluble
Petroleum Ether (Hot)		soluble

Specifications

The specifications for genuine pure candelilla wax are as follows:[1]

Scope

1. These specifications cover the following types and grades:

Mexican Designation		*American Designation*	*Color*
PRIME CRUDE	=	PRIME CRUDE	Tan to Brown
		U.S.A. PURE REFINED	Light Yellow to Tan

Composition and Properties

The wax shall be an original virgin product made directly from the wax coating on the reeds of the plant. (Principally from the *Euphorbia Antisiphilitica*, the *Euphorbia Cerifera*, and the *Pedilanthus Pavonis*.)

It shall be free of extenders or admixtures of other substances and shall conform to the following requirements and shall be so guaranteed by the seller.

---

[1] American Wax Importers and Refiners Assn., Inc.

		Prime Crude	U.S.A. Pure Refined
Melting Point—	minimum	68.5°C–155.3°F	68.5°C–155.3°F
	maximum	72.5°C–162.5°F	72.5°C–162.5°F
Flash Point—minimum		241°C–465°F	241°C–465°F
Volatile Matter (moisture include)—maximum percent		1.5	Negligible
Insoluble Impurities—maximum percent		1.0	
Acid Number—	minimum	12.0	12.0
	maximum	22.0	22.0
Saponification Number—	minimum	43.0	43.0
	maximum	65.0	65.0
Paraffinic Hydrocarbons—maximum percent		45.0	45.0
Acetone Soluble Resinous Matter—percent at 15°C—	minimum	15.0	15.0
	maximum	28.0	28.0
Benzene Solubility—maximum percent at 15°C		40.0	40.0

CARNAUBA WAX

Carnauba Wax is exuded by the leaves of the Brazilian "Tree of Life" (*Copernica Cerifera*) to conserve the moisture within the tree and leaves. This Carnauba Palm grows in the northern and northeastern parts of Brazil along the river banks, valleys, and lagoons where the soil is dark and fertile. The tree needs very little water to grow, but is very prolific and after about fifty years attains a height of forty to fifty feet. The natives in the surrounding area use the various products of the tree for many necessities in their living; hence the name "Tree of Life." Of interest is the fact that only in Northern Brazil does the Carnauba palm produce wax. The tree also grows in other parts of Brazil and adjacent countries in South America, Ceylon, and Equatorial Africa, but owing to the irregular rainy seasons of these places the trees fail to provide themselves with the wax.

Production

The leaves containing the wax are cut off the tree during the period from September to March. They are left in the sun several days to dry and then the wax is removed by beating the shriveled leaves. This wax is in a fine powdery form which is melted and strained in a primitive fashion. After cooling, this wax is broken into lumps which have a yellow to dark brownish green color. The color and quality are governed by the age of the leaves and the care used in processing the wax. In addition to color, the wax is also graded according to area of origin—Parahyba, Piaui, Ceara, and Bahia, principally.

Composition

Carnauba wax contains varying amounts of water. The refined grades contain 0.5 to 1.5%. The chalky grade usually contains 5% and some samples have tested up to 11%.

Table 4 : 5

COMPOSITION OF CARNAUBA WAX

Alkyl Wax Acid Esters	84–85%
Wax Acids (Free)	3–3.5
Alcohols	2–3
Lactides	2–3
Hydrocarbons	1.5–3
Resins (alcohol-soluble)	4–6
Moisture, Inorganic Content	0.5–1

Grades Available

Flor

Prime No. 1 Yellow

Medium or No. 2 Yellow

Cauhype (Cauipe)

Gorda Clara, Light Fatty, or No. 2 North Country

Fatty Grey or No. 3 North Country

Chalky

These crude grades are processed in the United States to meet consumers specific requirements.

Crude	Powdered
Refined	Lumps and Flakes
Refined	Powdered
Refined	Molded

Table 4 : 6

PROPERTIES OF CARNAUBA WAX

Melting Point	180–187°F (82.5–86°C)
Flash Point	Minimum 570°F
Specific Gravity at 25°C	0.996–0.998
Acid Number	2–10
Ester Number	75–85
Saponification Number	78–88
Iodine Number	7–14
Unsaponifiable Matter %	50–55
Acetyl Number	51–60
Color	yellow, green, gray, brown
Refractive Index, 60°C	1.463
Dielectric Constant	2.67–4.20
Effective A.C. Conductivity	250–310
Volume Resistivity	0.5–4

Properties

Carnauba wax is amorphous, hard, tough, and lustrous. It has a pleasing odor; it breaks with a clean fracture.

It is saponified by strong alkalies and is the hardest, highest-melting, natural commercial wax, except for some crude grades of ouricury wax. It is added to other waxes to increase their melting point, hardness, toughness, and luster and to decrease stickiness, plasticity and crystallizing tendencies.

Carnauba wax is compatible with all vegetable, animal, and mineral waxes, a large variety of natural and synthetic resins; with fatty acids, glycerides, and hydrocarbons.

Table 4:7

YIELDS AND PROPERTIES OF FRACTIONS FROM SAPONIFICATION OF CARNAUBA WAX[1]

	Carnauba wax	*Unsaponifiable material*	*Acids*
Weight, grams	5.0	2.55	2.52
Yield, %	—	51.0	50.0
Yield, % (12, 13)	—	54.0	43.0
M.p., °C	80–84.5	83.5–84.5	75–85
Acid[a]	0.07	0.00	0.98
Ester[a]	1.39	0.00	1.84
Hydroxyl[a]	0.69	2.43	0.44
Total moles[a]	1.29	2.73	1.81

[a] Conc. in moles/kg.

Table 4:8

SOLUBILITY OF CARNAUBA WAX

(25–30°C)

Solvent	*g/100cc*
Acetone	0.324
Benzene	0.518
Chloroform	1.690
Ethyl Acetate	0.375
Ethyl Alcohol (95%)	0.141
Ethyl Ether	0.421
Methanol	0.179
Naphtha (High Flash)	0.220
Turpentine	0.440
Xylene	0.610

Addition of large percentages of cumarone resins to carnauba wax increases its resistance to fracture (toughness) and makes it stringy when melted.

The melting point of carnauba wax is raised 3°C by the addition of 1% Acrawax. Its flexibility is increased by the addition of 2% or more of beeswax, microcrystalline paraffin wax, or oleic acid.

Carnauba wax, dissolved in mineral oil, wets pigments and other insoluble particles quite well—better than most oils. Although a carnauba wax-oil mixture has good wetting properties, it sets fast (in thin layers) and does not penetrate rapidly into paper, textiles, etc. Solutions of carnauba wax precipitate more readily than those of other natural waxes on addition of nonsolvents. Evaporation of solutions of carnauba wax yields noncontinuous powdery layers rather than films.

[1] T. W. Findley, *J. Am. Oil Chem. Soc.* **30**, 291–6 (1955).

Properties of Brazilian Grades of Carnauba Wax

In Brazil, seven types of carnauba wax are available: 1) flower, 2) first, 3) medium clear, 4) medium purple, 5) "cauipe," 6) sandy, 7) fat.

Of these, the best known are:

1) *Flower Wax*

This is made in small quantities, seldom appearing in export lists; it is the best quality, with hardly any impurities. Its color is almost the same as that of the yolk of an egg.

	%		
Wax	99.32	Iodine Number	22.6
Moisture	0.48	Melting Point	84°C
Impurities	0.20		

2) *First Wax*

It varies in color from dark to light with an insignificant percentage of impurities.

	%		
Wax	99.65	Iodine Number	24.3
Moisture	0.35	Melting Point	83°C
Impurities	0.00		

3) *Medium Wax*

This is generally of a uniform color, varying from light yellow to light grey, according to the age of the tree and the treatments.

	%		
Wax	99.53	Iodine Number	21.6
Moisture	0.42	Melting Point	83°C
Impurities	0.00		

6) *Sandy Wax*

This wax is light to dark grey and rough to the touch.

	%		
Wax	97.16	Iodine Number	16.7
Moisture	1.09	Melting Point	86°C
Impurities	1.75		

7) *Fat Wax*

This variety is the darkest variety, almost black, containing a large percentage of impurities. Taken from the palm leaves, it is melted to drive off moisture.

	%		
Wax	96.91	Iodine Number	16.00
Moisture	0.59	Melting Point	85°C
Impurities	2.50		

Uses

Carnauba wax is a preferential ingredient in all the following:

1. Buffiing of Wooden Ware
2. Candles
3. Casting for Precision
4. Cosmetics: Creams, Lipsticks
5. Finishes for Leather
6. Glazing: Candies, Gum, Pills, Paper
7. Inks for Carbon Paper
8. Lubricants: Greases, Mold Release, Phonograph Records
9. Medicinals: Salves, Ointments
10. Polishes: Shoe, Furniture, Leather, Automobile, Floor, Fruit, Vegetables
11. Protective Coatings: Varnishes, Lacquers, Enamels

Specifications

The specifications for genuine pure carnauba wax are as follows:[1]

Scope

These specifications cover the following types and grades:

Type	*Brazilian Designation*		*American Designation*
Type 1	Flor	=	Flor
	Primeira	=	Prime or No. 1 Yellow
Type 2	Mediana	=	Medium or No. 2 Yellow
Type 2A	Roxa	=	Roxa or No. 3 Yellow
Type 3	Cauipe	=	Cauhype
	Gorda Clara	=	Gorda Clara Light Fatty No. 2 North Country
Type 4	Gordarosa	=	Fatty Gray or No. 3 North Country
Type 5	Arenosa	=	Chalky

Color

The color of types 1, 2, 2A, and 3 CRUDE shall be no darker than that of the official color card of The American Wax Importers and Refiners Ass'n., unless otherwise agreed upon by the seller and the purchaser.

Composition and Properties

The wax shall be an original virgin product made directly from the powder obtained from the leaves of the Carnauba Palm (*Coperniea Cerifera*). It shall be free of extenders or admixtures of other substances and shall conform to the following requirements and shall be so guaranteed by the seller.

---

[1] American Wax Importers and Refiners Assn., Inc.

	Crude			U.S.A. Pure Refined	
	Yellow Types 1, 2, 2A	*Fatty Types 3 & 4*	*Chalky Type 5*	*Yellow Types 1, 2, 2A*	*Fatty Types 3 & 4*
Melting Point—minimum	83°C 181.4°F	82.5°C 180.5°F	82.5°C 180.5°F	83°C 181.4°F	82.5°C 180.5°F
Flash Point—minimum	310°C 590°F	299°C 570°F	299°C 570°F	310°C 590°F	299°C 570°F
Volatile Matter (moisture included)—max. percent	2.0	1.5	6.0	negligible	negligible
Insoluble Impurities—maximum percent	1.0	2.0	1.5	negligible	negligible
Acid Number—minimum	2.0	4.0	4.0	2.0	4.0
Acid Number—maximum	6.0	10.0	10.0	6.0	10.0
Saponification Number—minimum	78.0	78.0	78.0	78.0	78.0
Saponification Number—maximum	88.0	88.0	88.0	88.0	88.0
Paraffinic Hydrocarbons—maximum percent	2.0	2.0	2.0	2.0	2.0
Acetone Soluble Resinous Matter—max. percent at 15°C	5.0	3.5	3.5	5.0	3.5
Benzene Solubility—maximum percent at 25°C	8.0	8.0	8.0	8.0	8.0

Adulterations of Carnauba Wax[1]

To detect possible adulterants, the Cleveland open-cup flash-point tester was used. This proved to be a very handy tool. The melting points of combinations of ingredients with carnauba wax could be artificially raised by heating with ethyl cellulose. The flash point would be lowered.

If the open-cup flash-point test[2] is followed carefully, reproducible results can be obtained. Care should be taken to have no drafts in the room which would dilute the fumes before they ignite. The sample should be heated at a constant rate of 9–11°F a minute. These are the common sources of error.

A list of the highest and lowest flash-point tests of carnauba wax of various grades is given in the following tables:

Table 4:9

Flash Points of Carnauba Waxes Tested Prior to 1942

Carnauba Wax	*Flash Point, °F*
No. 1 Yellow	600–620
No. 2 Yellow	600–610
No. 3 N. C. Refined	585–595

Carnauba waxes tested since 1942 have shown a tendency to lower flash points as shown in Table 4:10.

[1] Bowers, *Soap and Sanitary Chemicals*, 113 (1949).

[2] *Standard Methods of Analysis of the Oil Chemists Society*, p. 41.

Table 4:10

FLASH POINTS OF CARNAUBA WAXES TESTED AFTER 1942

Carnauba Wax	*Flash Point, °F*
No. 3 N. C. Crude	575
No. 3 N. C. Refined	572
No. 3 N. C. Refined	572
No. 3 N. C. Refined	509
No. 3 N. C. Refined	536
No. 3 N. C. Refined	543
No. 3 N. C. Refined	Sputter 445 to 575 flashed
No. 3 N. C. Refined	536
No. 3 N. C. Refined	554
No. 3 N. C. Refined	558
No. 3 N. C. Refined	568
No. 3 N. C. Refined	542
No. 3 N. C. Refined	554
No. 3 N. C. Refined	546
No. 3 N. C. Refined	564
No. 3 N. C. Refined	570

The low flash points in carnauba wax may be due to adulteration or to different methods of collecting and refining. In some formulas, these waxes do not make a stable emulsion, regardless of the usual formula adjustments. In other wax formulas, with more efficient emulsifiers, emulsions are formed, but the resulting waxes are sticky and, when stored at 120°F for a few days, set up into jells. These results are shown in Tables 4:11 and 4:12. Table 4:11 shows the stickiness of two lots of wax of low flash points compared to waxes of high flash point made by the same formula.

Table 4:11

VARIATION OF TACK OF CARNAUBA WAX WITH FLASH POINT

Carnauba Wax	*Pounds Pull*	*Temperature, °F*	*Humidity, %*
536	14	74	40
543	12	74	40
590	7	74	40
595	7	74	40

The method of measuring tack is as follows. Linoleum strips are coated with two coats of wax. The wax is dried overnight. A clean woven wool pad of polisher's felt is placed on the wax. A block of wood covers the felt. The block contains an eye screw. A 20-lb block of lead is placed on the wood cover. A 15-lb capacity spring scale is attached to the eye screw. Two minutes are allowed to lapse between putting on the weight and checking the pull in pounds necessary to move the felt pad.

The figures in Table 4:12 shows some correlation between the shelf life and flash point. The same formula was used in each case. The expected life of this formula at 120°F is 1 year or more.

Table 4:12

RELATION BETWEEN FLASH POINT AND SHELF LIFE

Flash Point of Wax, °F	*Shelf Life at 120°F*
536	2 days
545	6 days
509	14 days
572	185 days
590	12 months

A correlation between both tack and shelf life and low flash points would depend on the ingredients causing the low flash point. These ingredients may vary slightly.

Table 4:13 shows the flash points of various wax blends made with carnauba wax.

Table 4:13

FLASH POINTS OF WAX BLENDS CONTAINING CARNAUBA WAX

			Flash Point, °F
Refined Carnauba No. 3 N.C., alone			590
Refined Carnauba No. 3 N. C. with	5%	candelilla	575
	100%	candelilla	490
	5%	135° paraffin	527
	10%	180° paraffin	570
	10%	oxidized petroleum wax	582
	20%	oxidized petroleum wax	572
	100%	oxidized petroleum wax	539
	10%	Piccolyte S-100	543
	25%	Piccolyte S-100	503
	25%	Durez 219	540
	10%	Wood I rosin	510

Table 4:14

BENZENE INSOLUBLES, SPECIFIC GRAVITY AT 25°C, AND CAPILLARY—TUBE MELTING POINTS OF WAXES

	% Benzene Insoluble	*Specific Gravity*		*Melting Point °F*	
		as Received	*Extracted*	*as Received*	*Extracted*
1. No. 3, Refined Carnauba Wax N.C.	0.18	0.9912	0.9916	180.5	183.0
2. No. 3, Crude Carnauba Wax N.C.	1.60	1.0097	0.9979	181.2	180.1
3. Refined Ouricury Wax	0.005	1.0025	1.0040	181.0	179.5
4. Crude Ouricury Wax	4.54	1.0321	1.0085	181.0	181.0

COTTON WAX

Cotton wax is obtained from the so-called green cotton by solvent extraction. It is a hard, yellow, high-melting wax.

Table 4 : 15

PROPERTIES OF COTTON WAX

Melting Point	83–85°C	(76–80°C)
Specific Gravity	0.902	(0.9760–1.005)
Saponification Number	2.5	(50–76)
Acid Number	6.7	(21–28)
Iodine Number (Hanus)	12.75	(20–27)
Acetyl Number	—	(48–83)
Unsaponifiable Matter	—	(57–68%)
Acetyl Number of Unsaponifiable Matter	—	(115–124)

Table 4 : 16

COMPOSITION OF SEMIREFINED COTTON WAX

Alkyl-Wax Acid Esters	46–48%
Acids (Free)	25–26
Sterols	10–12
Glycerides	1–3
Hydrocarbons	7–9
Moisture & Carbohydrates	2–3

ESPARTO WAX

Esparto Wax (also called Fiber Wax and Reed Wax) is found as a coating on esparto grass which grows in South Spain and North Africa.

During the process of making fine grades of paper, in Scotland, the grass is threshed and the wax is removed in the form of dust. The wax is then recovered from the dust by extraction, cooled in cakes, and broken into lumps.

Properties

Esparto wax is a hard, brittle, non-tacky wax of brownish color.

It is somewhat similar to candelilla wax but is less brittle and harder. When melted, it smells like decomposed straw. It has a granular structure, but much less pronounced than that of carnauba wax. When melted and cooled slowly, the surface is very smooth and flat. It contracts much less than carnauba wax. The purified commercial grade has an average melting point of 75.3°C.

Esparto (Fiber) wax contains about 65–70% hydrocarbon of the formula $C_{31}H_{64}$.

It gives good wetting and penetrating properties in oil compositions and softer films than carnauba wax. It is a better solvent for basic dyes than carnauba. It can be plasticized by the addition of small amounts of amorphous waxes or oils.

Table 4 : 17

PROPERTIES OF ESPARTO WAX

Melting Point	162–165°F, (72–74°C)
Flash Point	480–500°F
Specific Gravity at 25°C	0.983–0.989
Acid Number	28–33
Ester Number	33.6–36.9
Saponification Number	61.6–69.4
Unsaponifiables, percent	42–48.5
Iodine Number	8–16
Ash	0.4–0.6
Color	brown

Esparto wax is compatible with vegetable, animal, and mineral waxes; fatty acids and glycerides; hydrocarbons, and a large variety of natural and synthetic resins.

Uses

Esparto wax is used in carbon paper inks, leather finishes, polishes, and plastics. It imparts smoothness to various compositions.

FIR WAX

So-called *Fir wax* is contained primarily in the corky fractions of the bark of the Douglas Fir tree. Typical old growth bark will contain about 5% wax as measured by carbon tetrachloride extraction.

Douglas fir wax, although it is generally classified as a natural wax, more properly belongs in the category of a manufactured wax. Its characteristics depend heavily on the method of extraction from the bark. For example, a wax offered several years ago, based on benzene extraction of the bark, was a soft, dark colored, low-melting product. A wax produced on pilot scale by another company, by heptane extraction, was a very hard, pleasant smelling, moderately high-melting product.

If one defined Douglas fir wax as only the aliphatic ester content of the bark, such as might be produced by urea adduction of a less pure product, it would have approximately the following characteristics:

Acid Value	15–25
Saponification Value	120–130
Melting Point, °F	162–165
Hardness	1.5
Color	Pale Yellow

Practically, a wax of the type described exists only as a laboratory sample. More commonly the wax contains a greater or lesser percentage of non-adducting materials, principally aromatic acids with smaller amounts of triterpenoid materials and sterols. The aromatic acids can be present either in free or esterified form. One analysis showed lignoceryl alcohol (20%), lignoceric acid (60%), and ferulic acid (20%).

The characteristics of the these waxes can vary widely. Some of them give beautiful emulsions with a minimum of soap, whereas others would only curdle. In some forms the Douglas Fir waxes form excellent pastes. Other types show very poor solvent-binding power.

Properties of Fir Wax

FIRWAX 102

Acid Value	25–30
Saponification Value	145
Melting Point, Fisher-Johns, °C	65–67
Penetration, mm × 10/77°F/100 g	2.5
Specific Gravity	1.025
Color	Dark Brown

Firwax 120 will have properties similar to the above. In this grade, intended for investment casting, ash and shrinkage are also controlled. Typical values are as follows:

Volumetric Shrinkage, M.P. + 10°C to 25°C, %	10.8
Ash, %	0.005

In addition to low shrinkage and ash content, Firwax 120 contributes good hardness at moderate melting point, freedom from cavitation and post-shrinkage, and good compatibility with commonly used resins.

Firwax 102 which is used in carbon inks, shows excellent dye solubilizing properties and pigment dispersibility. It is not outstanding by itself in retaining oil; however, it can be easily modified by ozoceritic waxes to formulate carbon inks of excellent intensity, good oil- and dye-retention, low viscosity, and good stability of vicsosity. Both black and blue inks of very low cost and good quality have been made.

FLAX WAX

Flax wax is a hard brittle wax obtained from the stems of the flax plant. It is chiefly composed of aliphatic fatty acid esters and hydrocarbons.

Table 4 : 18

PROPERTIES OF FLAX WAX

Melting Point	67–70°C	(67.3–69.5°C)
Specific Gravity	0.971–0.985	(0.936–0.985)
Saponification Number	78–83	(78.4–83.7)
Iodine Number (Hanus)	21–28	(21.6–28.8)
Acid Number	18–23	(17.5–23.8)
Unsaponifiable Matter	70–80%	(20%)
After Distillation with Superheated Steam		
Saponification Number	70	
Acid Number	33	
Iodine Number	18	

JAPAN WAX

Japan wax is derived from the protective coating on the berry kernels from several varieties of the Japanese sumac trees. The wax is prepared by crushing the aged berries and then extracting by means of pressure. To refine, the wax is again melted, filtered, and bleached by sunlight and/or chemical bleach. The wax is usually molded into square cakes and packed in wooden cases.

Composition

Chemically, Japan wax is a fat (glyceride) and is composed of palmitin, palmitic, and other high-molecular-weight fatty acids and their glycerides.

It contains 5.2–6.3% (sometimes 7.1%) dibasic acids of which one of the formula $C_{23}H_{44}O_4$ is predominant. Dibasic acids occur in Japan wax as mixed glycerides with oleic acid.

The composition has been given essentially as follows:

Glycerides	93–7%
Fatty Acids (Free)	3.8–5.5
Alcohols (Free)	1.2–1.6

The dibasic acids of Japan wax have been determined to be *n*-alkanedioic acids having 16, 18, 19, 20, 21, 22, 23, 24, and 26 carbon atoms in the percentages of 1.2, 40, <0.1, 35.5., 0.7, 53.5, 0.4, 3.5, and 1.2% respectively.[1]

Grades Available

There are many grades of genuine Japan waxes available which are distinguished by brand names.

Properties

Japan wax is a pale cream colored wax with a gummy feeling. Actually it resembles a fat more than a wax.

Japan wax is not hard but tough, malleable, and sticky. It has a tallowy odor and becomes rancid, effloresces, and darkens on aging. It breaks easily and the broken surfaces are dull.

When Japan wax is melted and allowed to solidify, its specific gravity and melting point are lowered. On aging, however, these constants tend to return to the normal standard values.

It is soluble in hot alcohol, carbon disulfide, chloroform, ether, benzene, petroleum ether, carbon tetrachloride, isopropyl ether, naphtha, pyridine, tetralin, toluene, xylene, trichloroethylene, turpentine. It is insoluble in water and cold alcohol.

It is readily saponified by alkali to an emulsion.

---

[1] J. A. Lambarton, *Australia J. Chem.*, **14**, 323 (1961).

Table 4:19

PROPERTIES OF JAPAN WAX

Melting Point		122–133°F (50–56°C)
Flash Point		385–400°F
Specific Gravity at 25°C		0.975–0.984
Acid Number		6–20
Ester Number		210–225
Saponification Number		217–237
Unsaponifiables, percent		2–4
Iodine Number		4–15
Acetyl Number	27–31.2	—
Ratio Number	—	(10–11)
Hehner Number	—	(89–91)
Reichert-Meissl Number	—	(1–3)
Ash	0.020–0.08%	—
Refractive Index, 60°C	1.450–1.4560	—
Dielectric Constant	3.1–3.2	—
Effective A.C. Conductivity	33–34	—
Volume Resistivity	0.4–0.8	—

Japan wax is compatible with beeswax, cocoa butter, and glycerides. It is often adulterated with varying amounts of local vegetable oils.

Uses

Japan wax is used in buffing compounds for metal, candles, cosmetics, crayons and pencils, finishes for textiles and leather, lubricants, matches, packaging of foods, pharmaceuticals, polishes for furniture and leather, soaps, and as a common substitute for beeswax.

OURICURY WAX

Ouricury wax is exuded from the leaves of the Ouricury Palm (*Syagrus Coronata*), which grows in Bahia, Brazil. This palm which is found in dense forests, and often reaches a height of eighteen to thirty feet, furnishes the natives with several of their needs.

Ouricury wax is very similar to carnauba in character, but is available only in several shades of brown. The brown color varies with the care used in processing the wax.

Preparation

Unlike the carnauba palm, the ouricury tree produces only sparse amounts of wax. This wax is more difficult to gather because it can be obtained only by careful scraping of the leaves with sharp instruments. Because of this method of extraction, the wax may contain up to 25% foreign matter which is difficult to separate. Several methods are employed to purify the wax. After being melted and cooled, it is broken into lumps, graded, and priced on the basis of absolute insolubles and moisture present.

Composition

The analysis of ouricury wax is essentially as follows:[1]

Hydrocarbons	1.3%
Simple Esters	23.5
Hydroxy Mono-Esters	22.4
Hydroxy Di-Esters	17.2
Hydroxy-Acidic-Poly-Esters	5.4
Free Acids	8.7
Free Alcohols	3.0
Resins	14.8
Moisture	1.4
Ash	0.4

Grades Available

Crude—with various percentages of impurities

Refined—powdered, lumps, and flakes.

Foreign matter must be removed by filtering. Double-refined ouricury wax contains a much smaller amount of resin and foreign matter.

Properties

Ouricury wax is a hard, brittle, and lustrous wax. It contains a high percentage of resins. It differs from other vegetable waxes in that there is a lag of about 12° between its melting and setting points. Most vegetable waxes show a lag of 5–6° only.

Table 4 : 20

PROPERTIES OF OURICURY WAX

Melting Point	179–183°F (81.5–84°C)
Flash Point	540–580°F
Specific Gravity at 25°C	0.970–1.050
Acid Number	8–20
Ester Number	75–85
Saponification Number	70–100
Iodine Number	6–8
Unsaponifiable matter %	50–55
Ash	0.55%

Ouricury wax is compatible with all vegetable, animal, and mineral waxes; resins; fatty acids and glycerides; hydrocarbons (in certain proportions). When melted with other waxes, a resinous precipitate is formed.

Uses

Ouricury wax is used in many products as a substitute for carnauba wax where color is not an important factor. One of its principal uses is in the carbon paper industry; other applications are in shoe polishes, and as lubricants for mold release.

---

[1] L. J. N. Cole, *J. Am. Oil Chem. Soc.*, **37**, 359–63 (1960).

Specifications[1]

The specifications for genuine pure ouricury wax are as follows:

Scope

These specifications cover the following types and grades:

Brazilian Designation		*American Designation*	*Color*
CRUDE—ORIGINAL	=	CRUDE—ORIGINAL	Dark Greenish Brown to Dark Greenish Black
—		U.S.A. PURE REFINED	Light Tan to Dark Brown

Composition and Properties

The Brazilian Crude Ouricury shall be an original virgin product made directly from the powder obtained from the leaves of the Ouricury Palm (*Syagrus Coronata*) and shall be free of extenders or admixtures of other substances and shall conform to the following requirements and shall be so guaranteed by the seller.

	BRAZILIAN CRUDE	U.S.A. PURE REFINED
Melting Point—minimum	82.5°C–180.5°F	82.5°C–180.5°F
Flash Point—minimum	277°C–530°F	277°C–530°F
Volatile Matter(moisture included—maximum percent	1.0	Negligible
Insoluble Impurities—maximum percent	3.0	Negligible
Acid Number—minimum	8.0	8.0
Acid Number—maximum	18.0	18.0
Saponification Number—minimum	80.0	80.0
Saponification Number—maximum	105.0	105.0
Paraffinic Hydrocarbons—maximum percent	3.0	3.0
Acetone Soluble Resinous Matter—maximum percent at 15°C	9.0	9.0
Benzene Solubility—maximum percent at 25°C	18.0	18.0

PALM WAXES

The term, *Palm Wax*, when applied to the commercial grades, refers to particular types of blends of true waxes, which have been compounded for specific purposes. These waxes are available in a variety of grades.

Grades Available[1]

No. 1 No. 2 No. 693.

The Melting points of the various grades are as follows:

No. 1	170–175°F (77.0–79.4°C)
No. 2	162–167°F (72.2–75.0°C)
No. 693	171–175°F (77.2–79.4°C)

Cuban Palm Wax

This wax compares favorably with carnauba wax. It is harder and more friable than the latter. It is now being produced in Brazil on a small scale and is harvested and refined like carnauba wax.

[1] American Wax Importers and Refiners Assn., Inc.

Table 4:21

PROPERTIES OF CUBAN PALM WAX

Acid Value	6–16
Saponification Value	70–101
Melting Point	82–86
Hydroxyl Value	55–59
Penetration 100 g/55 cc	< 1
Hydrocarbons	1%
Solubility in Acetone (25°C)	0.9–1.5%

Compatibility

Palm waxes are compatible with all vegetable waxes, all animal waxes, resins, fatty acids, glycerides, and hydrocarbons.

Uses

Palm wax is ideal for incorporating in the inks on one-time carbon paper; and also as the necessary lubricant and moisture-repellent in the manufacture of rope and twine.

Other applications are in polishes for floors, automobiles, furniture, and leather.

RICE BRAN WAX

Rice Bran Wax is quite hard, dry, slightly crystalline and varies in color from tan to light brown.

Properties

Melting point	169–180°F (76 to 82°C)
Flash point	Minimum 520° F.
Acid Number	1–10
Ester Number	73–110
Saponification Number	75–120
Iodine	Maximum 10
Unsaponifiable Matter	39–57

Compatability:

Vegetable, Animal and Mineral Waxes
Fatty acids
Glycerides
Hydrocarbons

(Rice Bran Wax is F.D.A. approved for the end uses indicated. Regulations 121.1059, 121.1098)

RICE-OIL WAX

Rice-oil wax is prepared commercially by hydrogenation of crude rice oil at high pressure and temperature. The oil can be dewaxed by evacuated unglazed porcelain cylinders covered with filter cloth, or by a De Laval and basket-type centrifuge.

The crude wax is purified by washing repeatedly with methyl alcohol, acetone, ether, and finally chloroform, to yield a substance that is an ester mainly composed of tetracosanoic acid and myricyl alcohol with small amounts of saturated acids and some unsaturated acids.

The commercial rice wax melts at 67–69°C and has a saponification number of 142–144. The purified wax melts at 75–76°C and has a saponification number of 72.8, iodine number of 12.6 and 48.71% unsaponifiable matter.

SUGAR CANE WAX

Source

Crude cane wax is extracted from cachaza, a by-product in the manufacture of raw sugar. Cachaza is the filter-press mud separated by sedimentation and filtration from the cane juice that is squeezed from the ripe cane. It consists of the suspended particles and the precipitate formed by the action of heat and chemical treatment on the colloid, and is removed in the clarification of the cane juice. The precipitate contains the cane wax that was originally on the surface of the cane stalk before it was crushed.

Preparation

The cachaza is removed from the cane juice, after chemical treatment and heating, by sedimentation and filtration.

When the filter cake is removed from the filters, it is mixed with a solvent which dissolves the came wax as well as other soluble substances. The solvent is then evaporated, recovered, and reused. The mass residue is waxy and contains about 50% sugar-cane wax, 30% sugar-cane fatty oil and 20% resinous matter.

In a refining process, the crude sugar cane wax is dispersed in a solvent which readily dissolves the fatty-oil fraction and leaves the hard wax and resinous matter as a suspended solid. This slurry is filtered to remove the soluble fatty-oil fraction which is recovered as a green semifluid paste.

The sugar-cane wax and the resinous matter from which the sugar cane fatty oil has been removed are now reslurried with additional solvent and heated until the cane wax melts. At this temperature, the resinous matter separates from the wax and is removed by decantation. The light layer containing the sugar cane wax is then evaporated to remove all the solvent, leaving the refined cane wax which is cast into pans.

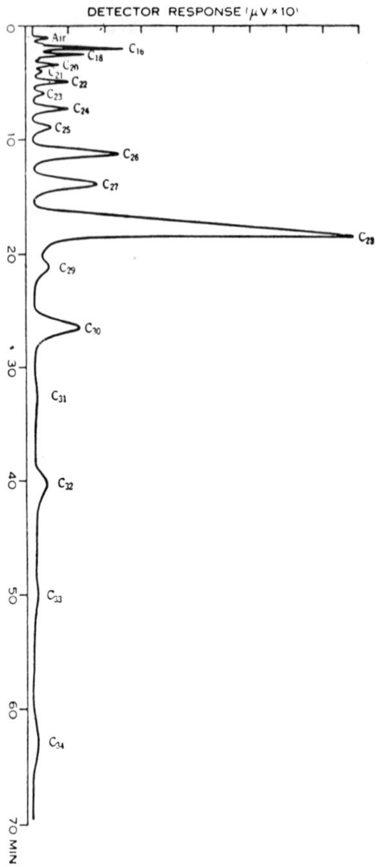

FIG. 4:1. CHROMATOGRAM OF HYDROCARBONS DERIVED BY SAPONIFICATION FROM SUGAR CANE CUTICLE WAX.

TEMPERATURE OF COLUMN, 270°C; RATE OF FLOW, 29.8 ML/MIN 600 MM PRESSURE; CHARGE, 1 μ 1.

Table 4:22

COMPOSITION OF SEMI REFINED SUGAR CANE WAX

(M.P. 67°C)

	%
Esters	78–82
Acids (Free Fatty)	14
Alcohols	6–7
Hydrocarbons	3–5
Inorganic Matter	0.5–1

Table 4:23

HYDROCARBONS DERIVED FROM CONSTITUENTS OF SUGAR CANE CUTICLE-WAX (WT. PER CENT.)

Carbon Number	*Natural Hydro-carbons*	*Alcohols*	*Alcohols (from saponifica-tion)*	*Acids (from saponifica-tion)*	*Aldehydes (from semi-carbazones)*	*Aldehydes (from oxidation to acids)*
14, 15						
16				2.9		
17	5.4 (14–19)			Trace		0.6 (14–19)
18				1.8		
19				0.2		
20	0.4			1.2		—
21	0.4			0.5		—
22	0.4			1.9		—
23	0.8			0.6		—
24	1.1	0.6	0.6	2.5	0.1	0.4
25	7.0	0.4	0.2	1.9	0.1	0.3
26	4.9	15.0	13.1	8.2	6.6	8.1
27	55.7	4.6	2.7	8.0	1.3	2.1
28	2.9	72.1	73.0	50.5	66.5	65.9
29	12.8	1.7	1.5	2.4	1.5	2.7
30	2.2	3.5	6.3	8.1	13.8	10.7
31	4.4	0.9	1.0	1.0	1.0	2.2
32	—	1.2	1.6	4.4	5.3	3.9
33	1.6			1.4	—	—
34				2.5	3.8	3.1

CONSTITUENTS OF SUGAR-CANE CUTICLE WAX

Class	*Weight (%)*	*Class*	*Weight (%)*
Aldehydes	50 (approx.)	Acids	15;* 7–8†
Alcohols	25–27	Hydrocarbons	8–9

* Yield from saponification, includes acids produced in Cannizzaro reaction.
† Calculated from acid and saponification values assuming average chain-length of 28.

Constituents of Sugar Cane Wax[1]

Sugar cane cuticle wax previously found to consist of long-chain aldehydes, alcohols, acids, and hydrocarbons (Lamberton and Redcliffe, 1960) has been re-examined by gas chromatography to identify and estimate the individual constituents. The aldehydes, alcohols, and acids, all straight chain, are principally of even carbon number in each instance with the C_{28} member predominant, but appreciable amounts of odd carbon homologues are also present. The hydrocarbons are also of odd and even carbon number but with the odd members in greater amounts and *n*-heptacosane as the major hydrocarbon constituent. (Figure 4:1).

Properties of Sugar Cane Wax

The characteristics of cane wax are similar to those of the other valuable hard natural vegetable waxes. The supply of this wax which could be obtained from the world cane crop would far exceed that of any other vegetable wax.

The wide variation in the constants is due to varying grades of purity, the higher figures referring to the pure and bleached wax.

Table 4:24

ELECTRICAL PROPERTIES OF DARK HARD SUGAR CANE WAXES AND OTHER COMMERCIAL WAXES

Wax	*Frequency kc/sec.*	*Dielectric Constant*	*Power Factor*	*A.C. Conductivity*[1] *mho-cm*$^{-1}$	*Volume Resistivity ohm–cm*
Louisiana	1.0	2.93	103×10^{-4}	0.17×10^{-10}	
Sugar Cane	7.5	2.87	124×10^{-4}	1.5×10^{-10}	
	100.0	2.82	120×10^{-4}	19.0×10^{-10}	
					6×10^{15}
Cuban	1.0	2.65	96×10^{-4}	0.14×10^{-10}	
Sugar Cane	7.5	2.61	118×10^{-4}	1.3×10^{-10}	
	100.0	2.56	131×10^{-4}	19.0×10^{-10}	
					2×10^{15}
		*		*	*
Carnauba No. 2,					
N.C.	All	2.66–2.83		$4.9–\ 5.3 \times 10^{-12}$	$60–120 \times 10^{13}$
Candelilla	samples	2.38–2.49		$6.0–\ 6.4 \times 10^{-12}$	$>290 \times 10^{13}$
Crude Montan	measured	2.62–2.67		$29.0–\ 31.0 \times 10^{-12}$	$>290 \times 10^{13}$
Crude Beeswax	at	2.87–2.88		$45.0–\ 48.0 \times 10^{-12}$	$5–\ 12 \times 10^{13}$
Chinese	1 kc/s	2.94–2.98		$125.0–137.0 \times 10^{-12}$	$1–\ 2 \times 10^{13}$
Yellow Ceresin		2.16–2.24		$<0.37 \times 10^{-12}$	$>290 \times 10^{13}$
Refined Paraffin		2.21–2.24		$<0.37 \times 10^{-12}$	$>290 \times 10^{13}$

[1] Kranz, Z. H. *et al.*, *Australian J. Chem.*, **13**, 498–505 (1960).
* Initial values.

Duplicane 517

General Properties

Melting Point	165–170°C
Penetration	1 maximum
Color	brown
Acid Number	20–30
Saponification Number	65–75
Oil-retention Penetration	42 maximum

Solubility in Various Solvents

Wt. of Wax	*Solvent*	*Cloud point, °C**
10%	Ethanol	Insoluble
	Butanol	56°C
	Acetone	Insoluble
	Benzol	66°C
	Ethyl ether	partially soluble
	Ethyl acetate	44°C
	Carbon tetrachloride	28.5°C
	Kerosene	40°C
	V M and P naphtha	38°C
	Turpentine	34°C
	Hydrogenated naphtha	32.5°C

* The wax is completely soluble above the temperature shown.

Warwick Waxes

Name of Wax	*M.P.* ASTM D-127-30	*Penetration* 100/77°/ 5 sec	*Color* N.P.A.	*Acid Number*	*Saponification Value*	*Type*
Cane Wax 500	170–176	2 Max.	Tan	26–32	75–95	Veg. Wax
Cane Wax 511	176–180	2 Max.	Brown	62–68	130–140	Veg. Wax
Cane Wax 517–711	176–180	2 Max.	Black	11–14	75–85	Veg. Wax
Cane Wax 700	171–177	2 Max.	Light Brown	26–32	70–90	Veg. Wax

UCUHUBA WAX

This wax (also called Ucuhuba Butter, Ucuhuba Resin, Ucuhuba Tallow) is obtained from the ucuhuba plant, which grows in Brazil. It is of a yellowish color and of the consistency of butter. Its composition is as follows:

	%
Moisture	19.380
Albuminous Substances	11.409
Ether Extract	10.000
Cellulose	3.162
Ash	5.520
Nonnitrogenous Substance	50.529
Phosphorus (P_2O_5)	1.552

Properties

Melting Point	116–120°F (46–49°C)
Flash Point	470–490°F
Specific Gravity at 25°C	0.977–0.982
Acid Number	3–5
Ester Number	205–215
Saponification Number	208–217
Unsaponifiables, percent	0.3–0.5
Iodine Number (Hanus)	2–10

COCOA BUTTER

Cocoa butter (also balled cacao Butter, Oil of Theobroma) is a fatty wax-like material obtained from roasted cocoa beans, principally as a by-product of chocolate manufacture.

It is a yellowish-white, greasy solid of a faint, agreeable odor and a bland, chocolaty taste. It is brittle below 25°C. It consists principally of glycerides of stearic, palmitic, oleic, lauric acids, and other lower fatty acids.

Properties

Melting Point	30–35°C
Specific Gravity, 100–25°C	0.858–0.864
Saponification Number	188–195
Iodine Number	35–40
Refractive Index, 40°C	1.4537–1.4578
Reichert-Meissl Number	0.1
Polenske Number	0.2

Cocoa butter is soluble in benzene, chloroform, ether and hot anhydrous alcohol.

Uses: In confectionery, cosmetics, pharmaceutical ointments, suppositories.

A propylene glycol diester of mixed higher fatty acids having a melting point of 33°C is a good substitute for cocoa butter.[1]

[1] W. G. Alsop, U.S. Patent 2,993,063, July 18, 1961.

Chapter 5

ANIMAL WAXES

BEESWAX

Origin

Beeswax, also called White Wax (*Cera Alba*) and Yellow Wax (*Cera Flava*),[1] is secreted by the honey bee for building its combs. The combs are melted and the beeswax strained, or boiled with water and then skimmed off, to obtain the crude beeswax. Crude beeswax varies in color from a deep brown to a light yellow. The color depends upon the type of flower from which the bees have gathered pollen, the amount of natural contaminants such as propolis, and the age and treatment of the combs. Generally, the light yellow crudes are of better quality because they bleach more readily than the darker crudes.

Preparation

In refining, the crude beeswax is remelted and thoroughly washed, impurities are either skimmed off the top or are allowed to settle out. The refined beeswax is cast into approximately 1-oz squares, 1-lb cakes, or 25-lb slabs, ready for market. Refined beeswax may be further decolorized and cleaned by filtration with diatomaceous earth and carbon.

Wax to be bleached is chemically treated with oxidizing chemicals such as the permanganates or bichromates and either shredded mechanically, to expose a maximum surface for sun bleaching, or bleached by the sun in the liquid state. At the end of the sun-bleaching process, the beeswax is washed, given a cleaning filtration, and cast into discs, cakes, or slabs, ready for use.

Composition

Chemically, beeswax is composed of myricyl palmitate, cerotic and homologous acids, with small amounts of hydrocarbons, cholesterol esters, and ceryl alcohols. Pollen and propolis are the two chief colorants and contaminants in nature. The presence of propolis (bee glue) in undue amounts makes bleaching difficult. The free fatty-acid content is an important factor in emulsifiability. It is easily saponified with alkalies.

---

[1] Roy A. Grout, Dadant & Sons, Hamilton, Ill.

Table 5:1

COMPOSITION OF BEESWAX

Spectroscopic and gas chromatograpic[1] examination of beeswax after saponification shows it to consist of:

Hydrocarbons	16%
Monohydric Alcohols	31
Diols	3
Acids	31
Hydroxy-Acids	13
Balance[2]	6

[1] D. T. Downing, *Australian J. Chem.*, **14**, 253 (1961).
[2] Pigments, propolis, etc.

Table 5:2

COMPOSITION OF YELLOW BEESWAX

Esters of Wax Acids	70.9%
Cholesteryl Esters of Fatty Acids	1.1
Lactones	0.6
Flavones (Coloring Matter)	0.3
Free Alcohols	1–1.25
Free Wax Acids	13.5–14.5
Hydrocarbons	10.4–13.6
Moisture & Foreign Matter	1–2

Grades Available

Crude, Bleachable, and Non-Bleachable: lumps and blocks.

Yellow Refined, U. S. P.: 1 lb cakes and slabs.

Pure U. S. P.: white sunbleached—squares 1 lb cakes and slabs.

Properties

Natural Beeswax is amorphoid and it varies in color from a deep brown to a light taffy shade. The wax has a distinctive honey odor, an aromatic taste, and it does not adhere to the teeth or become pasty when chewed. It is plastic when kneaded by hand, but it is hard and it breaks with a snap when it is bent in the cold. It breaks with a flaky granular surface.

Table 5:3

CONSTANTS OF DIFFERENT GRADES OF BEESWAX

	Specific Gravity at 15°C	*Melting Point °C*	*Acid No.*	*Saponification No.*	*Ester No.*	*Ratio No. (ester No. divided by acid No.)*	*Unsaponifiables %*
Yellow Beeswax (Cera Flava)	0.958–0.970	62–64.0	17–23	87.0–97	70–80.0	3.3–4.0	50–56
White Beeswax (Cera Alba)	0.958–0.970	62–64.0	18–24	90.0–102	70–80.0	3.3–4.0	50–56
Extraction Beeswax (Unbleached)	0.953–0.957	61–62.5	23–27	92.0–95	66–70.5	2.4–3.0	50–56
Extraction Beeswax (Bleached)	0.970–0.984	69–72.5	22–30	91.5–104	69–77.5	2.5–3.3	50–56

When beeswax is heated to 150–250°C, the acid number decreases whereas the ester and saponification numbers increase; the melting point also is raised. Similar changes take place when beeswax is held in the liquid state at lower temperatures for periods of time. Continued heating causes re-esterification or the formation of estolide with decrease of ester and saponification numbers.

Beeswax is sparingly soluble in cold alcohol, completely soluble in chloroform, ether, and in fixed and volatile oils. It is partly soluble in cold benzene and cold carbon disulfide, and is completely soluble in these liquids at about 30°C. It mixes with fats, oils, waxes, and resins when melted with them.

Compatibility

Beeswax is compatible with vegetable, animal, and mineral waxes, tallows, fatty acids, glycerides, hydrocarbons, fatty alcohols; almost all other waxes and oils.

Beeswax mixed with ceresin and some aromatic resins, or with microcrystalline wax, is used for preparing bee comb foundation for rearing bees in Europe.

Specifications

The U.S.P. specifications for beeswax are as follows:

Melting Point	142–149°F (61–65°C)
Flash Point	468–482°F
Specific Gravity at 25°C	0.950–0.960
Acid Number	16–23
Ester Number	72–79
Ester-Acid Ratio	3.6–4.3
Saponification Number	88–102
Iodine Number	8–11
Unsaponifiable Matter %	52–55
Acetyl Number	15.1
Color	white, yellow, brown
Refractive Index, 75°C	1.4398–1.4451
Odor	honey
Dielectric Constant	3.1–3.3
Effective A.C. Conductivity	70–86
Volume Resistivity	0.9–1.5

Beeswax is widely used in:

Candles
Confectionery
Cosmetics: Creams, Lotions, Pomades
Emulsions
Engraving
Finishes for Leather, Textile, Wood, Paper
Lithography
Medicinals: Ointments, Salves
Modeling
Polishes: Furniture, Wood, Leather
Precision Castings

Adulteration of Beeswax

Beeswax is sometimes adulterated with various materials. Tallow, stearic acid, Japan wax, rosin, paraffin, microcrystalline wax, and ceresin are common adulterants. The purity of beeswax is tested by the ratio of the ester value to the acid value. Even this is not reliable if taken by itself. Waxlike mixtures can be made which, although free from beeswax, have a normal ratio number, as for example, a mixture obtained by melting 37.5 parts of Japan wax, 6.5 parts of stearic acid, and 6.5 parts of paraffin or ceresin. In buying beeswax, therefore, the safest course is to buy from a reliable and recognized firm.

CHINESE WAX (INSECT WAX)

Chinese wax is deposited on the branches of a species of ash tree, in China, by certain insects. The wax is scraped off and refined by melting in hot water, skimming off, remelting, filtering, and bleaching, if necessary.

Chinese wax is yellowish white, fairly hard, shiny, fibrous, and crystalline.

Chemically, Chinese wax is composed of ceryl cerotate and minor amounts of other esters.

Table 5:4

Composition of Chinese (Insect) Wax

	%
Aliphatic Monocarboxylic Acid Esters	93–95
Monoethanoid Alcohols, Free	<1
Wax Acids, Free	0.5–1
Hydrocarbons	2–3
Resins	1–0.5

Table 5:5

Properties of Chinese Wax

Melting Point	149–176°F
Specific Gravity	0.926–0.970
Acid Number	3–13
Saponification Number	80–92
Unsaponifiable Matter	49–50%
Iodine Number	1.5–2
Color	yellow-white
Odor	very faint
Taste	very faint
Refractive Index, 40°C	1.4566
Dielectric Constant	3.65–3.90
Effective A.C. Conductivity	190–210
Volume Resistivity	0.04–0.18

Chinese wax is soluble in alcohol and ether (slightly); in benzene, isopropyl ether, toluene, xylene, trichloroethylene (completely).

Uses

In candles, furniture polishes, and coatings for paper and cloth.

SHELLAC WAX

Shellac wax is a by-product in the production of alcoholic solutions of shellac. It precipitates out of such solutions as a haze or sludge and is filtered off and dried.

Composition

The composition of deresinated shellac wax has been reported as follows:

Wax Acid Esters	80–82%
Wax Acids (Free)	10–14
Wax Alcohols (Free)	<1
Hydrocarbons	2–6
Lac Resins	2–4

Properties

The commercial product is a hard, tough, brown, lustrous wax containing some shellac and impurities. It approaches carnauba wax in hardness and luster.

Table 5:6

PROPERTIES OF SHELLAC WAX

Melting Point	74–78°C	(74–80°C)	82°C
Specific Gravity	0.971–0.980	(0.974–0.982)	0.972 (20°C)
Acid Number	12–16	(10.9–25)	1.2
Saponification Number	120–126	(79–85)	46.7
Unsaponifiable Matter	72–76%	(68–72%)	—
Ratio Number	—	(7–8.6)	—
Ester Number	—	—	45.5
Iodine Number	—	—	1.25
Volatile Matter at 105°C	—	—	0.2–0.4
Ash	—	—	0.12

It is used as a component of insulating material.

SPERMACETI WAX

Origin

Spermaceti (*Cetin*, *Cetaceum*), is a wax obtained from the head cavities (the "case") and the blubber of the sperm whale. Of 30,000 lb of oil which one sperm whale can yield, the "case" furnishes 2500 to 3500 lb with the remainder coming from the blubber.

Preparation

Spermaceti wax is purified from the oil in the following manner: The oil is chilled at 32°F for several days before being inserted in a hydraulic press. The press cake is crude wax and is subjected to hydraulic pressure twice, first at 50–60°F, and then at 80°F to drive out the remaining oils. To produce white crystals, the cake is further refined by boiling with dilute caustic soda, separating, and then by washing with hot water until free from alkali.

Composition

Spermaceti wax consists mainly of cetyl palmitate.

The composition is essentially as follows:[1]

Esters	95.4%
Alcohols	2.6
Acids	1.2

Grades Available

Spermaceti U.S.P.
- 1 lb Cakes
- 65 lb Blocks

Properties

Refined spermaceti wax forms lustrous white masses of long, glistening white crystals. The crystals are quite brittle so that the wax can easily be powdered. The wax is without either taste or odor.

Spermaceti is partially saponified by aqueous alkalies and forms emulsions.

It is soluble in boiling alcohol, ether, chloroform, carbon disulfide, fixed and volatile oils.

Table 5:7

PROPERTIES OF SPERMACETI

Melting Point	113–120°F (45–49°C)	
Flash Point	470–480°F	
Specific Gravity at 25°C	0.940–0.946	
Acid Number	0.0–0.5	
Ester Number	116–125	
Saponification Number	116–125	
Unsaponifiables, percent	45–50	
Iodine Number	3.0 (max.)	
Acetyl Number	2.7	—
Color	white (translucent)	
Refractive Index, 60°C	1.440	—
Dielectric Constant	6–18	—
Effective A.C. Conductivity	> 500	—.
Volume Resistivity	0.0006–0.0003	—
Viscosity at 100°C	—	6.7–7.4)
Fatty Acids	—	(49–54)

Spermaceti is compatible with beeswax, ethyl cellulose, and stearic acid.

Specifications

The U.S.P. specifications for Spermaceti are essentially as given in Table 5:7.

Uses

Spermaceti does not compete with such waxes as beeswax, carnauba, or the mineral waxes. It is used chiefly for cosmetics, in creams, lotions, soaps; for medicinals, in ointments, and salves; and for candles.

---

[1] L. J. N. Cole, *J. Amer. Oil Chem. Soc.*, **37**, 359–63 (1960).

Moby Dick Wax 90

Moby Dick Wax 90 is derived from sperm oil. It is the solid wax portion, composed chiefly of the more saturated fatty alcohol esters of fatty acids, but it also has about 25% unsaturated compounds, with only one double bond. As the iodine number is near 50 it can be sulfonated the same as tallow, and it will produce a product having many advantages. It does not oxidize or turn rancid; it will be a superb lubricant, and it will make stable emulsions.

If this wax is saponified, the alcohols are freed. They may be removed with solvents by distillation; or used as superfatting or coupling agents. By blending with mahogany soaps, excellent rust-proofing products are made. At room temperature, Wax 90 has the consistency of lard.

COMPOSITION (AVERAGE)

Chain Length	*Fatty Acids*	*Alcohols*
C12	5.0%	—
C14	10.0	1.0
C16	20.0	25.0
C18	15.0	17.0
C20	7.0	—

Specifications for Moby Dick Wax 90

Melting Point	80°F to 110°F (27 to 43°C)
Color	10 Gardner
Free Fatty Acid	2.0% Maximum
Acid Number	4.0 Maximum
Iodine Number	25 to 50
Saponification Number	123 to 133
Unsaponifiable (Fatty Alcohols)	40 to 47%
Shipment	400 pound steel drums or tank cars

WOOL WAX

Wool wax, often called wool fat, wool grease, or *dégras*, is extracted from crude sheep's wool. When it is refined the product is called *lanolin*. The raw wool is first scoured with water and a detergent. By centrifuging, the wool grease is obtained. This is refined by aqueous alkali to remove free fatty acids and foreign matter. It is then bleached, deodorized, and dried:

Table 5:8 HYDROCARBONS IN WAXES[1]

	Total Hydrocarbons	*Saturated Paraffins*	*Unbranched Paraffins*	*Branched Paraffins*	*Olefins*	*Non-Aliphatic Hydrocarbons*
Candelilla Wax	50	48.2	48.2	0.5	2.5	—
Beeswax	15	11.9	11.2	0.7	3.1	—
Paraffin Wax (petroleum)	91.2	88.7	45.3	43.4*	2.5	8.8
Paraffin Wax (Brown Coal)	—	86.6	—	13.4*	—	—

* Including Naphthenes.

[1] F. Spengler, *Fette, Seifen, Anstrich*, **59**, 607 (1957).

Table 5:9

CALCULATED COMPOSITIONS OF FRACTIONS FROM WAXES AND OF THE UNHYDROLYZED WAXES[1]

Wax fraction	*Hydro-carbons*	*Free alcohols*	*Free acids*	*Esters*	*Hydroxy esters*	*Acid esters*	*Diesters*	*Acid diesters*	*Hydroxy diesters*
	All values in moles %[2]								
150° Beeswax	54	2	28	17	0	0	0	0	0
250° Beeswax	0	0	0	76	16	8	0	0	0
Residue beeswax[3]	0	0	0	0	0	0	67	5	12
Beeswax[4]	23	1	12	45	8	4	6	1	1
150° Caranda wax	5	75	9	11	0	0	0	0	0
250° Caranda wax	0	16	4	69	11	0	0	0	0
Residue Caranda wax	0	0	0	0	13	3	83	0	0
Caranda wax	1	26	4	44	9	1	16	0	0
150° Crude candelilla	69	19	10	2	0	0	0	0	0
Residue crude candelilla	0	0	0	1	29	37	33	0	0
Crude candelilla wax	46	13	7	2	8	10	9	0	0
Frn. 1 (150°) ref. cand.	80	12	6	2	0	0	0	0	0
Frn. 2 (150°) ref. cand.	31	17	44	8	0	0	0	0	0
Residue refined candelilla	0	13	6	58	23	0	0	0	0
Refined candelilla wax	57	14	7	21	8	0	0	0	0
150° Ouricury wax	27	24	15	35	0	0	0	0	0
250° Ouricury wax	0	22	0	16	45	17	0	0	0
Residue ouricury wax[5]	0	0	0	0	0	0	8	0	25
Ouricury wax	7	—	—	—	—	—	—	—	—
Unsap. carnauba wax	11	89	0	0	0	0	0	0	0
Acids carnauba wax	0	0	0	44[6]	0	54[7]	2[8]	0	0
Carnauba wax	12	0	0	10	53	5	19	0	0

[1] T. W. Findley, *J. Am. Oil Chem. Soc.*, **30**, 291–6 (1955).

[2] Totals for some of the waxes are not exactly 100% because of molecular weight change of a few per cent during distillation.

[3] 13% Triester.

[4] 1% Triester.

[5] 29% Acid hydroxy diester and 38% hydroxy triester.

[6] Lactone.

[7] Includes 24% acid hydroxy ester.

[8] Lactide.

Table 5:10

ORIGIN AND USE OF NATURAL WAXES

Wax	*Type*	*Source*	*Locality*	*Original Color*	*Hardness*	*Commercial Grades*
Beeswax	Animal	*Apis mellifica*	Worldwide	Light yellow to dark greenish brown	Fairly brittle, but plastic when warm	Crude; yellow refined: bleached white
Candelilla	Vegetable	*Euphorbia antisyphilitica*	Mexico and southern Texas	Dark brown, or brownish to greenish	harder than beeswax, not so hard as carnauba	Technical
Carnauba	Vegetable	*Corypha cerifera*	Brazil and South American countries	Sulphur-yellow or yellowish green	Very hard and brittle	No. 1 yellow; No. 2 yellow; No. 2 north country; No. 3 north country; No. 3 north country refined; No. 3 chalky
Ceresin	Mineral	Purified ozocerite from lignite		Yellow or white		White; yellow; orange, sold according to melting-point range, 130–133°, 135–138°, 140–145°, 150–160°, 160–170°F
Insect Wax or Chinese Wax	Insect	*Coccus ceriferus*	Western China	White, opaque, fibrous		
Japan	Vegetable	*Rhus succedanea, Rhus vernicifera*	China, western provinces of Japan	Greenish, pale yellow or light brown	Hard and brittle	Kitagumi and chickusan
Montan	Mineral	Lignite	California, Germany Czechoslovakia	Dark brown, brownish black	Hard and brittle; conchoidal fracture	Crude and bleached
Ozocerite	Mineral	Lignite	Austria, Egypt, Poland, Russia, Ukraine, Yugoslavia, Texas, Utah	Pure yellow to dark brown, greenish brown, or dark amber	Varies from quite soft to as hard as gypsum	Crude; natural yellow; white
Spermaceti	Animal	*Physeter macrocephalus*		White, crystalline and glistening	Very brittle, readily powdered	Blocks, cakes, technical, U.S.P.

Table 5:10 (Continued)

Wax	Containers and Weights in pounds	Special Features	Uses	other names
Beeswax	barrels 300, kegs 100, cases and bags	distinctive odor	manufacture of wax polishes, furniture polishes, wood polishes and finishes; candles; transparent papers; pomades, toilet preparations, cosmetics; food products, chewing gums, sweetmeats; adhesive compositions; dressing and polishing leather; modeling fruits and flowers; base for plasters, etc., engraving, lithography; sizing and finishing textiles, in medicine	
Candelilla	bags 150–200	distinctive odor	manufacture of leather dressing, furniture polish, shoe polish; candles; cements; varnishes; sealing wax; electrical insulating compositions; phonograph records; paper size; celluloid, rubber; waterproofing and insectproofing containers; in paint removers; soft wax stiffener; substitute for carnauba and beeswax	
Carnauba	bags and cases of varying weights	characteristic odor of new-mown hay on warming. Takes a fine hard gloss on rubbing with a cloth	manufacture of shoe and furniture polishes; lacquers, varnishes; phonographic records, carbon-paper coatings; hardening candles; leather finishes; electrical-insulating compositions; water-proofing textiles, wood, etc.; inks and ink compositions; mold lubricant; substitute for beeswax	Brazil wax
Ceresin	bags	plasticity	manufacture of candles; shoe, floor, and leather polishes; antifouling paints; wood polishes and fillers; incandescent gas mantles; paper size; waxed paper; lubricating compositions; waxes; figures and forms; toys, dolls; bottles for hydrofluoric acid; electrical insulations; general size; impregnating and preserving agent; substitute for beeswax in crayons, perfume pastes, pomades, cosmetics, salves, and other pharmaceutical products; matrix compositions in printing; printing forms; rubber mixtures; waterproofing textiles, textile dressing	cerin, cerosin, earth wax, mineral wax, purified ozokerite

Table 5:10 (Continued)

Wax	Containers and Weights in pounds	Special Features	Uses	Other Names
Insect Wax, or Chinese Wax	bags and barrels	fibrous crystalline structure	manufacture of candles; leather and furniture polishes; treating silk and cotton fabrics; sizing and glazing papers	insect wax, tree wax, vegetable spermaceti, wrongly called Japanese wax
Japan	cases 224, and other sizes	effloresces; tallowy odor	manufacture of furniture, shoe, and floor polishes; candles, wax matches; leather polishing and finishing compositions; substitute for beeswax	Japan tallow, vegetable wax of Japan; sumac wax
Montan	bags of various sizes	lustrous appearance; bituminous odor on friction or warming	manufacture of polishing preparations, paints, varnishes, tarry roofing compositions, waterproofing paints; shoe creams and polishes; phonograph records; insulating compositions; candles; soaps, adhesive pastes; finishing leather goods; increasing hardness of fat compositions; paper sizing; to raise the melting point of softer waxes; substitute for carnauba and beeswax	lignite wax
Ozokerite	bags of various sizes; stands 80–100	plasticity; mixture of hydrocarbons	manufacture of paints, varnishes, wood fillers, wood polishes; candles; lithographic and printing inks; dolls, toys; carbon paper, shoe polishes, incandescent gas mantles; pomades, cosmetics, pharmaceutical ointments; matrices in galvano-plastic work; crayons, waxed paper and cloth, linen and cotton sizing preparations; electrical insulating; leather finishing; lubricant, and sealing-wax compositions; process engraving and lithography; electrotyper's wax; rubber filler; substitute for beeswax, carnauba, ceresin	mineral wax, fossil wax, native paraffin ozacerite
Spermaceti	cases 50–60	odor and taste free when pure	manufacture of emulsions, cerates; candles, soaps, sweet-meats, candies, confectionery; cosmetics, ointments, pomades, toilet preparations; finishing textiles; in medicine	cetaceum

Table 5:11
PROPERTIES OF WAXES

Wax	Refractive Index	Specific Gravity	Melting Point °C.	Melting Point °F.	Direct Acid Number
Beeswax	1.4398–1.4451 at 75°C., 167°F.	0.959–0.975	62–70		16.8–20.6
Candelilla	1.4555 at 71.5°C., 160.7°F.	0.969–0.993	65–69		16
Carnauba	1.463 at 60°C., 140°F.	0.990–0.999	83–91		4–8
Ceresin	1.4415–1.4464	0.88–0.92	54–77	130–170	0
Insect Wax, or Chinese Wax		0.926–0.970	65–80	149–176	13
Japan	1.450 at 60°C., 140°F.	0.975–0.984	50–56		6, usually not over 20
Montan		crude 1.0	crude 76–92 distilled 72–77 refined 77–84	169–198 162–171 171–183	crude 25 distilled 73–85 refined 15–20
Ozokerite	1.4415–1.4464	0.85–0.95	58–100	136–212	0
Spermaceti	1.440 at 60°C., 104°F.	0.905–0.960	41–49		0.5–2.8, not usually over 0.5

Table 5:11 (Continued)

Wax	Fatty Acids, %	Saponification Number	Unsaponifiables, %	Ratio Number*	Iodine Number	Acetyl Number
Beeswax	46.77	88–96	52–56	3.6–3.8	8.3–11	15.2
Candelilla	29.0	46–65	74.0	4.7	14.0–36.8	
Carnauba	47.95	78–88	54–55	31	13.5	55
Ceresin	0	< 2	100	0	7–9	
Insect Wax, or Chinese Wax	51.5	80–93	49.5	29.3	1.4–2	
Japan	4–12	217–237	1.1–1.6	11–35	4–15	27–31.2
Montan	distilled 56–64 refined 11–15	crude 58 distilled 75–89 refined 70–80	crude 3.58 distilled 30–45	crude 3–3.5 distilled 0–0.3 refined 3–3.5	crude 16 distilled 10–15	
Ozokerite	0	0	100	0	7.8	
Spermaceti	52–53	121–135	51–54	124	2.6–3.8	2.6

* $\frac{\text{Saponification No.} - \text{Acid No.}}{\text{Acid No.}}$

FATS

The higher fats and fatty acids are waxy in character and, therefore, are included here.

Table 5:12
TALLOW FATTY ACIDS
(Proctor & Gamble)

	T-10*		T-11		T-22		T-44		T-55	
	Typical	*Limit*	*Typical*	*Limit*	*Typical*	*Limit*	*Typical*	*Limit*	*Typical*	*Limit*
Saponification Value	206	202-208	205	202-208	206	202-209	206	197-212	207	195-212
Acid Value	205	200	204	200 min.	205	200 min.	206	195-212		
Iodine Value	0.8	1.0 max.	42	38-45	54	40-65	60	45-80	73	50-80
Titer, °C.	60	58-62	45	44-47	41	38-45	41	35-45	38	35-42
Unsaponifiables, %	0.2	1.0 max.	0.4	1.5 max.	0.5	1.5 max.	2.1	4.0 max.	6.3	10.0 max.
Moisture, %	0.04	0.5 max.	0.1	0.3 max.	0.1	0.5 max.	0.1	0.5 max.	0.1	0.5 max.
Color—Lov. 5¼″ Red	0.5	1.0 max.	1	1.5 max.	5	20 max.				
Gardner (1963)							6	10 max.	8	10 max.
Total Polyunsaturates, %			0.5	0.8 max.						
Average Molecular Weight	272		274		272		272		271	
Composition (GLC%)										
Saturated Acids										
C_{12} Lauric			0.5		0.5		1.2		1.6	
C_{14} Myristic	3.5		3.5		3		2.9		1.3	
C_{15} Pentadecanoic	0.5		0.5		0.5		0.3			
C_{16} Palmitic	28.5		26		25		23		18	
C_{17} Margaric	1		1		1		1			
C_{18} Stearic	65		28		18		18		15	
Unsaturated Acids										
C_{14} Myristoleic			0.5		1		0.4			
C_{16} Palmitoleic	0.5		2.5		3		2.6		0.5	
C_{16} (2=) Hexadecadienoic			0.1		0.5		0.3			
C_{18} Oleic	1		37		43		43		43	
C_{18} (2=) Linoleic			0.4		4		7		16	

Table 5:13

PALM FATTY ACIDS

Typical	*Limit*
211	207–216
210	207–216
54	Approx. 55
45	43–48
0.1	0.5 max.
0.3	2.0 max.
4	9 max.
265	
0.1	
0.2	
2	
2	
47	
0.5	
3	
34	
11	
0.2	

Part 2

SYNTHETIC WAXES

Chapter 6

FATTY ALCOHOLS AND ACIDS

CETYL ALCOHOL

Commercial cetyl alcohol is obtained by "splitting" sperm oil or spermaceti, or by the reduction of cetyl palmitate or palmitic acid. Synthesis from petroleum hydrocarbons will soon be on a commercial production basis.

Properties

Melting Point	45–50°C
Specific Gravity, 15°C	0.810
Boiling Point	334°C
Iodine Number	<3
Acid Number	0
Saponification Number	0
Unsaponifiable Matter	98–100%
Ash	0
Refractive Index, 60°C	1.4345
Color	white
Form	crystalline
Heat of Fusion (cal/mol)	8900

Cetyl alcohol is now obtainable as an odorless and tasteless product that is stable to alkali, acid, light, and air. It is a good solvent for basic dyes.

Soluble in: alcohol, benzol, carbon disulfide, ether, glycol ethers, fats, mineral, and vegetable oils.

Uses of Cetyl Alcohol

Cetyl alcohol is an emulsifier for water-in-oil type of emulsions. It causes mechanical adherence of the particles of water and when it is thus incorporated in viscous substances, such as petrolatum or hydrogenated oils, it will tend to stiffen the mixture, yielding an emulsion containing a larger percentage of water than would otherwise be possible.

Under synthetic waxes are included wax-like materials of distinct chemical composition, e.g., fatty alcohols, acid esters, amines, amides, imides, nitriles, terphenyls, chlorinated and fluorinated hydrocarbons, polyolefins, and chemically-reacted natural and synthetic materials.

In this section are also included materials not truly synthesized but extracted from natural oils and fats with or without chemical treatment, e.g., stearic acid and glycerides.

Examples

Petrolatum alone will only take up 9–15% water. The addition of about 4% cetyl alcohol will permit the absorption of 39–52% water.

A mixture of 4% cetyl alcohol, 10% lanolin, and 86% petrolatum is able to take up 82–108% water.

1% cetyl alcohol more than doubles the water absorbing power of hydrogenated peanut oil, 75% water being taken up without cetyl alcohol and 170% being absorbed with it. The addition of 3% cetyl alcohol to lard increases the amount of water absorbed from 8–245%.

Emulsions of the water-in-oil type, with large proportions of water, are less stable than those containing less water, but cetyl alcohol usually has a beneficial effect on the stability of these emulsions when it is used as an addition to the standard formulae, as it will act as an auxiliary emulsifying agent.

The proportions of cetyl alcohol to be incorporated in various types of water-in-oil emulsions will, as a rule, range from 1 to 5%, depending on the individual formula.

Since it acts to promote water-in-oil type emulsions with inert substances, cetyl alcohol alone cannot produce the reverse, i.e., oil-in-water emulsions. However, its oil-soluble hydrocarbon radical, when used with standard emulsifying agents, such as soaps, lauryl sulfates and wetting agents will give body and stability to oil-in-water emulsions, in general, and creams and lotions, in particular.

Example

A mixture of equal parts of mineral oil and water, containing about 1% soap, will soon separate into two more or less distinct layers; the addition of about 2% of cetyl alcohol (dissolved in the oil) will make a thick, creamy emulsion which will show only slight separation, on long standing.

Although the stabilizing action is most marked in lotions, a small amount of cetyl alcohol will definitely stabilize brushless shaves, vanishing creams, etc. Approximately 1% is usually sufficient and here again, it may be stated that cetyl alcohol acts as an auxiliary emulsifying agent.

Because of its ability to become hydrated and to assist the hydration of other substances, larger amounts of cetyl alcohol may make a cream softer and more translucent. An addition of 0.5–1% will hardly alter the physical appearance of the product, but will still increase stability. Properly incorporated, cetyl alcohol will also cause a cosmetic cream to rub out more smoothly on the skin, as a result of the finer dispersion of ingredients.

LANETTE WAX

Lanette wax is a mixture of cetyl and stearyl alcohols.

Lanette wax SX consists of cetyl and stearyl alcohols, with a neutral emulsifier.

Lanette wax ester is a mixture of palmitic acid esters of cetyl and stearyl alcohols.

Lanette wax extra consists of higher fatty alcohols and saponifiable free fatty acid.

Properties

Melting Point	50°C
Specific Gravity	0.81
Acid Number	0
Saponification Number	0
Ester number	0
Acetyl Number	180–190

Soluble in: alcohol, benzene, ether.

Lanette Wax, with soap, sulfonated oils, aromatic sulfonic acids, or other emulsifiers, acts as a stabilizer for emulsions. Lanette Wax Extra with diluted alkali or soap, acts as an auxiliary emulsifier, particularly for paraffin- and ceresin-wax emulsions.

Lanette Wax SX is neutral, melts at about 60°C and is, by itself, an emulsifying agent. It disperses in hot water. On cooling, this dispersion thickens or gels. Secondary emulsifiers or stabilizers, such as glue, gelatin, methyl cellulose, or gum arabic, are used with it to increase the stability of emulsions. If Lanette Wax SX is heated above 100°C, it tends to lose its dispersing properties. Small amounts of this wax are said to give better tensile strength and elasticity to varnish films. Certain drug and cosmetic emulsions, made with this wax, are more stable to weak acids, such as boric, carbolic, and salicylic. It is also more stable to mild oxidizing and reducing agents than soap.

STEARYL ALCOHOL

Stearyl alcohol (octadecanol) is a white solid waxlike product; mixture of fatty alcohols, stearyl alcohol predominating. A technical grade (Stenol) has properties as follows:

Melting Point	54–57°C
Specific Gravity	approx. 0.85
Distillation Range	300–375°C
Hydroxyl Number	approx. 215
Calculated Molecular Weight	approx. 260
Iodine Number	<2

It is practically free from acids and esters.

POLAWAX[1]

Polawax is a mixture of reaction products of stearyl alcohol and ethylene oxide. It is a creamy wax-like solid.

It is the original of the newer forms of the self-emulsifying combination now known, even officially, by the simple term *Emulsifying Wax.*

[1] Croda, Inc.

Table 6:1

CHARACTERISTICS OF MONO-UNSATURATED FATTY ALCOHOLS[1]

Alcohols	*Predominant Chain-Length*	*Iodine Value*	*Acid Value Max*	*Sap. Value Max*	*Acetyl Value*	*Hydroxyl Value*	*Color Lovibond* $5\frac{1}{4}$" *Max.*	*Cloud Point* °C	*Melting Point* Closed Tube °C	*Titer* °C	*Calculated Mol. Wt.*	*Av. Viscosity* SSU/210°F	*Specific Gravity* 25/25°C	*Boiling Range* 760 mm. °C	*Unsaturated Content %*
Oleyl-Stearyl															
Adol 42	45% C_{18} Oleyl	40–60	1	3	181–188	210–218	15Y/1.5R	29–35	31–36	31–35	257–270	40	0.860	315–345	48
Oleyl															
Adol 85 (cosmetic and pharmaceutical)	80% C_{18} Oleyl	85–95	1	2	178–189	205–220	5Y/0.5R	−7 to −1			256–274	41	0.840	282–349	80
Adol 32(comm)	45% C_{18}	75–85	1	3	181–196	210–230	15Y/1.5R	14–17			244–267	39	0.845	305–370[2]	75
Adol 34 (comm)	40% C_{18}	60–75	1	3	181–196	210–230	15Y/1.5R	18–30			244-267	39	0.845	300–370[2]	61
Erucyl															
Adol 22 (tech)	80% C_{22} Erucyl	75–90	1	3	152–165	172–188	25Y/2.5R	22–30			298–326	42	0.849	334–376	98
Ricinoleyl															
Adol 40 (tech)	88% C_{18} 1,12-Diol	85–95	1	3	277–290	350–370	10Y/1R	−18 to −7			270–284	51	0.876	245–328	99
Linoleyl-Linolenyl															
Unadol 40 (linoleyl)	50% $C_{18}(2=)$	115–135	1	3	178–185	205–215	25Y/2.5R	13–17			261–274	35	0.855		82
Unadol 90 (linolenyl)	50% $C_{18}(3=)$	165–185	1	3	178–185	205–215	25Y/2.5R	7–13			261–274	39	0.864		85
Arachidonyl-Clupanodonyl															
Unadol 21 (pilot plant production)	50% C_{18}-C_{22} Unsaturated	175–200	1	3		200–215	25Y/2.5R	13–18			261–280				

[1] From Archer-Daniels-Midland Co.

[2] 90% N. F. Method

Compositions shown are based on gas chromatographic data. They are approximate and may vary ±10 % of the values given depending upon the raw materials available.

Characteristics of Saturated Fatty Alcohols

Alcohols	*Predominant Chain-Length*	*Iodine Value Max.*	*Acid Value Max.*	*Sap. Value, Max.*	*Acetyl Value*	*Hydroxyl Value*	*Color Lovibond 5¼″ Max.*	*Cloud Point °C*	*Melting Point Closed Tube °C*	*Titer °C*	*Calculated Mol. Wt.*	*Av. Viscosity SSU/210°F*	*Sp. Gravity 25/25°C*	*Boiling Range 760 mm °C*
Ocetyl-Decyl														
Adol 9	55% C_8	3	1	3	284–300	361–387	5Y/0.5R	−20 to −15	−20 to −14	−20 to −14	145–156	32	0.826	200–275
Lauryl														
Adol 12 (high purity)	90% C_{12}	0.5	0.5	2	241–245	295–300	5Y/0.5R	16–21	17–23	17–22	187–190	36	0.832	245–274
Adol 16 (high purity)	80% C_{12}	0.5	0.5	2	238–243	290–297	5Y/0.5R	17–22	18–23	18–22	189–194	36	0.832	240–295
Adol 14 (tech)	65% C_{12}	0.5	0.5	3	228–238	275–290	5Y/0.5R	17–22	18–23	18–22	193–204	36	0.833	240–309
Adol 13 (comm)	55% C_{12}	0.5	0.5	3	223–233	268–283	5Y/0.5R	17–22	18–23	18–22	198–209	36	0.834	240–316
Myristyl														
Adol 18 (high purity)	90% C_{14}	1.5	0.5	2	216–220	258–263	5Y/0.5R	33–37	34–38	34–37	214–225	36	0.795[2]	255–308
Cetyl														
Adol 52 (N.F.)	90% C_{16}	1.5	1	2	194–197	227–231	10Y/1R	46–50	47–51	47–50	244–249	39	0.815[3]	316–336[5]
Adol 54 (tech)	60% C_{16}	3	1	2	187–193	218–225	10Y/1R	45–50	46–51	46–50	250–258	40	0.816[2]	310–345[5]
Steryl														
Adol 62 (U.S.P.)[1]	80% C_{18}	2	1	1	177–183	204–211	10Y/1R	52–59	53–60	53–59	266–275	42	0.817[2]	337–360
Adol 64 (tech)	70% C_{18}	3	1	2	178–186	206–216	10Y/1R	50–52	51–54	51–53	260–272	44	0.817[2]	320–362
Adol 65 (comm)	65% C_{18}	5	1	3	174–193	204–226	20Y/2R	43–52	44–53	44–52	248–275	40	0.816[2]	305–355
Adol 63 (cetyl-stearyl)	60% C_{18}	3	1	3	180–188	208–218	10Y/1R	46–51	47–53	47–52	257–270	44	0.816[2]	312–344
Arachidyl-Behenyl														
Adol 71 (tech)	85% C_{20}-C_{22}	6	1	3	152–165	172–188	10Y/1R	55–64	56–66	56–65	298–327	43	0.799[4]	342–380
Adol 67 (comm)	55% C_{20}-C_{22}	4	1	3	158–270	180–195	10Y/1R	53–57	54–59	54–58	288–312	43	0.800[4]	330–380
Behenyl														
Adol 60	80% C_{22}	3	1	3	155–167	175–190	10Y/1R	59–63	60–65	60–64	295–320	46	0.793[4]	342–377
12-Hydroxyl Stearyl														
Adol 45	86% 12-Hydroxy-Stearyl	10	1	3	271–287	340–365	10Y/2R	64–75	65–77	65–76	263–282	57	0.839[4]	315–365

[1] Also available to special specifications with up to 90% C_{18}.
[2] 60/25°C
[3] 65/25°C
[4] 100/25°C
[5] 90% N.F. Method
Compositions shown are based on gas chromatographic data. They are approximate and may vary ±10 of the values given depending upon the raw materials available.

It differs significantly from the older type of self-emulsifying wax in that its oil-in-water-producing properties do not derive from the presence of auxiliary anionic surface-active agents of the soap, sulfonated oil or sulfated fatty alcohol, or from alkalies or gums—a most important point.

Polawax is non-ionizing in character, and from this aspect alone, it is considerably in advance of previous products, and it will be of especial interest in the making of ointments or creams that must be autoclaved. Penicillin creams prepared by autoclave sterilizing of a 10 per cent Polawax emulsion and incorporation of 10 units per gram show activity of penicillin after at least eight days at room temperature.

Some chemical and physical properties of Polawax are as follows:

Iodine value (Wijs)	0.2–2.4
Drop Point °C	47–51
pH value (3% sol.)	5.4–7
Saponification value mg	7.8–15
Acid value mg	0.8–1.8

Unlike some emulsifying waxes, Polawax does not deteriorate on heating at 150°C–152°C for two hours, other than slight alteration in color. Under these conditions, it loses 2–3 per cent in weight with slight hardening of the wax.

Heating 30 per cent Polawax in water in an autoclave at 10 lb pressure for 45 minutes gives emulsions of which the pH value moves down only two points from 6.6. to 6.4.

STEARIC ACID

Commercial stearic acid is not a chemically pure compound, but a mixture of fatty acids in which stearic acid predominates.

Stearic acids are chiefly produced from saponified and distilled animal fatty acids. These fatty acids are usually composed of approximately 60% liquid and 40% solid acid. The bulk of the liquid acids is separated from the solid acids by hydraulic pressing. From the first pressing single-pressed stearic acid is obtained, from the second pressing, double-pressed stearic acids, and from the third pressing, triple pressed stearic acids. Each successive pressing removes additional liquid fatty acids, thus increasing the purity and quality of the resulting solid stearic acid. The liquid acids present are unsaturated and the solid acids, saturated.

The stearic acid of commerce is frequently called stearin (stearine). This is a misnomer as stearin is a glyceride of stearic acid.

Soluble in: Alcohol, ether, chloroform, carbon bisulfide, carbon tetrachloride, methanol, pyridine.

The acetyl number should be close to 0.0. There should be very little spread between iodine and thiocyanogen numbers as no doubly unsaturated acids are present.

Table 6:3

POLYGLYCOL WAXES

CARBOWAX Polyethylene Glycol *Data determined on typical commercial material*	*Formula Molecular Weight/ Range*	*Apparent Specific Gravity, 20/20°C.*	*Freezing Range, °C.*	*Water Solubility, 20°C., % by Wt.*	*Viscosity, Centistokes at 210°F.*	*Comparative Hygroscopicity (Glycerine=100)*	*Surface Tension, 25°C., Dynes/ Cm.*	*Refractive Index $n_D25°C.$*	*Pounds Per Gal. at 20°C.*	*Flash Point, °F.(b)*
CARBOWAX Polyethylene Glycol 1500 (g)	500-600	1.151(d)	38 to 41	73	13-18	35	(c)	(c)	9.17 55°C.	430
CARBOWAX Polyethylene Glycol 1540	1,300-1,600	1.0910	43 to 46	70	25-32	30	(c)	(c)	9.17 55°C.	510
CARBOWAX Polyethylene Glycol 4000	3,000-3,700	1.204	53 to 56	62	80-95	..	(c)	(c)	...	515
CARBOWAX Polyethylene Glycol 6000	6,000-7,500	1.207	60 to 63	50, appr.	700-900	..	(c)	(c)	...	520
CARBOWAX Polyethylene Glycol 20,000 Linear	18,000-19,000	1.215	56	...	5,960	..	(c)	(c)	Solid	550
Polyethylene Glycol Compound 20M	15,000, appr.	1.207	50 to 55	50, appr.	96,000 cps.	..	52 (f)	(c)	...	>465

Table 6:4

PROPERTIES AND COMPOSITION OF COMMERCIAL STEARIC ACIDS

	Single-Pressed	*Double-Pressed*	*Triple-Pressed*
Titer	52.2–53.0°C	53.5–53.9°C	54.6–55.2°C
Melting Point (Open Tube)	52.8–53.6°C	54.1–54.5°C	55.2–55.8°C
Iodine Number (Wijs)	10–14	7–9	2–4
Free Fatty Acids (as Oleic)	103–104%	104–105%	104–105%
Acid Number	205–207	207–209	207–209
Saponification Number	207–209	208–210	208–210
Unsaponifiable Matter	0.50–0.60%	0.20–0.30%	0.10–0.20%
Average Molecular Weight	272–274	269–272	269–272
Specific Gravity 60°/15°	0.853	0.855	0.857
Flash Point	400°F	400°F	405°F
Fire Point	430°F	425°F	425°F
Unsaturated Content	12–16%	8–10%	2–4%
Lovibond Color (Max.)	16 Y–3.0 R	4 Y–1.0 R	2 Y–0.5 R
Lovibond Color (Min.)	5 Y–0.6 R	2 Y–0.2 R	1 Y–0.1 R
Myristic Acid Composition	1–3%	1–3%	1–3%
Palmitic Acid	47–48%	49–50%	46–47%
Stearic Acid	28–33%	34–37%	46–48%
Palmitoleic Acid	trace	trace	nil
Oleic Acid	12–16%	8–10%	2–4%

Properties of Pure Stearic Acid

Melting Point	69.3°C
Iodine Number	0.0
Acid Number	197.5
Refractive Index, 80°C	1.4300
Specific Gravity 69.3°C	0.847
Boiling Point	232°C (15 mm)
Molecular Weight	284.28
Dipole Moment, 25°C	0.77

Uses: Cosmetics, ointments, polishes, soaps, candles, waterproofing, sun-checking compounds (rubber), emulsifiers, textile finishes, lubricants, greases, color solvents, crayons, plastics, soldering fluxes.

Stearic acid, produced by hydrogenation of fish oils, is used for many technical purposes. It is marketed in a number of grades having the following properties:

Titer °C	*Acid No.*	*Iodine No.*
52	190	6
53	195	25
57	192	13
58	192	8

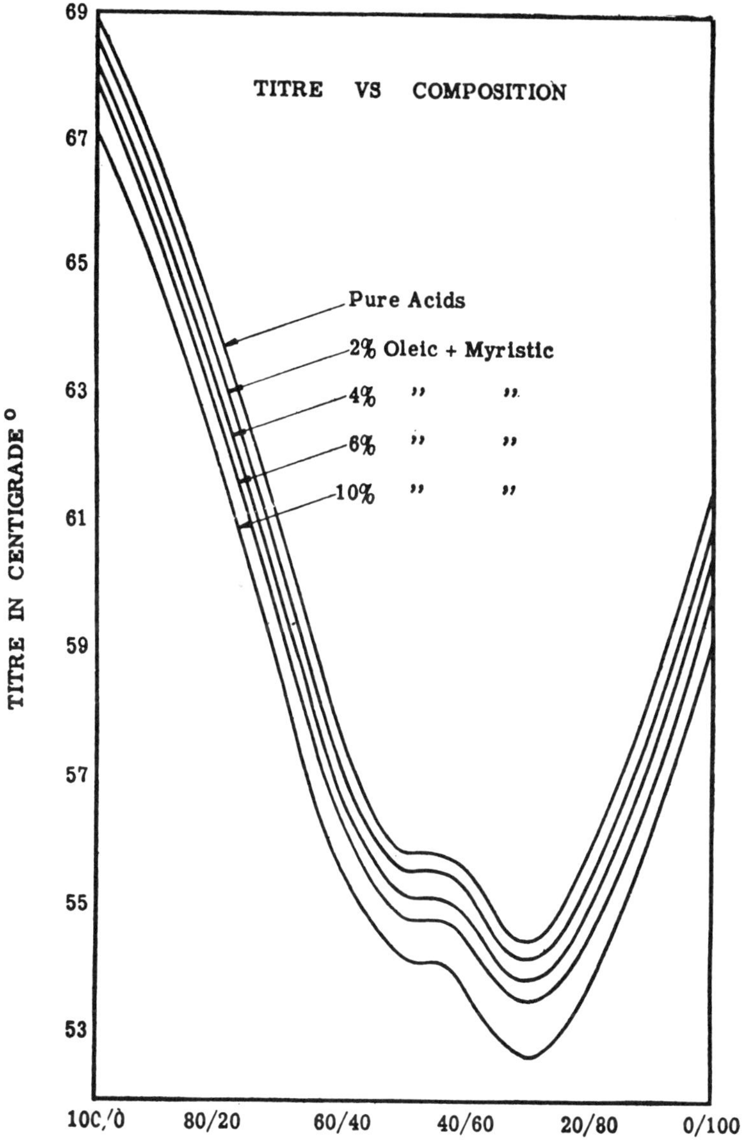

Fig. 6:1. Titre of Pure Stearic and Palmitic Acid Blends.

PALMITIC ACID

Palmitic acid occurs naturally as the glyceride in various vegetable and animal fats and oils. It is a hard, crystalline substance lying between myristic and stearic acids in the saturated fatty acid series.

Table 6:5

COMPARISON OF THE PROPERTIES OF COMMERCIAL AND PURE PALMITIC ACIDS

	Commercial Palmitic Acid	*Pure Palmitic Acid*
Chemical Formula	—	$C_{15}H_{31}COOH$
Mean Molecular Weight	258.0	256.25
Melting Point	58.0°C	62.85°C
Iodine Number (Wijs)	3.0	none
Neutralization Number	216.0	219.0
Unsaponifiable Matter	0.5%	none
Dipole Moment, 25°C	—	0.77; 1.75[1]
Color	white	white
Odor	trace	none
Composition:		
Palmitic Acid	90.0%	—
Oleic Acid	4.0%	—
Stearic Acid	6.0%	—
Approximate Weight per U.S. Gallon	—	7.5% lb.

Uses: The metallic salts of palmitic acid are of particular interest in paints, varnishes, lacquers, inks, etc. Aluminum palmitate is one of the best thickening, suspension, and flattening agents known. Zinc and calcium palmitate also find many applications in this field.

By mixing palmitic acid with myristic acid or with lauric acid, low-titer, completely saturated fatty acid mixtures can be obtained which are particularly suitable for preparing oxygen-stable greases.

MYRISTIC ACID

Myristic acid is found as the glyceride in vegetable oils and fats, in varying, sometimes rather small, percentages; e.g., in peanut oil, coconut oil, and some less accessible raw materials. By splitting the raw vegetable oils and subjecting the fatty acids to a fractional vacuum steam distillation, sometimes combined with fractional crystallization, filtration, and deodorization, a fairly pure myristic acid is obtained.

Myristic acid is a hard, crystalline, fatty acid, ranging in color from white to light brown, according to purity. The faint odor of the technical grades is very agreeable, somewhat resembling that of ionone or sweet peas. Special requirements sometimes call for a practically odorless or colorless product, which can be obtained at an increased cost.

---

[1] Bhattacharya, *Indian J. Phys.* **16**, 396 (1942).

Table 6:6

COMPARISON OF THE PROPERTIES OF PURE AND TECHNICAL MYRISTIC ACIDS

	Pure	*Technical*
Formula	$C_{13}H_{27}COOH$	—
Purity	100	Approx. 80% Myristic Acid
Acid Number	246.1	237–243
Saponification Number	246.1	240–246
Iodine Number (Hanus)	0	7–13
Titer	—	42–46
Melting Point	53.8°C	—
Specific Gravity	0.8622 54°/4°C 0.8584 60°/4°C	0.869 at 50°C
Refractive Index	$n_D^{60} = 1.43075$ $n_D^{76} = 1.4218$	—
Molecular Refraction	68.77	—
Dipole Moment, 25°C	0.77	—
Combustion Heat	cal/g=9.34; cal/mol=2084.4	—
Boiling Range	196.5°C at 15 mm 250.0 at 100 mm	ca. 220–230 at 20–30 mm
Color	white	light straw
Odor	none	characteristic, mild
Weight per Gallon	—	ca. 7.25 lb

Technical myristic acid is composed of:

	Approximate %
Lauric Acid	10
Myristic Acid	80
Palmitic-Stearic Acid	5
Oleic Acid	5

Soluble in: Acetone, benzene, carbon tetrachloride, hexane, methyl alcohol, butyl alcohol, methyl Cellusolve, butyl Cellosolve, *o*-dichlorobenzene, etc., and other hydrocarbons, alcohols, and solvents of similar composition. It is insoluble in water.

Grades of Distilled Mixed Fatty Acids

Mixed fatty acids are offered in many grades:

Type	*Titer, °C*	*Free Fatty Acids*	*Iodine No.*
Animal No. 1 Grade	40–42	100–102	50–60
Animal No. 2 Grade	38–39	99–100	50–60
Animal No. 3 Grade	35–37	98–99	60–70
Cottonseed	34–37	98–100	90–100
Coconut	22–25	128–130	14–16
Myristic	41–44	118–121	8–10
Soybean	22–26	96–100	130

Solubility of Fatty Acids

Fatty acids are, in general, more soluble than the glycerides. They are very soluble in ether and soluble in chloroform, carbon tetrachloride, carbon disulfide, aromatic hydrocarbons, paraffin hydrocarbons, acetone and hot alcohol.

The glycerides are also very soluble in ether, but almost insoluble in alcohol.

Table 6:7

Solidification Points of Binary Fatty-Acid Mixtures[1]

Palmitic-Stearic Mixture			Stearic-Arachidic Mixture		
C_{16} Mol. %	C_{18} Mol. %	Sol. Pt. °C. (Corr.)	C_{18} Mol. %	C_{20} Mol. %	Sol. Pt. °C. (Corr.)
0.00	100.00	68.85	0.00	100.00	74.35
2.81	97.19	68.15	2.80	97.20	73.65
6.05	93.95	67.20	5.85	94.15	72.80
9.77	90.23	66.20	7.65	92.35	72.30
14.00	86.00	65.00	13.03	86.97	70.80
18.64	81.36	63.80	14.09	85.91	70.55
18.85	81.15	63.72	16.71	83.29	69.80
24.33	75.67	62.20	21.49	78.51	68.30
31.29	68.71	60.10	23.46	76.54	68.00
34.16	65.84	59.20	26.76	73.24	67.00
38.85	61.15	57.70	28.90	71.10	66.50
39.82	60.18	57.28	30.74	69.26	65.95
42.33	57.67	56.70	35.03	64.97	64.80
44.19	55.81	56.52	38.24	61.76	63.90
45.65	54.35	56.25	39.16	60.84	63.75
47.73	52.27	56.00	42.03	57.97	63.25
49.54	50.46	56.02	43.38	56.62	63.15
50.26	49.74	56.00	46.69	53.31	62.75
52.87	47.13	56.00	46.72	53.28	62.70
56.11	43.89	55.85	48.69	51.31	62.50
58.65	41.35	55.70	50.46	49.54	62.30
62.24	37.76	55.45	52.00	48.00	62.20
65.00	35.00	54.90	54.05	45.95	62.15
65.36	34.64	54.85	55.82	44.18	62.15
68.06	31.94	54.45	58.30	41.70	62.05
68.30	31.70	54.42	60.60	39.40	61.85
68.98	31.02	54.39	62.78	37.22	61.72

[1] Schuette and Vogel, *Oil Soap,* July 7, 156 (1940).

Table 6:7 (continued)

Palmitic-Stearic Mixture			Stearic-Arachidic Mixture		
C_{16} Mol. %	C_{18} Mol. %	Sol. Pt. °C. (Corr.)	C_{18} Mol. %	C_{20} Mol. %	Sol. Pt. °C. (Corr.)
70.80	29.20	54.21	63.99	36.01	61.50
71.72	28.28	54.20	66.60	33.40	61.25
73.32	26.68	54.20	69.76	30.24	61.10
73.40	26.60	54.20	72.19	27.81	61.00
74.34	25.66	54.20	74.59	25.41	60.90
76.48	23.52	54.35	76.20	23.80	60.98
77.65	22.35	54.60	76.47	23.53	61.00
80.26	19.74	55.15	79.09	20.91	61.60
85.16	14.84	56.60	85.40	14.60	63.35
90.01	9.99	58.20	88.68	11.32	64.18
94.30	5.70	59.70	88.79	11.21	64.23
97.19	2.81	60.85	93.50	6.50	66.10
100.00	0.0	62.22	97.49	2.51	67.75
			100.00	0.00	68.85

Solidification points of fatty acids containing water are lowered as follows:[1]

Water	Solidification Point, °C.
Myristic Acid	
0.00	54.01
2.12	53.18
Palmitic Acid	
0.00	62.41
1.50	61.79
Stearic Acid	
0.00	69.20
1.28	68.74

Table 6:8

SOLUBILITY OF WATER IN MELTED FATTY ACIDS

	% Water	Solution at °C.
Myristic	1.70	53.2
Palmitic	1.25	61.8
Stearic	1.02	92.4

[1] Hoerr, Pool, Ralston, *Oil Soap,* **19,** 126 (1942).

Table 6:9

Refractive Indices (n_D^t) of Normal Saturated Fatty Acids[1]

Acid	20.0°C.	25.0°C.	30.0°C.	40.0°C.	45.0°C.	50.0°C.
Caproic	1.4170	1.4150	1.4132	1.4095		1.4054
Enanthic	1.4230	1.4209	1.4192	1.4155		1.4114
Caprylic	1.4280	1.4260	1.4243	1.4205		1.4167
Pelargonic	1.4322	1.4301	1.4287	1.4250		1.4210
Capric				1.4288		1.4279
Hendecanoic				1.4319		1.4248
Lauric					1.4323	1.4304
Tridecanoic						1.4328

Acid	55.0°C.	60.0°C.	65.0°C.	70.0°C.	80.0°C.
Caproic		1.4012		1.3972	1.3931
Enanthic		1.4073		1.4037	1.3993
Caprylic		1.4125		1.4089	1.4049
Pelargonic		1.4171		1.4132	1.4092
Capric		1.4210		1.4169	1.4130
Hendecanoic		1.4240		1.4202	1.4164
Lauric	1.4288	1.4267	1.4250	1.4230	1.4191
Tridecanoic	1.4310	1.4290	1.4272	1.4252	1.4215
Myristic	1.4329	1.4310	1.4291	1.4273	1.4236
Pentadecanoic	1.4348	1.4329	1.4310	1.4292	1.4254
Palmitic			1.4328	1.4309	1.4272
Margaric			1.4340	1.4324	1.4287
Stearic				1.4337	1.4299

Table 6:10

Densities of Normal Saturated Fatty Acids at 80°C.

Acid	d_4^t	Acid	d_4^t
Caproic	0.8751	Tridecanoic	0.8458
Enanthic	0.8670	Myristic	0.8439
Caprylic	0.8615	Pentadecanoic	0.8423
Pelargonic	0.8570	Palmitic	0.8414
Capric	0.8531	Margaric	0.8396
Hendecanoic	0.8505	Stearic	0.8390
Lauric	0.8477		

[1] A. Dorinson, M. R. McCorkle, A. W. Ralston. *J. Am. Chem. Soc.* **64,** 2739 (1942).

Table 6:11

Composition and Unsaturation of Commercial Fatty Acids

Average Percentage of Saturated Acids in	Capric, Caproic and Caprylic	Lauric	Myristic	Palmitic	Stearic	Higher Saturated
Coconut Fatty Acid, Distilled	ca. 13.5	ca. 45.5	ca. 20	ca. 15	ca. 5	—
Cottonseed Fatty Acid, Distilled	—	—	—	20–26	2–3	0.5–1.0
Cotton, High-Titre Fatty Acid, Distilled	—	—	ca. 1	40–45	4–5	—
Chinawood Fatty Acid, Split	Total about 2–3%					—
Helioline-D Fatty Acid, Distilled	—	—	—	4.5–9	0.5–1	—
Hempseed Fatty Acid, Distilled	—	—	—	10–12		?
Isoline Fatty Acid, Distilled	Total about 3–4%					—
Lauric Acid (Technical)	ca. 15	ca. 58	ca. 20	ca. 9	ca. 1	—
Linseed Fatty Acid, Distilled	—	—	—	2.7–3.8	5.4–7.7	—
Marine (Raw) Fatty Acid, Split	Total ca. 15–25%					—
Myristic Acid, Distilled	—	ca. 20	ca. 65	ca. 5		—
Palm Fatty Acid, Split	—	—	1–2	35–43	4–5	—
Perilla Fatty Acid, Distilled	Total about 6.5–8.0—mostly palmitic					—
Rapeseed Fatty Acid, Distilled	—	—	0–1.5	1–2		1–3
Ricinoleic Fatty Acid, Split	Total about 3–5%					—
Sardine Fatty Acid, Distilled	Total about 10–20%—mostly palmitic					—
Soybean Fatty Acid, Distilled	—	—	—	6.5–10	4.2–6.3	1–2
Sunflower Fatty Acid, Distilled	—	—	—	3–4	2.5–3.5	1–2

Table 6:11 (Continued)

Average Percentage of Unsaturated Acids in	Ricin-oleic	Oleic	Linolic	9/11 Octadeca-dienic	Lino-lenic	Elaeo-stearic	Clupan-donic
Coconut Fatty Acid, Distilled	—	ca. 6	3–4	—	—	—	—
Cottonseed Fatty Acid, Distilled	—	35–45	30–40	—	—	—	—
Cotton, High-Titre Fatty Acid, Distilled	—	22–25	25–30	—	—	—	—
Chinawood Fatty Acid, Split	—	5–15	—	—	—	80–85	—
Helioline-D Fatty Acid, Distilled	—	45–55	40–45	—	—	—	—
Hempseed Fatty Acid, Distilled	—	12–15	50–55	—	20–25	—	—
Isoline Fatty Acid, Distilled	—	6–9	40–45	40–45	—	—	—
Lauric Acid (Technical)	—	ca. 4	—	—	—	—	—
Linseed Fatty Acid, Distilled	—	6–12	35–47	—	37–42	—	—
Marine (Raw) Fatty Acid, Split	—	C16 30(?)	?	Containing jecoleic and therapic acid (?)		—	+
Myristic Acid, Distilled	—	ca. 10	—	—	—	—	—
Palm Fatty Acid, Split	—	40–50	8–12	—	—	—	—
Perilla Fatty Acid, Distilled	—	4.0–10.5	33–44	—	44–49	—	—
Rapeseed Fatty Acid, Distilled	—	30–40	10–20	Erucic 45–50	1–4	—	—
Ricinoleic Fatty Acid, Split	84–87	7–10	1–3	—	—	—	—
Sardine Fatty Acid, Distilled	—	?	?	Mostly jecoric (?) acid			+
Soybean Fatty Acid, Distilled	—	22–29	50–55	—	7–10	—	—
Sunflower Fatty Acid, Distilled	—	36–40	52–56	—	—	—	—

Table 6:11 (Continued)

Constants	Acetyl No.	Degree of Unsaturation: Thio-cyanide No.	Iodine No. Hanus (1 hour)	Iodine No. Wijs (¾ hour)	Iodine No. Wijs (3 days) R-Rossman Method	Hexa-bromide No.
Coconut Fatty Acid, Distilled	—	5.5–12	8–13	—	—	—
Cottonseed Fatty Acid, Distilled	—	65–70	90–100	—	—	—
Cotton, High-Titre Fatty Acid, Distilled	—	40–50	60–75	—	—	—
Chinawood Fatty Acid, Split	—	85–90	—	<160	ca. 230–R 240–R	—
Helioline-D Fatty Acid, Distilled	—	80–88	120–140	—	—	—
Hempseed Fatty Acid, Distilled	—	85–100	155–165	—	—	4–17
Isoline Fatty Acid, Distilled	—	?	155–165	135–145	170–178–W	—
Lauric Acid (Technical)	—	2–4(?)	4–6	—	—	—
Linseed Fatty Acid, Distilled	—	Below 120	184–190	—	198–W	ca. 40
Marine (Raw) Fatty Acid, Split	?	105–110	160–170	—	—	?
Myristic Acid, Distilled	—	3–6(?)	5–10	—	—	—
Palm Fatty Acid, Split	—	40–50	45–60	—	—	—
Perilla Fatty Acid, Distilled	—	>130	185–200	—	—	>60
Rapeseed Fatty Acid, Distilled	—	80–83	100–109	—	—	4–7
Ricinoleic Fatty Acid, Split	153–156	86–93	87–90	—	—	—
Sardine Fatty Acid, Distilled	?	?	110–130	—	—	?
Soybean Fatty Acid, Distilled	—	80–85	128–138	—	—	5–11
Sunflower Fatty Acid, Distilled	—	78–88	125–135	—	—	—

Table 6:12

Physical Constants of Pure Fatty Chemicals[6]

Systematic Name	Common Name	Empirical Formula	Molecular Weight	Acid Value	Melting Point °C	Double Bonds	Boiling Point °C 760 mm	Specific Gravity	Moles Olefin Per KG
Saturated Acids									
Octanoic	Caprylic	$C_8H_{16}O_2$	144.21	389	16.5		240	$0.905^{25/4}$	
Decanoic	Capric	$C_{10}H_{20}O_2$	172.26	326	31.3		270	$0.893^{32/4}$	
Dodecanoic	Lauric	$C_{12}H_{24}O_2$	200.31	280	43.6		299	$0.868^{50/4}$	
Tetradecanoic	Myristic	$C_{14}H_{28}O_2$	228.37	246	53.8		326	$0.853^{70/4}$	
Hexadecanoic	Palmitic	$C_{16}H_{32}O_2$	256.42	219	62.9		352	$0.841^{80/4}$	
Octadecanoic	Stearic	$C_{18}H_{36}O_2$	284.47	198	69.9		376	$0.848^{70/4}$	
Eicosanoic	Arachidic	$C_{20}H_{40}O_2$	312.52	179	75.2				
Docosanoic	Behenic	$C_{22}H_{44}O_2$	340.58	165	80.2		306/60 mm[2]		
12-Hydroxy Octadecanoic	Hydroxy Stearic	$C_{18}H_{36}O_3$	300.47	187	85.0				
Unsaturated Acids									
9-Dodecenoic	Lauroleic	$C_{12}H_{22}O_2$	198.30	285		1	164–165[2]	$0.913^{20/4}$[2]	
9-Tetradecenoic	Myristoleic	$C_{14}H_{26}O_2$	226.35	248		1			
9-Hexadecenoic	Palmitoleic	$C_{16}H_{30}O_2$	254.40	221	−1[2]	1			
9-Octadecenoic	Oleic	$C_{18}H_{34}O_2$	282.46	199	14	1	234/15 mm[2]	$0.895^{20/4}$[2]	
9-Eicosanoic	Gadoleic	$C_{20}H_{38}O_2$	310.50	181	24.5	1			
13-Docosenoic	Erucic	$C_{22}H_{42}O_2$	338.56	166	33.5[2]	1	241–243/5 mm[2]	$0.860^{57.1/4}$[5]	
12-Hydroxy-9-octadecenoic	Ricinoleic	$C_{18}H_{34}O_3$	298.46	188		1		$0.940^{27.4/4}$[2]	
9, 12-Octadecadienoic	Linoleic	$C_{18}H_{32}O_2$	280.44	200	−5[5]	2		$0.903^{20/4}$[5]	
9, 12, 15-Octadecatrienoic	Linolenic	$C_{18}H_{30}O_2$	278.42	201	−11[5]	3		$0.914^{20/4}$[5]	
9, 11, 13-Octadecatrienoic	Eleostearic	$C_{18}H_{30}O_2$	278.42	201	49[4] (α)	3			
5, 8, 11, 14-Eicosatetraenoic	Arachidonic	$C_{20}H_{32}O_2$	304.46	185		4			
4, 8, 12, 15, 19-Docosapentaenoic	Clupanodonic	$C_{22}H_{34}O_2$	330.49	170		5			
Saturated Alcohols									
Hexanol	Caproyl	$C_6H_{13}OH$	102.17		−51.6[1]		157.2[2]	$0.819^{20/4}$[2]	
Octanol	Caprylyl	$C_8H_{17}OH$	130.23		−16.3[1]		194.5[2]	$0.825^{20/4}$[2]	
Decanol	Capryl	$C_{10}H_{21}OH$	158.27		7.0[1]		120/12 mm[2]	$0.830^{20/4}$[2]	
Dodecanol	Lauryl	$C_{12}H_{25}OH$	186.32		24.0[1]		150/20 mm[2]	$0.831^{24/4}$[2]	
Tetradecanol	Myristyl	$C_{14}H_{29}OH$	214.38		38.0[1]		170–173/20 mm[2]	$0.824^{38/4}$[2]	
Hexadecanol	Cetyl	$C_{16}H_{33}OH$	242.43		49.3[1]		190/15 mm[2]	$0.811^{60/4}$[2]	
Octadecanol	Stearyl	$C_{18}H_{37}OH$	270.48		58.5[1]		210/15 mm[2]	$0.812^{59/4}$[2]	
Octadecan-1, 12-diol	Hydroxy Stearyl	$C_{18}H_{36}(OH)_2$	286.48						
Eicosanol	Arachidyl	$C_{20}H_{41}OH$	298.53		65.5[2]		220/3 mm[2]		
Docosanol	Behenyl	$C_{22}H_{45}OH$	326.59		70.6[2]		180/0.22 mm[2]		
[illegible]		[illegible]	[illegible]				210/0.4 mm[2]		

Table 6:12 (continued)

Unsaturated Alcohols								
9-Dodecen-1-ol	Lauroleyl	$C_{12}H_{23}OH$	184.31		1			
9-Tetradecen-1-ol	Myristoleyl	$C_{14}H_{27}OH$	212.36		1			
9-Hexadecen-1-ol	Palmitoleyl	$C_{16}H_{31}OH$	240.41		1			
9-Octadecen-1-ol	Oleyl	$C_{18}H_{35}OH$	268.46		1	208–210/15 mm[2] (*cis*)		
9-Eicosen-1-ol	Gadoleyl	$C_{20}H_{39}OH$	296.51		1			
13-Docosen-1-ol	Erucyl	$C_{22}H_{43}OH$	324.57		1	240–242/10 mm[2] (*cis*)		
9-12-Octadecadien-1-ol	Linoleyl	$C_{18}H_{33}OH$	266.45	−5 to −2[2]	2			
9-12-15-Octadecatrien-1-ol	Linolenyl	$C_{18}H_{31}OH$	264.43		3			
9-Octadecen-1, 12-diol	Ricinoleyl	$C_{18}H_{34}(OH)_2$	284.47		1			
5, 8, 11, 14-Eicosatetraen-1-ol	Arachidonyl	$C_{20}H_{33}OH$	290.31		4			
4, 8, 12, 15, 19-Docosapenten-1-ol	Clupanodonyl	$C_{22}H_{35}OH$	316.51		5			
Alpha Olefins								
1-Dodecene	α-Dodecylene	$C_{12}H_{24}$	168.31	−35.3[3](FP)	1		0.760	5.95
1-Tetradecene	α-Tetradecylene	$C_{14}H_{28}$	196.36	−12.9[3](FP)	1		0.769	5.10
1-Hexadecene	Cetene	$C_{16}H_{32}$	224.16	4.1[3] (FP)	1		0.778	4.47
1-Octadecene		$C_{18}H_{36}$	252.47	18	1		0.787	3.97
1-Eicosene		$C_{20}H_{40}$	280.52		1		0.796*	3.57
1-Docosene		$C_{22}H_{44}$	308.57		1		0.805*	3.25
Saturated Hydrocarbons								
Dodecane		$C_{12}H_{26}$	170.33	−9.6[3](FP)		215	0.776 $^{0/4}$	
Tetradecane		$C_{14}H_{30}$	198.38	5.9[3](FP)		253	0.765	
Hexadecane	Cetane	$C_{16}H_{34}$	226.44	18.2[3]		288	0.775 $^{20/4}$	
Octadecane		$C_{18}H_{38}$	254.49	28.2[3]		317	0.777 $^{28/4}$	
Eicosane		$C_{20}H_{42}$	282.54	38		205	0.778 $^{37/4}$	
Docosane		$C_{22}H_{46}$	310.59	44.4		317	0.778 $^{44/4}$	

1 International Critical Tables.
2 Ralston, A. W., "Fatty Acids and Their Derivatives".
* Extrapolated liquid density.
3 *Analytical Chemistry*, **27**, p. 411 (1955).
4 Eckey, E. W., "Vegetable Fats and Oils".
5 Markley, K. S., "Fatty Acids."
6 Archer-Daniels-Midland Co.

Table 6:13

Solubility of Fatty Chemicals

Solubilities of alcohols are expressed in grams of alcohol per 100 grams of solvent.	*Isopropyl Alcohol*		*Benzene*		*Trichloro-ethylene*		*Ethyl Ether*		*Acetone*		*Turpentine*		*V.M.P. Naphtha*		*Kerosene*		*Light Mineral Oil*	
	25°C	50°C	25°C	50°C	25°C	50°C	25°C	50°C	25°C	50°C	25°C	50°C	25°C	50°C	25°C	50°C	25°C	50°C
Saturated Alcohols								Boiling Point of Diethyl Ether—34.5°C										
Adol 18	100	100	100	100	100	100	100		100	100	100	100	100	100	100	100	20	100
Adol 52	100	100	100	100	100	100	100		60	100	80	100	80	100	36	100	1	80
Adol 54	80	100	100	100	80	100	100		70	100	70	100	90	100	80	100	2	100
Adol 63	100	100	100	100	60	100	100		30	100	60	100	60	100	20	100	2	100
Adol 65	100	100	100	100	60	100	100		30	100	60	100	60	100	20	100	2	100
Adol 62	40	100	22	100	18	100	60		8	100	24	100	14	100	6	100	1	80
Adol 64	100	100	60	100	36	100	80		10	100	36	100	30	100	8	100	2	100
Adol 67	2	100	20	100	28	100	36		1.8	100	14	100	6	100	2	100	0.8	40
Adol 71	1.4	100	16	100	12	100	22		0.8	100	8	100	2	100	1.4	100	0.4	36
Adol 60	1	100	4.0	100	6	100	12		0.6	100	6	100	2	100	1.2	60	0.4	20
Adol 45	16	100	1.6	100	1.6	100	12		8	100	2	36	0.2	1	0.2	0.6	0	0.6
Unsaturated Alcohols																		
Adol 34	100	100	100	100	100	100	100		100	100	100	100	100	100	100	100	100	100
Adol 32	100	100	100	100	100	100	100		100	100	100	100	100	100	100	100	100	100
Adol 85	100	100	100	100	100	100	100		100	100	100	100	100	100	100	100	100	100
Unadol 90	100	100	100	100	100	100	100		100	100	100	100	100	100	100	100	100	100
Solubilities below are expressed in grams of the Chemifat per 100 *grams of* solution.																		
Spermoil Products																		
Spermafol 45	9	MS	52	MS	32	MS	29		8	MS	34	MS	36	MS	29	MS	9	MS
Spermafol 52	4	99	43	99	29	MS	14		5	99	24	MS	25	MS	23	MS	5	MS
Cosmol 1000	4	MS	48	MS	29	MS	37		7	MS	25	MS	29	MS	23	MS	5	MS
Spermaceti Wax	1	85	37	100	23	MS	17		1	100	10	MS	21	MS	4	MS	<1	3

MS—Mutually Soluble.

Table 6:13 (continued)

SOLUBILITY OF FATTY CHEMICALS

Solubilities below are expressed in grams of the Chemifat per 100 grams of solution.								Boiling Point of Diethyl Ether—34.5°C										
Hydrofol Acids																		
Hydrofol Acid 630	MS	MS	MS	MS	MS	MS	MS		MS	MS	MS	MS	MS	MS	23	MS	6	MS
Hydrofol Acid 46	35	MS	48	MS	40	MS	49		27	MS	29	MS	35	MS	25	MS	3	MS
Hydrofol Acid 51	33	99	35	99	33	94	38		19	99	22	98	23	99	25	99	4	98
Snodotte Acid	19	86	26	92	25	83	34		14	92	11	87	15	86	9	87	2	69
Hydrofol Acid 150	14	65	12	78	16	63	26		7	64	7	54	6	56	4	55	1	24
Distal 42	MS	MS	MS	MS	MS	MS	MS		21	MS	MS	MS	MS	MS	25	MS	8	MS
Hydrofol Acid 405	15	76	19	81	20	75	29		10	79	12	78	10	77	6	72	1	48
Hydrofol Acid 410	15	76	19	81	20	75	29		10	79	12	73	10	77	6	72	1	43
Hydrofol Acid 420	20	87	29	91	26	84	33		11	86	16	74	15	86	10	84	3	68
Hydrofol Acid AB	10	71	12	76	14	58	25		5	67	7	55	5	57	3	55	<1	23
Hydrofol Acid 200	19	51	5	75	6	31	15		11	64	1	15	<0.5	4	<0.5	2	<0.5	<0.5
Hydrofol Acid 565 (Erucic)	MS	MS	MS	MS	MS	MS	MS		88	MS	MS	MS	MS	MS	MS	MS	MS	MS
Hydrofol Acid 560 (Behenic)	1	17	3	36	4	24	6		2	26	2	21	16	39	<1	14	<1	4
Hydrofol Glycerides																		
Hydrofol Glyceride 42	5	MS	53	MS	60	MS	42		9	MS	50	MS	44	MS	36	MS	15	MS
Hydrofol Glyceride 45	3	MS	45	MS	50	MS	30		8	MS	41	MS	36	MS	29	MS	13	MS
Hydrofol Glyceride 50-51	<1	MS	37	MS	32	MS	22		2	MS	40	MS	31	MS	19	MS	6	MS
Hydrofol Glyceride T-57-L	1	9	20	87	36	MS	4		<1	63	19	MS	7	MS	4	MS	1	MS
Hydrofol Glyceride T-57-N	<1	5	15	86	32	MS	5		2	80	17	MS	7	MS	3	MS	1	MS
Hydrofol Glyceride T-57	<1	5	15	86	32	MS	5		2	80	17	MS	7	MS	3	MS	1	MS
Hydrofol Glyceride 150	<1	3	5	74	11	86	2		1	16	4	65	2	47	1	59	<1	41
Hydrofol Glyceride 200	<1	12	4	24	12	45	4		1	18	<0.5	10	<0.5	0.5	<0.5	<0.5	<0.5	<0.5

MS—Mutually Soluble.

Table 6:14

COMPATIBILITY FATTY ACIDS WITH MINERAL WAXES
Compatibilities are expressed in grams of fatty acid per 100 grams of *blend*

	Crude German Montan	*Microcrystalline Wax*	*Single Bleached Ozckerite*		*Crude German Montan*	*Microcrystalline Wax*	*Single Bleached Ozckerite*
Hydrofol Acid 51	C	C	C	Hydrofol Glyceride 50-51	C	C	C
Hydrofol Acid 150	C	C	C	Hydrofol Glyceride 150	C	C	C
Hydrofol Acid 200	C	79	71	Hydrofol Glyceride T-57-L	C	C	C
Hydrofol Acid 405	C	C	C	Hydrofol Glyceride 200	C	85	78
Hydrofol Acid 410	C	C	C	Cosmol 1000	C	C	C
Hydrofol Acid 66	C	C	C	Spermafol 52	C	C	C
Hydrofol Acid AB	C	C	C	45° N.W. Spermoil	C	C	C

Table 6:15

Average Fatty Acid Content and Constants of Fats and Oils

	Chemical Formula	Coconut	Palm Kernel	Tallow	Lard	Palm	Olive	Cottonseed	Corn	Peanut	Soybean	Sunflower	Linseed	Perilla	Castor	Tung	Whale	Menhaden	Sardine	Herring
SATURATED ACIDS																				
CAPROIC	$C_6H_{12}O_2$	0.2	Trace																	
CAPRYLIC	$C_8H_{16}O_2$	8.0	3.0																	
CAPRIC	$C_{10}H_{20}O_2$	7.0	6.0																	
LAURIC	$C_{12}H_{24}O_2$	48.0	50.0																	
MYRISTIC	$C_{14}H_{28}O_2$	17.5	15.0	2.0	1.0	1.0	Trace	0.5									8.0	7.0	5.0	7.0
PALMITIC	$C_{16}H_{32}O_2$	8.8	7.5	30.0	26.0	42.5	9.0	21.0	7.5	7.0	6.5	3.5	5.0	7.5		4.0	11.0	16.0	14.0	8.0
STEARIC	$C_{18}H_{36}O_2$	2.0	1.5	21.0	11.5	4.0	2.3	2.0	3.5	5.0	4.5	3.0	3.5	Trace	2.0	1.5	2.5	1.0	3.0	Trace
ARACHIDIC	$C_{20}H_{40}O_2$						0.2	Trace	0.5	4.0	0.7	0.6	Trace							
BEHENIC	$C_{22}H_{44}O_2$																			
LIGNOCERIC	$C_{24}H_{48}O_2$					Trace			0.2	3.0	Trace	0.4								
UNSATURATED ACIDS																				
MYRISTOLEIC	$C_{14}H_{26}O_2$																1.5	Trace	Trace	Trace
PALMITOLEIC	$C_{16}H_{30}O_2$																17.0	17.0	12.0	18.0
OLEIC	$C_{18}H_{34}O_2$	6.0	16.0	45.0	58.0	43.0	82.5	33.0	46.3	60.0	33.5	34.0	5.0	8.0	8.6	15.0	34.0	27.0	10.0	9.0
LINOLEIC	$C_{18}H_{32}O_2$	2.5	1.0	2.0	3.5	9.5	6.0	43.5	42.0	21.0	52.5	58.5	61.5	38.0	3.5		9.0	Trace	15.0	13.0
LINOLENIC	$C_{18}H_{30}O_2$										2.3		25.0	46.5			Trace			
ELAEOSTEARIC	$C_{18}H_{30}O_2$															79.5				
RICINOLEIC	$C_{18}H_{34}O_3$														85.9					
C_{20} UNSATURATED	$C_{20}H_{2(20-x)}O_2$																5.0	20.0	22.0	20.0
C_{22} UNSATURATED	$C_{22}H_{2(22-x)}O_2$																12.0	12.0	19.0	25.0
CONSTANTS																				
SAPONIFICATION VALUE		251-264	240-250	196-200	195-200	196-206	185-200	192-200	188-193	185-192	189-194	189-194	189-196	188-197	175-183	189-195	185-195	189-193	189-193	179-194
IODINE NUMBER		8-10	16-23	35-44	50-69	48-58	74-94	100-115	116-130	83-95	124-148	120-136	179-204	185-206	82-86	160-170	110-136	148-185	170-190	130-144
TITER—°C		20-23	20-23	37-46	36-43	38-47	18-25	32-38	18-20	28-32	20-21	17-20	19-21	12-17	—	37-38	22-24	31-33	28-34	23-27

Table 6:16

WAX–TYPE ACIDS AND WAX–TYPE HIGHER ALCOHOLS

Name	Formula	Melting Point °C.	Specific Gravity at 15°C.	Soluble in	Occurrence
Cerotic Acid	$CH_3(CH_2)_{24}COOH$	77.8	0.836 at 79°C.	warm alcohol	free in beeswax, montan wax, carnauba; also as cerotate in insect wax and carnauba
Montanic Acid	$CH_3(CH_2)_{26}COOH$	83.0		methyl alcohol	free in montan wax
Melissic Acid	$CH_3(CH_2)_{27}COOH$	91.0			free in beeswax and montan wax
Palmitic Acid	$C_{16}H_{32}O_2$	62.2	0.846	alcohol; ether	as tripalmitin in Japan wax; as cetyl palmitate in spermaceti; as myricyl palmitate in beeswax
Lauric Acid	$C_{12}H_{24}O_2$	43.5			as laurin in Japan wax
Cetyl Alcohol	$C_{16}H_{33}OH$	50.0	0.810	alcohol; ether; benzol	as cetyl palmitate in spermaceti
Octadecyl Alcohol	$C_{18}H_{37}OH$	59.0		alcohol; ether	spermaceti
Ceryl Alcohol	$C_{26}H_{53}OH$	79.0			as ceryl cerate in Chinese insect wax
Myricyl Alcohol	$C_{31}H_{63}OH$	88.0		alcohol; ether	as myricyl palmitate in beeswax and carnauba
Anonymous Alcohol	$C_{24}H_{48}(OH)_2$	103.0			carnauba wax

Chapter 7

FATTY ACID ESTERS AND GLYCERIDES

GLYCERYL STEARATES

The commercial, waxlike glyceryl esters are glyceryl tristearate, glyceryl distearate and glyceryl monostearate. The first contains as impurities small amounts of oleic, linoleic, and linolenic esters. The second contains varying amounts of monostearate and the third, varying amounts of distearate. The mono- and distearates (water-dispersible grades only) contain varying amounts of soap.

Glyceryl Monostearate S (Aldo 28)

Properties

Color and Form	cream, waxlike solid
Melting Point	57-61°C.(capillary tube)
Odor	faint fatty
pH (3% aqueous dispersion at 25°C.)	9-10
Specific Gravity (20°/20°C.)	0.97

Soluble in: Hot alcohol, hot hydrocarbons; dispersible in hot water.

Uses

Protective coatings for edible hygroscopic powders, etc. Edible powders, crystals, and tablets, which absorb moisture from the air, have a tendency to lump, cake, and decompose, on standing. A solution of Aldo 28 in hot alcohol, or other suitable solvent, is sprayed on the powder which can be tumbled in a heated barrel until all the solvent has evaporated. This leaves the product coated with a thin, protective film of Aldo 28 which disperses when the product is thrown into water. "Sun-checking" can be prevented by formulating polybutene-rubber combinations with Aldo 28, which not only improves resistance to light, but also acts as a plasticizer and increases slightly the rate of vulcanization. By using glyceryl tristearate, the compound will become more rigid, but will still be flexible. Development of surface tack on aging can be eliminated from butadiene compounds (Vistanex), by compounding with Aldo 28. Apparently, the glyceryl monostearate sweats out of the compound at a

Table 7:1

PROPERTIES OF GLYCERYL MONOSTEARATE

Product	*Grade*	*Form & Color at 25°C*	*Specific Gravity 25°C*	*M.P. or Solid Pt. °C*	*Acid Value*	*Free Fatty Acid %*	*Saponification Value*	*Iodine Value*	*pH of 5% Aqueous Dispersion*
Glycerol Mono Stearate	Aldo ®28	White beads	0.97	56–61	<10	<5	135–145	<5	9.0–10.0
Glyceryl Mono Stearate	Aldo ®33	White beads	0.97	57–61	< 5	<2.5	160–165	<5	7.0–8.0
Glyceryl Mono Stearate	Aldo ®72	White beads	0.97	54–58	< 5	<2.5	165–175	<6	5.5–6.5

Table 7:2

SOLUBILITY OF GLYCERYL MONOSTEARATE

Product	*Grade*	*Water*	*Methanol*	*Ethanol*	*Acetone*	*Ethyl Acetate*	*Toluol*	*Naphtha*	*Mineral Oil*	*Vegetable Oil*
Glyceryl Mono Stearate	Aldo ®28	DH	SH	SH	SXH	SXH	SH	SH	SH	SH
Glyceryl Mono Stearate	Aldo ®33	DH	SH	SH	SXH	PSH	SH	SH	SH	SH
Glyceryl Mono Stearate	Aldo ®72	DH	SH	SH	SXH	SXH	SH	SH	SH	SH

S: soluble or miscible in all proportions; ND: not or poorly dispersible; SH: soluble hot; D: dispersible; DH: dispersible hot; SX: clearly miscible in certain proportions; SXH: clearly miscible in certain proportions, hot; I: insoluble or very slightly soluble; PS: poorly soluble. Not completely miscible at any concentrations; PSH: partly soluble hot.

similar rate to the development of tackiness and the two properties compensate each other so that the surface presents neither a tacky nor a waxy appearance and feel.

Cosmetics: Emulsifying agent for the manufacture of greaseless creams.

Edible and Pharmaceutical Emulsions: For paste emulsions of edible oils and greaseless pharmaceutical ointments.

Lubricant: Integral lubricant for vinyls, particularly where a nontoxic edible agent is required; will not interfere with printing on vinyl sheeting; integral lubricant for ureaformaldehyde resins.

Table 7:3

MELTING POINTS (°C) OF FATTY ESTERS[1]

	Palmitate	*Stearate*	*Montanate*
Ethyl	23.5	33.6	64–65
Propyl	20.4	30.5	63.5
Butyl	16.9	27.5	—
Amyl	19.4	30.0	—
Isoamyl	12.5	23.0	—
Octyl	22.5	31.8	—
Cetyl	51.6	56.5	—

Table 7:4

PENTAERYTHRITOL ESTERS, MELTING POINTS OF (°C)[2]

	TETRA ESTERS			DI-ESTERS		
	β	β′	α	β	β′	α
Laurate	52.0	46.8	37	52	48.8	35
Tridecylate	53.5	47	43	—	—	—
Myristate	62.5	56	48	62	58.0	46
Palmitate	71.0	65	58	70	66.0	55
Stearate	77.0	72	66	75	71.5	64

GLYCOL FATTY-ACID STEARATES

Ethylene glycol, diethylene glycol, and propylene glycol derivatives will be discussed in this section. They are essentially the monoesters obtained from the reaction of one molecule of glycol with one molecule of fatty acid. The straight esters are water-insoluble, but can be made water-dispersible by adding small amounts of soap which will yield modified esters.

Glycol esters are widely used in industry as emulsifiers, stabilizers, plasticizers, lubricants, and thickening agents.

Propylene glycol fatty-acid esters are nontoxic and edible. They are used in the food and pharmaceutical industries as emulsifiers and defoaming agents.

[1] G. S. Whitby, *J. Chem. Soc.*, 1458–65 (1926).
[2] T. Malkin, *Fette, Seifen, Anstrichr.*, **61**, 868 (1959).

MYVEROL Distilled Monoglycerides

Myverol 18-40 is prepared from prime steam lard. In addition to a very high monoester content, the product is bland and has excellent stability.

As is the case with other edible fats and oils, monoglycerides are subject to oxidation on exposure to air. To prevent such oxidation during extended storage, Myverol Distilled Monoglycerides contain small amounts of approved antioxidants.

Average Chemical and Physical Data

Monoester Content	90.0% (minimum)
Saponification Value	165
Iodine Value	50
Glycerol Content	1.5% (maximum)
F.F.A. (as oleic)	1.5% (maximum)
Specific Gravity	0.92 at 80°C.
Congeal Point	47°C (117°F)
Cloud Point	54°C (129°F)
Clear Point	60°C (140°F)

Myverol 18-15 is made from refined cottonseed oil.

Average Chemical and Physical Data

Monoester Content	90.0% (minimum)
Saponification Value	155–165
Iodine Value	85
Glycerol Content	1.0% (maximum)
F.F.A. (as oleic)	1.5% (maximum)
Specific Gravity	0.94 at 60°C
Congeal Point	46°C (115°F)
Clear Point	56°C (133°F)

Myverol 18-19 is made from safflower oil.

Average Chemical and Physical Properties

Monoester Content	90.0% (minimum)
Appearance	Cloudy liquid
Iodine Value	110–120
Glycerol Content	1.0% (maximum)
F.F.A. Content (as oleic)	2.0% (maximum)
Congeal Point	25°–28°C (77°–82°F)
Clear Point (supercools readily)	39°–42°C (102°–108°F)

MYVACET Distilled Acetylated Monoglycerides

These acetylated monoglycerides are unique in their sharp melting points, non-greasy nature, film-forming characteristics, oxidative stability, solvent power, and emulsifying and plasticizing ability. They can improve the shelf life, texture, appearance, and stability of a variety of foods and industrial products.

Made by a molecular distillation process, Myvacet products are bland and odor free. They hydrolyze in water at a rate comparable to triglyceride fats. Their oxidative stability is many times that of the parent fats. This remarkable stability against rancidity is an important feature in the use of Distilled Acetylated Monoglycerides as food coatings and components.

Distilled Acetylated Monoglycerides are intermediate in polarity between triglycerides and monoglycerides and melt at lower temperatures. These two properties make them more soluble in usual solvents than either the triglycerides or monoglycerides. One of these products, a liquid, is a good solvent for materials that may be slightly too polar to dissolve in petroleum or glyceride oils.

The solid types of Myvacet Distilled Acetylated Monoglycerides form flexible films at room temperatures which retain their flexibility at low temperatures. The films are good barriers against oxygen and water vapor, comparing favorably with coated cellophane. They are relatively permeable to carbon dioxide and may be useful where a selective barrier is desired. Such films can be cast directly on the food product to form skin-tight coatings to afford greater protection than is now obtainable with common "wrap" type films.

Myvacet Distilled Acetylated Monoglyceride films can be strengthened by incorporation of small amounts of plastic materials such as cellulose acetate butyrate. Addition of 15% or more cellulose acetate butyrate to Myvacet Type 9-40 forms a strippable film coating.

Myvacet Type 5-00 is a non-fracturing, flexible, waxy solid. Melting above body temperature, it is a plastic product which has none of the "greasy feel" usually associated with fats. The oxidative stability has seldom, if ever, been approached by any other fat. It has unusually high solubility particularly in the more polar solvents. It is completely miscible with worm 80% aqueous ethanol, about 5–10% of the Myvacet product crystallizing out at room temperature.

Since natural raw materials (hydrogenated lard) vary somewhat in properties, Myvacet Type 5-00 reflects that variation. The properties below were observed, and are reasonably typical of the product.

Typical Physical Properties

Physical state	Creamy white, waxy solid
Melting point	41°–43°C (106°–109°F)
Clear Point	55°C (131°F) (approx.)
Refractive index	1.444 (50°C : 122°F)
Viscosity	30 cps (50°C : 122°F)
Specific gravity at	
50°C (122°F)	0.94 (liquid)
20°C (68°F)	1.00 (solid)

Solubility at 20°C: Major portion soluble in most common organic solvents, but some crystals form even at low concentrations.

Typical Chemical Properties

Iodine value	1
Saponification no.	296 (approx.)
% Monoglyceride	14 (approx.)
Peroxide value (PV)	0.5
Stability (PV after 96 hours aeration at 98°C)	0.5
Reichert-Meissl value	90 (approx.)
Acid value	Less than 4

Myvacet Type 7-00

Typical Physical Properties

Physical state	Waxy solid
Color-appearance	creamy white
Melting point	37°–40°C (99°–104°F)
Refractive index at 50°C	1.442
Viscosity	22 cps (50°C : 122°F)

Typical Chemical Properties

Iodine value	1
Saponification no.	320 (approx.)
% Monoglyceride	7 (approx.)
Peroxide value (PV)	0.5
Stability (PV after 96 hours aeration at 98°C)	0.5
Reichert-Meissl value	105 (approx.)
Acid value	Less than 6

Myvacet 7-00 has a tendency to crystallize fractionally on slow cooling. It is recommended that this product be melted completely before removing samples.

Myvacet Type 7-15

Typical Physical Properties

Physical state	Creamy white, waxy solid
Melting point	32–35°C (90°–95°F)
Refractive index at 40°C	1.447
Viscosity at	
100°C	10 cps
50°C	25 cps
Specific gravity at	
100°C	0.95
50°C	0.96

Typical Chemical Properties

Iodine value	14
Saponification value	345
% monoglyceride	8
Peroxide value (PV)	0.5
Stability (PV after 120 hours aeration at 100°C)	1.0
Reichert-Meissl value	103
Acid value	Less than 6

Myvacet Type 9-40 (Acetylated Lard)

Typical Physical Properties

Physical state	clear, almost colorless liquid
Congeal point	8°C (approx.) (46°F)
Refractive index at	
50°C (122°F)	1.443
40°C (104°F)	1.447
Viscosity at	
121°C (250°F)	7 cps
50°C (122°F)	19 cps
20°C (68°F)	50 cps
Specific gravity at	
50°C (122°F)	0.96
20°C (68°F)	0.99

Solubility at 20°C. Soluble in all common organic solvents, 80% aqueous ethanol, and vegetable and mineral oils. Less than 4% in 70% alcohol.

Typical Chemical Properties

Iodine value	42
Saponification value	380
% Monoglyceride	0 to 2
Reichert-Meissl value	145
Acid value	Less than 4

Myvacet Type 9-85 (Acetylated Cottonseed Oil)

Typical Physical Properties

Physical State	Clear, very slightly yellow liquid
Congeal Point	5°C (41°F) (approx.)
Refractive Index	1.451 (40°C : 104°F)
Viscosity at	
20°C (68°F)	47 cps
50°C (122°F)	18 cps
Specific gravity at	
20°C (68°F)	0.98
50°C (122°F)	0.96

Table 7:5

PROPERTIES OF GLYCOL STEARATES

Product	*Grade*	*Form & Color at 25°C*	*Specific Gravity 25°C*	*M.P. or Solid. Pt. °C*	*Acid Value*	*Free Fatty Acid %*	*Saponi-fication Value*	*Iodine Value*	*pH of 5% Aqueous Dispersion*
Diethylene Glycol Stearate	Diglycol Stearate S	White beads	0.96	47–56	95–105	47–53	160–170	< 6	6.8-7.5
Ethylene Glycol Mono Stearate	S151	White flakes	0.96	57–60	< 6	< 3.0	188–201	< 8	3.0–5.0
Propylene Glycol Stearate	Aldo 25	Solid, cream	0.94	43–46	< 10	< 5	143–150	< 7	9.2–10.0
Propylene Glycol Mono Stearate	PGMS 64	White to cream beads	0.94	39–42	10 max.	5.0 max.	175–185	5 max.	6.0–8.0[1]

[1] 3% Aqueous dispersion.

Table 7:6

SOLUBILITY OF GLYCOL STEARATES[1]

Product	*Grade*	*Water*	*Methanol*	*Ethanol*	*Acetone*	*Ethyl Acetate*	*Toluol*	*Naphtha*	*Mineral Oil*	*Vegetable Oil*
Diethylene Glycol Stearate	Diglycol Stearate S	DH	SH	SH	SXH	SXH	SH	SH	SH	SH
Ethylene Glycol Mono Stearate	S151	ND	SH	SH	SH	SH	SH	SH	SH	SH
Propylene Glycol Mono Stearate	Aldo 25	DH	SH	SH	SH	PS	SH	SH	SH	SH
Propylene Glycol Mono Stearate	PGMS 64	DXH	SXH	SX	SXH	SXH	SH	SH	SH	SH

[1] See Table 7:2 for explanation of symbols.

Table 7:7

PROPERTIES OF GLYCOL ESTERS[1]

Compound		m.p. °C	R.I.	Solubility in Absolute Alcohol	
				0°C	15°C
Ethyleneglycol	Monostearate	58.5	1.4310	0.670	2.000
—	Distearate	75.0	1.4385	0.010	0.020
—	Monopalmitate	51.5	1.4411	1.620	10.000
—	Dipalmitate	68.7	1.4378	0.018	0.055
Propyleneglycol	Monostearate	59.5	1.4424	0.0211	0.0340
—	Distearate	76.5	1.4366	0.0012	0.0063
—	Monopalmitate	54.3	1.4405	0.0193	0.0907
—	Dipalmitate	68.8	1.4364	0.0052	0.0115
Trimethyleneglycol	Monostearate	60.5	1.4437	0.0143	0.0305
—	Distearate	64.7	1.4397	0.0013	0.0038
—	Dipalmitate	56.2	1.4374	0.0244	0.0517

Table 7:8

COMPARATIVE SOLUBILITIES OF 12-HYDROXYSTEARIC ACID ESTERS

cc of solvent to dissolve 1 g of ester at 25°C

Ester	*Ethanol*	*Acetone*	*Ligroin*	*Toluene*	*Minerol Oil*	*Cottonseed Oil*
C_1	3	5	20	2	—	20
C_2	2	3	10	1	40	20
C_3	2	3	4	1	40	20
C_4	1	1	2	1	30	10
C_5	1	2	2	1	30	20
C_6	2	3	2	1	30	20
C_8	10	5	5	2	50	30
C_{10}	50	20	20	3	—	40
C_{12}	150	50	50	5	—	50
C_{14}	200	80	50	5	—	50
C_{16}	—	450	200	20	—	—
C_{18}	—	—	450	50	—	—

[1] I. S. Howe, *Trans. Rev. Soc. Canada*, **12**, 13–18 (1918).

Table 7:9

PROPERTIES OF GLYCOL ESTERS OF 12-HYDROXYSTEARIC ACID[1,2]

Compound	*Hydroxyl Value*	*M.P. °C*
Ethyleneglycol Monoester	318.0	67.0–68.5
Propyleneglycol Monoester	316.0	63.0–66.0
Trimethyleneglycol Monoester	388.0	60.5–62.0
Ethyleneglycol Diester	178.0	90.0–92.0
Trimethyleneglycol Diester	176.0	81.3–82.5
Methyl	178.4	56.6–57.0
Ethyl	170.8	50.3–51.6
n-Propyl	163.8	48.3–49.5
n-Butyl	157.3	43.7–44.9
n-Amyl	151.4	45.0–46.0
n-Hexyl	145.9	46.1–47.4
n-Octyl	136.0	49.5–51.3
n-Decyl	127.3	56.0–57.2
n-Dodecyl (Lauryl)	119.7	60.0–61.5
n-Tetradecyl (Myristyl)	112.9	61.0–64.0
n-Hexadecyl (Cetyl)	106.9	66.5–69.5
n-Octadecyl (Stearyl)	101.5	76.0–76.5
1,2-Hydroxystearic Acid	186.7	81.0
Hydrogenated Castor Oil	152.3	86.0

[1] S. A. Bell, *J. Am. Pharm. Assn.*, **31**, 75 (1942).
[2] S. A. Bell, *J. Am. Pharm. Assn.*, **32**, 118 (1943).

POLYETHYLENE GLYCOL STEARATES

All these polymers dissolve in water to form clear solutions and are soluble in many organic solvents. Usually they do not hydrolyze or deteriorate. As their molecular weight *increases,* water solubility, vapor pressure, hygroscopicity, and solubility in organic solvents *decrease;* at the same time, freezing or melting range, specific gravity, flash point, and viscosity *increase.* By blending several members of the series, many intermediate combinations of these physical properties can be obtained.

These products are soluble in methanol, ethanol, acetone, and ethyl acetate. They are dispersible in hot water. Glycosperse TS-20 is dispersible in toluene and soluble in naphtha, mineral and vegetable oils. The others are insoluble or poorly soluble in the latter.

Table 7: 10

PROPERTIES OF SORBITOL STEARATES

Product	M.P. °C	Sp. Gr.	Free Fatty %	Acid No.	Saponification No.	Iodine No.	Odor	Color	Hardness	Solubility in Water	Alcohol	Naphtha	Mineral Oil
Sorbitol Distearate	49–51	0.96	15–17	30–34	168–178	<5	faint	cream	3	D	PS	SH	SH
Sorbitol Distearate (A)	42–43	0.96	17–21	38–42	175–185	<5	faint	cream	3	I	PS	SH	SH
Sorbitol Monostearate (Arlex G-909)	45–55	0.98	—	<5	—	—	—	—	—	DH	SH	SH	—

Table 7: 11

PROPERTIES OF SORBITAN FATTY ACID ESTERS

Product	Grade	Color & Form at 25° C.	Approx. Specific Gravity at 25° C.	Titer °C.	Acid No.	Saponification No.	Hydroxyl No.	Water Content %	Approx. Viscosity 25°C/CPS
Sorbitan-Mono Palmitate	Glycomul P	Tan Beads	—	45–47	7.5 max.	139–151	272–306	0–1.5	—
Sorbitan Mono Stearate	Glycomul S	Cream Beads	—	50–52	10 max.	146–158	236–259	0–1.5	—
Sorbitan Tri Stearate	Glycomul TS	Cream Beads	1.0	51	12–15	175–190	65–80	0–1.0	—

Table 7: 12

SOLUBILITY OF SORBITAN FATTY ACID ESTERS[1]

Product	Grade	Water	Methanol	Ethanol	Acetone	Ethyl Acetate	Toluol	Naphtha	Mineral Oil	Vegetable Oil
Sorbitan Mono Palmitate	Glycomul P	ND	SXH	PSH	PSH	SH	PSH	SXH	PSH	S
Sorbitan Mono Stearate	Glycomul S	ND	SXH	PSH	I	PSH	PSH	I	PSH	S
Sorbitan Tri Stearate	Glycomul TS	I	I	I	D	ND	ND	D	D	D

[1] See Table 7: 2 for explanation of symbols.

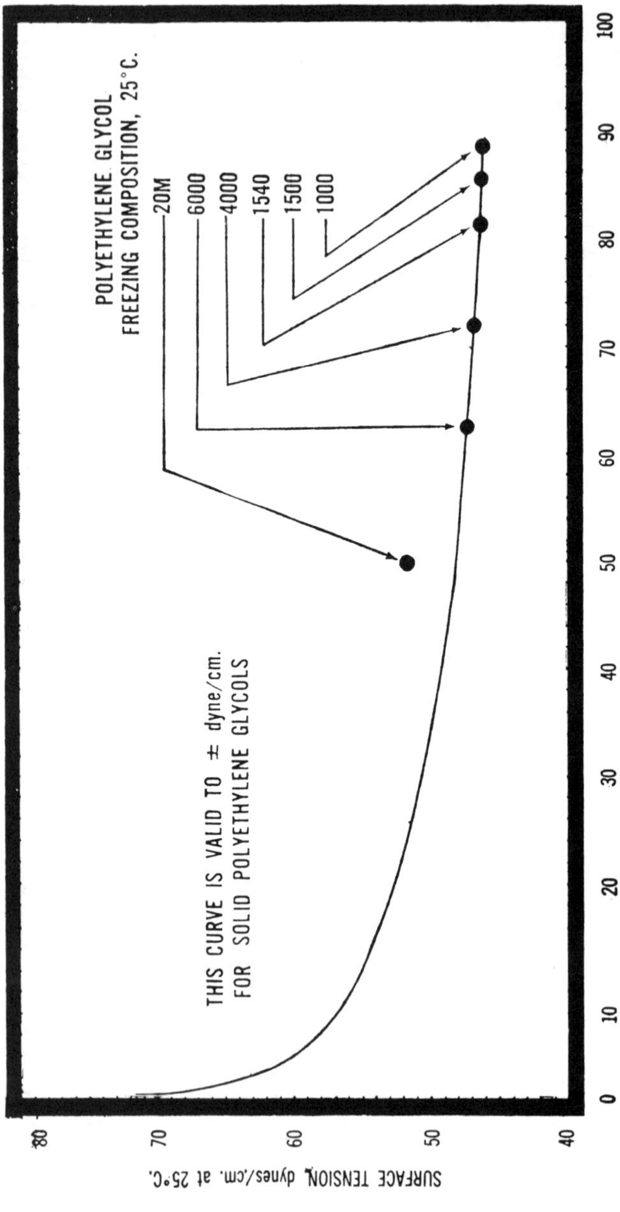

Fig. 7:1. Surface Tensions of Aqueous Solutions of Solid Polyethylene Glycols at 25°C.

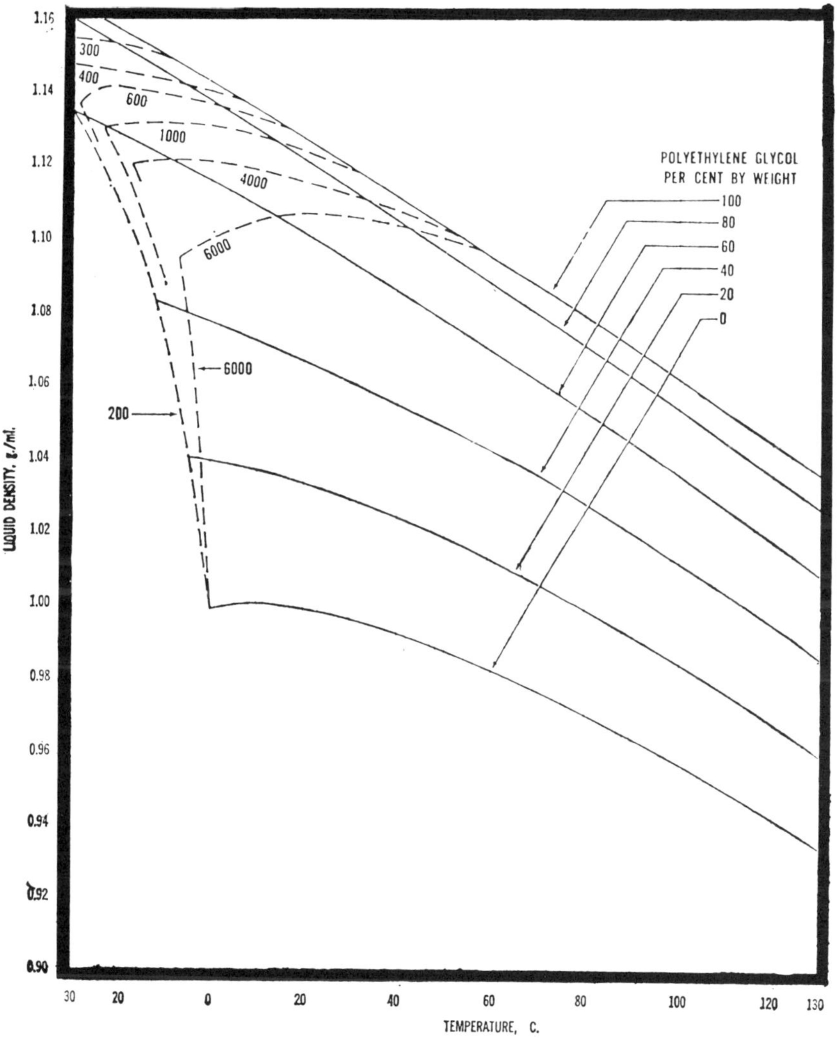

FIG. 7:2. LIQUID DENSITIES OF AQUEOUS SOLUTIONS OF LIQUID AND SOLID POLYETHYLENE GLYCOLS.

Table 7:13

PROPERTIES OF POLYETHYLENE GLYCOL STEARATES

Product	*Grade*	*Form & Color at 25°C*	*Specific Gravity 25°C*	*M.P. or Solid. Pt. °C*	*Acid Value*	*Free Fatty Acid %*	*Saponification Value*	*Iodine Value*	*pH of 5% Aqueous Dispersion*
Polyethylene Glycol 400 Mono Stearate	S 541	Soft solid, white	1.00	24–29	<10	<5.0	85–95	<4	3.0–4.5
Polyethylene Glycol 400 Di Stearate	S1009	Soft solid, cream	0.98	33–38	<10	<5.0	120–130	<5	3.0–5.0
Polyethylene Glycol 600 Mono Stearate	S1012	Soft solid, cream	1.01	33–38	<10	<5.0	68–75	<3	3.0–5.0
Polyethylene Glycol 600 Di Stearate	S1013	Waxy solid, cream	1.00	28–38	<10	<5.0	102–110	<5	3.0–5.0
Polyethylene Glycol 1000 Mono Stearate	S1016	Waxy solid, cream	1.02	40–45	<7	<3.5	45–52	<2	3.0–5.0
Polyethylene Glycol 1500 Mono Stearate	S1042	Waxy solid, cream	1.06	27–31	<8	<4.0	57–67	<2	3.0–5.0
Polyethylene Glycol 4000 Mono Stearate	S1004	Waxy solid, cream	1.10	54–61	<6	<3.0	17–22	<1	3.0–5.0

Table 7:14

SOLUBILITIES OF POLYETHYLENE GLYCOL STEARATES[1]

Product	*Grade*	*Water*	*Methanol*	*Ethanol*	*Acetone*	*Ethyl Acetate*	*Toluene*	*Naphtha*	*Mineral Oil*	*Vegetable Oil*
Polyethylene Glycol 400 Mono Stearate	S 541	DH	S	S	S	S	S	PSC, SH	PSC, SH	PSC, SH
Polyethylene Glycol 400 Di Stearate	S1009	DH	SH	S	PSH	S	S	S	S	SH
Polyethylene Glycol 600 Mono Stearate	S1012	SX	SH, PSC	S	S	S	PSH	I	I	S
Polyethylene Glycol 600 Di Stearate	S1013	DH	SH, PSC	S	I	S	PSH	SH	SH	S
Polyethylene Glycol 1000 Mono Stearate	S1016	SX	SH, PSC	S	PSH	S	S	I	I	SH
Polyethylene Glycol 1500 Mono Stearate .	S1042	SX	SH	SH	S	S	SH	PSC, SH	I	S
Polyethylene Glycol 4000 Mono Stearate	S1004	SH	SH	SH	SH	SH	SH	DC	PSH	SH

[1] See Table 7: 2 for explanation of symbols.
S: Soluble or miscible in all proportions; SX: Clearly miscible in certain proportions; PS: Poorly soluble, not completely miscible at any concentration; D: Dispersible; ND: Not dispersible; I: Insoluble; C: Cold; H: Hot.

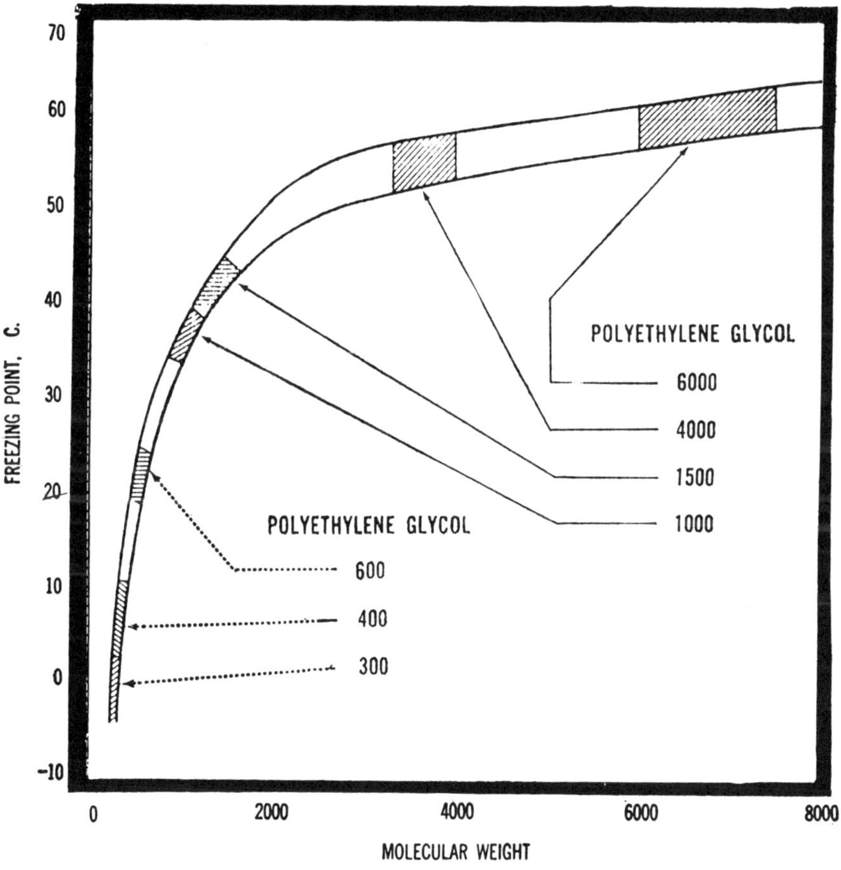

FIG. 7:3. FREEZING RANGES OF LIQUID AND SOLID POLYETHYLENE GLYCOLS PLOTTED AGAINST THEIR AVERAGE MOLECULAR WEIGHTS.

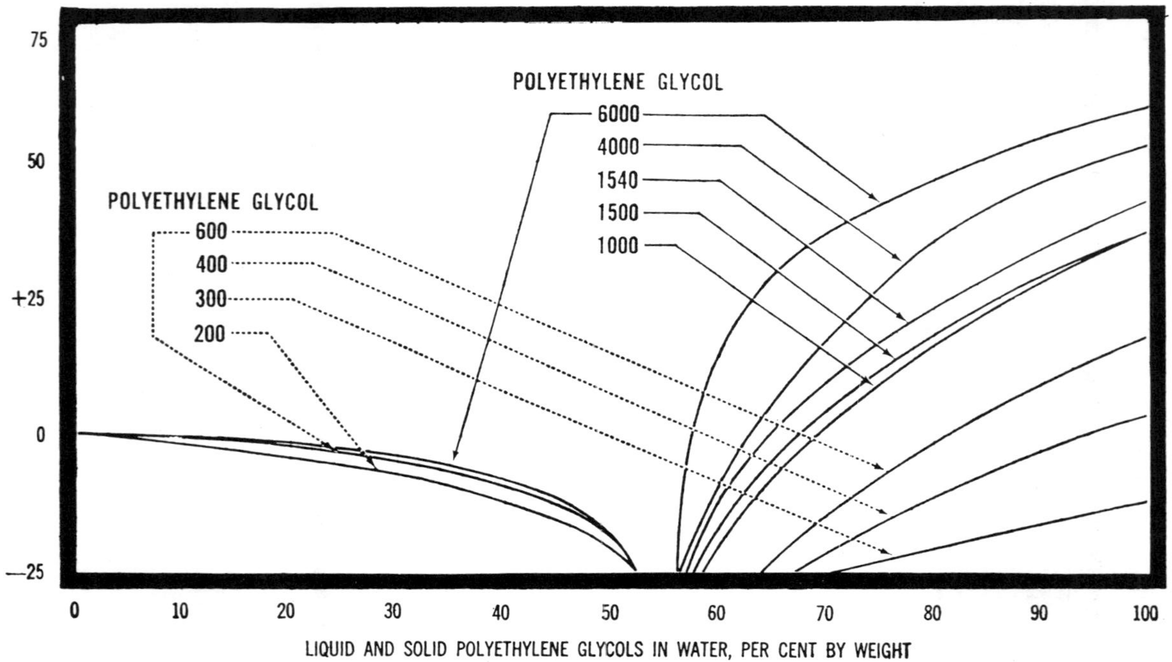

FIG. 7:4. PHASE DIAGRAMS OF AQUEOUS SOLUTIONS OF LIQUID AND SOLID POLYETHYLENE GLYCOLS.

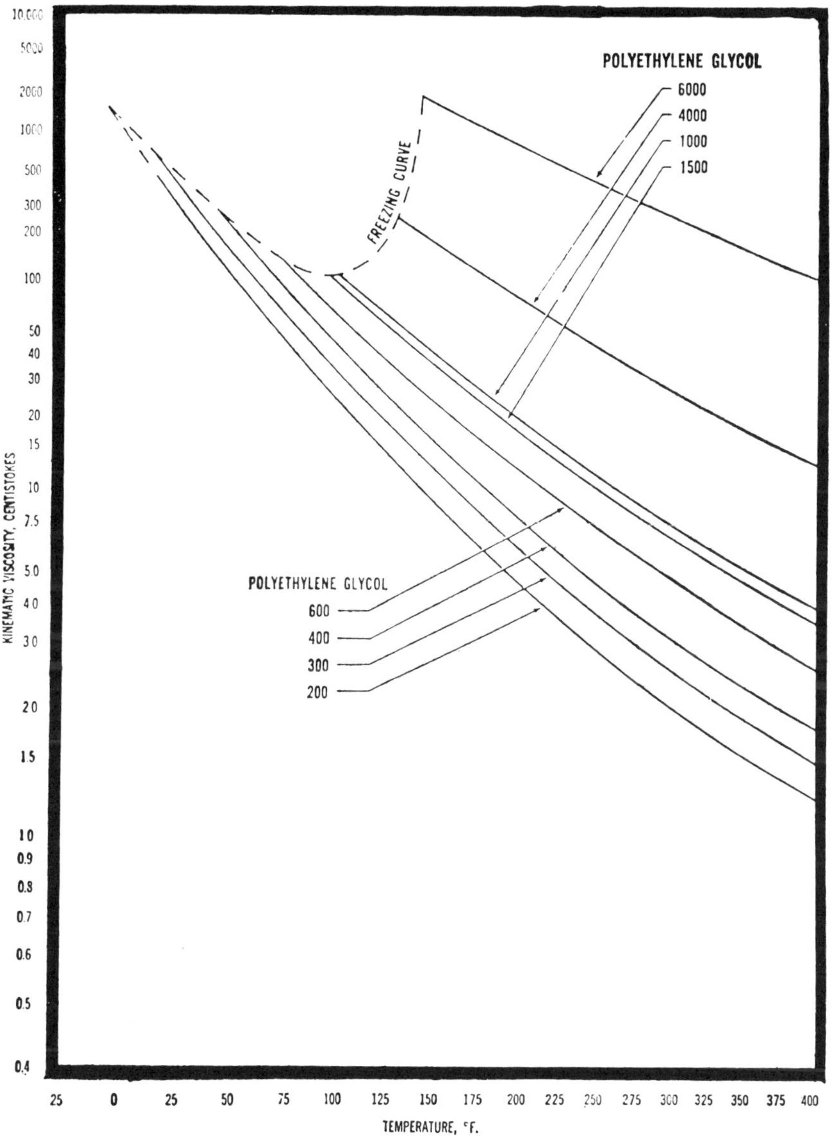

FIG. 7:5. KINEMATIC VISCOSITY OF POLYETHYLENE GLYCOLS.

Table 7:15

PROPERTIES OF POLYOXYETHYLENE FATTY ETHERS

Product	*Grade*	*Color & Form at 25°C*	*Approx. Sp. Gr. at 25°C*	*Titer °C*	*Acid No.*	*Saponification No.*	*Hydroxyl No.*	*Water Content %*
Ethylene Oxide (23 mols) lauryl alcohol ether	Glycosperse LA-23	White Wax	1.05	34	0–2.0	—	42–52	2.5–3.0
Ethylene Oxide (8 mols) stearate	SA-8 Glycosperse	Ivory Wax	1.0	28	0–1.0	87–97	85–100	2.5–3.0
Ethylene Oxide (9 mols) Stearate	SA-9 Glycosperse	White Solid	1.0	26	0–4.6	81–91	81–91	0–1.0
Ethylene Oxide (20 mols) tristearate	Glycosperse TS-20	Tan Wax	1.05	31	0–2.0	88–98	44–60	2.5–3.0
Ethyleneglycol monostearate	Pegosperse 50MS	White	—	58				
Diethyleneglycol monostearate	Pegosperse 100S	White	—	49				
Propyleneglycol monostearate	Pegosperse PS	Cream	—	46				

PROPERTIES OF SELECTED COMMERCIAL POLYGLYCOLS

"Carbowax"

Carbowax is the trademark for certain solid, water-soluble, high-molecular-weight (>500) polyethylene glycols that have a waxy appearance and feeling.

Compatibility of Carbowaxes

The compatibility of Carbowax 4000 is very limited, nitrocellulose and a few of the Bakelite resins being the only film-forming substances with which it is compatible in the proportions tested. The tests were run by adding a solution of Carbowax in alcohol, benzene, or water to a solution of the resin in an appropriate solvent and examining a pour of the mixture on a glass plate for evidences of non-homogeneity. In many cases, these examinations were supplemented by an examination of hot melts of the resin and Carbowax in the absence of solvent.

Toxicity

Carbowax polyethylene glycols are bland products possessing a very low order of toxicity. Studies of penetration on many of these products show that they do not penetrate the skin in harmful amounts. In fact, 20 milliliters per kilogram body weight, which is the maximum dosage that can be utilized on rabbits, failed to define an LD_{50} by skin penetration.

Single acute oral toxicity is extremely low and repeated feedings of 3 months to two years duration demonstrated a low order of toxicity.

Uses

Carbowaxes are extensively used as base in cosmetics and ointments, a binding agent, size, softener, humectant, lubricant, preservative, and adhesive, especially for textiles, paper, leather, and other porous materials. It offers possibilities as a plasticizer for such substances as casein, gelatin, glue, cork, special printing inks, and for cellulose transparent films since it will not diminish their transparency. Carbowax Compound 1500 can be used with excellent results as a lubricant for wire drawing and metal extrusion operations. A small amount of Carbowax Compound in water, applied as a lubricant to molds in which rubber parts are formed, causes a perfect, quick release when the mold is opened. It imparts an attractive finish to the molded rubber product, and the molds require less cleaning when this releasing agent is used. It can also be used as a water-soluble lubricant in warp sizes for cotton yarn; as a scrubbing liquid in air purification; and as an antisqueak fluid, with or without graphite. In addition, it shows promise in the formulation of carbon paper, because it is helpful as a wetting agent in wax mixtures; in belt dressings, in unvulcanized rubber compositions for self-sealing tires and tubes, and for transparentizing tracing-paper stock.

Table 7:16
PHYSICAL PROPERTIES OF CARBOWAXES

Compound	*1500*	*1540*	*4000*
Physical Appearance	Soft, waxy solid	Medium-hard, waxy solid	Hard, waxy solid
Density, g per cc at 20°C	1.151	1.15	1.204
Melting Range	34–37°C	40–45°C	50–55°C
Viscosity Range, Saybolt Universal at 210°F	60–90 sec	100–150 sec	400–600 sec
Flash Point, Open Cup	430°F	460°F	535°F
Freezing Points (°C) of Mixtures Containing, by Weight:			
90% Water	− 1.0	− 0.5	− 0.5
70% Water	− 3.0	− 3.0	− 4.5
50% Water	−15.0	−10.0	−10.0
30% Water	+16.0	+20.0	+30.0
10% Water	+35.0	+37.0	+49.0
Solubility in Water, % by Weight at:			
0°C	62	59	53
20°C	73	70	62
40°C	95	94	80
44°C	100	—	—
45°C	100	100	—
57°C	100	100	100
	Specifications		
Melting Point	not below 34°C		not below 50°C
Color, Pt-Co, 5% Aqueous Solution	not more than 100	—	not more than 50
Water Solubility, 5% Aqueous Solution	clear and substantially free from insoluble matter.		
Acidity, 5% Aqueous Solution	not more than 0.05%, calculated as acetic acid.		
Alkalinity, 5% Aqueous Solution	not more than 0.2%, calculated as sodium hydroxide.		

Carbowax Compound 4000 looks like paraffin wax, but is water soluble. It is not hygroscopic like Carbowax Compound 1500.

Table 7:17
COMPATIBILITY OF CARBOWAX 4000

	Ratio of Carbowax to Resin		
	5/1	1/1	1/5
Ethyl Cellulose	I	I	I
Methyl Cellulose	I	I	I
Polyvinyl Alcohol	I	I	I
Bakelite XR-3180		I	
Bakelite XR-4357		I	
Bakelite XJ-1434		I	
Bakelite XR-9396		I	
Bakelite XR-4357		I	
Bakelite XR-9392	I	C	C
Bakelite BR-3360	I	C	C
Bakelite BR-8900	I	I	C
Shellac		I	
Manila Loba B		I	
Carnauba Wax, No. 3 Refined		I	
Gum Damar		I	
Gum Accroides		I	
Gum Elemi		I	
Gum Mastic		I	
Batavia Damar		I	
Pale East India Macassar		I	
Gum Arabic	I	I	I
Raw Castor Oil	I	I	I
Tung Oil	I	I	I
White, Heat-bodied, Raw Linseed Oil	I	I	I
Lacquer Linseed Oil (A.D.M. No. 100)	I	I	I
Processed Linseed Oil (A.D.M.-O.K.O. M-17)	I	I	I
Paraffin Wax		I	
Ester Gum		I	
Rosin		I	
Algin S.P.		I	
Halowax No. 1012		I	
Uformite F-224		I	
Petrex No. 5		I	
Teglac Z-152		I	
Rezyl 1102		I	
Beeswax		I	
Mineral Oil		I	
Casein		I	I
Zein		I	
Polyvinyl Alcohol		I	
Nitrocellulose	I	SC	C

I = Incompatible
C = Compatible; completely homogeneous
SC = Partially compatible but with some evidences of inhomogeneity

Table 7:18

PROPERTIES OF CARBOWAX

Carbowax No.	*Average Molecular Weight*	*Specific Gravity, 20/20°C*	*Freezing Range, °C*	*Viscosity CKS, 210°F*	*Average Liquid Specific Heat cal/g/°C*
1000	950–1050	—	37.0–40.0	17.4	0.54 (a)
1500	500–600	1.151 (c)	38.0–40.1	13–18	0.54 (a)
1540	1300–1600	1.15	43.0–46.0	25–32	0.54 (a)
4000	3000–3700	1.204 (c)	53.0–56.0	75–85	0.55 (a)
6000	6000–7500	—	60.0–63.0	700–900	0.55 (a)
20M	15,000–20,000 (e)	—	50–55 (h)	57.3 (j)	0.59 (a)

Table 7:18 (continued)

Carbowax No.	*Heat of Fusion cal/g*	*Refractive Index 25 ^nD*	*Surface Tension 25°C dynes/cm*	*Heat of Combustion 25°C Btu/lb*	*Solubility in water, 20°C % by Wt*	*Comparative Hygroscopicity (Glycerol —100)*	*Vapor Pressure mm. Hg 100°C*	*Flash Point °F*(b)
1000	37	(i)	(i)	11,250	70	35	3.3×10^{-9}(d)	510
1500	37	(i)	(i)	11,090	73	35	—	430
1540	37	(i)	(i)	11,320	70	30	2.0×10^{-12}(d)	510
4000	43	(i)	(i)	11,350	62	1	(k)	515
6000	46	(i)	(i)	11,390	50	1	(k)	520
20M	41	(i)	52 (g)	—	60(f)	—	(k)	>465

(a) Mid-point of freezing range to 100° C.
(b) Cleveland open-cup tester.
(c) Density, g/ml at 20°C.
(d) From plot of vapor pressure vs. square root of molecular weight.
(e) Estimated
(f) With gel formation.
(g) 50 wt. per cent aqueous solution.
(h) Softening point.
(i) Solid at 25°C.
(j) 25 per cent aqueous solution.
(k) Less than 2.0×10^{-12}mm.

Polyethylene Glycol Compound 20M (Carbowax)

Polyethylene glycol compound 20M, which has an estimated molecular weight of 15,000 to 20,000, is synthesized by joining polyethylene glycol 6000 molecules with a diepoxide. A hard, translucent, waxy white solid, it is characterized by a much higher melt viscosity and aqueous solution viscosity than polyethylene glycol 6000. It is shipped in the form of white flakes. Non-hygroscopic and water-soluble, polyethylene glycol compound 20M is somewhat more flexible than the other solid polyethylene glycols. Water solutions show less penetration into porous surfaces because of their higher viscosities.

Polythylene glycol compound 20M has the high solubility in water of the other polyethylene glycols. Solubility in aromatic hydrocarbons is low at room temperature, but increases at higher temperatures. It is only slightly soluble in some aliphatic hydrocarbons. The softening point of 50–55°C is the same as that of polyethylene glycol 4000. Polyethylene glycol compound 20M, which has a viscosity of 450,000 Saybolt Seconds at 210°F, is compatible with the other solid polyethylene glycols and can be used to modify their viscosity characteristics and flexibility. Polyethylene glycol compound 20M is also compatible with nitrocellulose and methyl cellulose but only partly compatible with polyvinyl alcohol, carboxymethyl cellulose, sodium alginate, and Cellosize hydroxyethyl cellulose.

Table 7:19

GLYCOX (ACYLATED ETHOXYLATED POLYOLS)

	Glycox 6	*Glycox 9*	*Glycox 12*	*Glycox 15*
Appearance	Waxy Solid	Soft Solid	Soft Solid	Soft Solid
HLB Value	6	9	12	15
Color, Gardner	6 max.	6 max.	6 max.	6 max.
Odor	Bland	Bland	Bland	Bland
Melting Point, °C.	49–52	38–41	37–40	36–39
Saponification Value	141–145	109–113	80–84	50–54

Emulsion Properties

Stable emulsions can be prepared by adding the Glycox to a mixture of the oil and water phases. The oil-water composition is heated to 50°C and the melted Glycox added with agitation followed by cooling to 25°C with continued agitation.

Typical emulsion properties of a mineral oil-water system are as follows:

Hard Wax E-DM

This is a solvent-soluble, snow-white ester wax. Its characteristics have been determined as follows:

Penetration (100 g, 5 sec, 77°F)	1–2
Solidification Point (rotating thermometer), °C	52/53
Specific Gravity at 20°C	0.94
Acid Value	72–76
Saponification Value	76–80
Hydroxyl Value	0
Flash Point, C.O.C.	218°C
Fire Point, C.O.C.	218°C

Hard Wax E-DM is insoluble in water and ethyl alcohol. It is soluble in hydrocarbon solvents, except that its solubility is lower in the lowest boiling hydrocarbons, for example, petroleum ether.

Hard Wax E-DM is compatible in all proportions with all common waxes, and gum rosin.

Hard Wax E-DM is difficult to emulsify, except when dissolved in solvents.

Hoechst Waxes

I.G. Waxes and Gersthofen Waxes are now marketed under the name of Hoechst Wax. These waxes differ in chemical composition. Some contain glycerol, glycol, or polymerized ethylene esters of acids from montan wax and other sources; some contain metallic soaps and other modifiers.

BJ unbleached and **BJ bleached** are representatives of soft waxes (like beeswax). The color of the unbleached variety is brownish yellow, similar to that of beeswax; the bleached quality has an ivorylike color. They can be easily molded between the fingers at ordinary temperatures. They are soft and should be used for wax compositions which are required to be elastic and pliable. They are very easily saponified and emulsified and, therefore, suited primarily for the production of saponified and mixed-saponified preparations. They are also used in sizes, finishes, cosmetics, and similar products.

CR is similar to Wax Special in several respects, but it is a little more crystalline. It is very easy to saponify and can, therefore, be used for solid or liquid emulsions where the dark color is not a disadvantage.

E is a pale-yellow wax of crystalline structure. As it may be saponified and emulsified much more readily than O and OP, it is suited for the production of saponified and mixed-saponified preparations.

L has practically the same properties as S, but is slightly darker. It is, however, considerably harder than the S brand and, therefore, is used for polishing shoe trees and also for producing flatting pastes, stick polishes, etc.

N new is identical in properties with N, with the only difference that it yields emulsions of the oil-in-water type under normal working conditions. It is used for mixed saponified shoe and floor polishes.

O and **OP** are light-colored, solid waxes of high melting point. They have a conchoidal fracture and great hardness, are free from impurities, and always show similar properties, as indicated by their constants. Their outstanding property is their colloidal nature in the presence of organic solvents, as a result of which they take up and bind large quantities of these solvents. In this respect, they surpass, by far, the natural waxes. The gels and jellies produced from O and OP with solvents, such as turpentine and turpentine substitutes, give a high gloss and are, therefore, suited, with other waxes and waxy substances, for the manufacture of boot polishes, leather dressings, floor polishes, and similar articles.

S is a crystalline wax of very light color and great hardness. Owing to its crystalline structure, it hardens softer waxes or waxlike bodies, such as paraffin wax, ozocerite, etc. It yields polishes of very high luster and free from haziness. Owing to its high acid number it can form soaplike combinations and emulsions with alkalies. It is also excellently suited for dissolving color bases and, therefore, is used in the manufacture both of boot polishes and carbon papers.

Special Wax is chiefly intended for the production of black boot polishes. It makes these polishes resistant to variations of temperature, imparts a good surface, and forms surface rings, like carnauba wax, under suitable treatment.

V is a chemically inert product of high luster. It has an opaque, pure-white color. Its most pronounced property is an exceptional solubility in turpentine and turpentine substitutes. Wax V is not only soluble itself in high proportions in these solvents but also exerts a liquefying effect on other waxes which is maintained at low temperatures. It is suitable for the production of liquid floor polishes, either alone or in combination with the usual wax mixtures.

Solubility

The solubility of Hoechst waxes in organic solvents corresponds to the solubility of the natural waxes; in water they are insoluble, and at room temperature they are soluble in organic solvents only with comparative difficulty. Their solubility increases with rising temperature. Above the melting point they are completely compatible with solvents. The differences in solubility between the various grades of wax can be characterized by the temperature at which the same amount of any given wax is dissolved. Table 7:22 shows the temperatures (in Centigrade) at which about 10% of any given wax are dissolved. A small degree of turbidity caused by components soluble only with difficulty has, for this purpose, been neglected.

Table 7:20

SPECIFICATION DATA:

HOECHST WAXES	Drop Point Mod. ASTM D 566/49 °C		Density 20°C	Acid Value Mod. ASTM D 1386/55T	Sap. Value Mod. ASTM D 1387/55T	Color
Acid Waxes:						
S	78-84	172-183	1.00-1.02	135-155	155-175	light yellowish
L	80-85	176-185	1.00-1.02	120-140	140-160	yellowish
LP	78-84	172-183	1.00-1.02	115-130	135-155	yellowish
Ester Waxes:						
E	76-82	169-180	1.01-1.03	15-20	125-160	light yellowish
X 22	76-81	169-178	1.01-1.03	25-35	130-150	yellowish
F	74-80	165-176	0.97-0.99	6-10	85-105	light yellowish
X 77	75-80	167-176	0.97-0.99	8-12	85-105	yellowish
KP	81-86	178-187	1.01-1.03	20-30	130-150	brown
KPS	77-83	171-181	1.00-1.02	25-35	135-150	yellowish
KSL	80-86	176-187	1.00-1.02	25-35	120-145	yellowish
KSS	82-87	180-189	1.00-1.02	25-35	105-130	yellowish
BJ	78-83	172-183	0.97-0.99	17-25	130-150	light yellowish
RT	75-80	167-176	0.95-0.98	19-24	80-100	light yellowish
Ester Wax Containing an Emulsifier:						
NE	70-76	158-169	0.97-0.99	45-55	110-135	light yellowish
Partly Saponified Ester Waxes:						
OP	100-105	212-221	1.01-1.03	10-15	100-115	beige
X 55	98-103	208-217	1.01-1.03	10-15	90-110	yellowish
Special	97-107	207-225	1.00-1.02	13-18	85-105	brownish-black
O	100-105	212-221	1.01-1.03	10-15	100-115	beige
OM	92-97	198-207	1.00-1.02	20-25	110-125	beige
FL	92-97	198-207	0.99-1.01	34-45	80-100	yellowish
Waxes Not Based on Montan Wax:						
R 21	103-108	217-227	0.93-0.95	<10	<10	light yellowish
V	45-50	113-122	0.93-0.95	0	0	off-white
GL 3	88-93	190-199	1.01-1.04	20-30	0	brownish
PA 190	130-140	266-284	0.95-0.97	0	0	pale-grey

Test Methods

Drop Point — ASTM D 566/49 modified to provide for preparation of sample. Apparatus dimensions metric rather than English.

Acid Value — ASTM D 1386/55T
Saponification Value — ASTM D 1387/55T
} modified to provide for better solubility (xylene-ethanol in a 2:1 ratio is used instead of toluene-ethanol in 1:2 ratio).

*Because of the higher viscosity of the molten material, the waxes can be measured only if the speed of revolution of the thermometer is reduced to about 5-15 secs. per revolution.

Table 7:21 Properties of Hoechst Waxes

Hoechst Wax Grade	*Character, Appearance*	*Solv. Abs. Paste-Forming Properties*	*Solv. Binding Property*	*Glossy Effect*	*Emulsi-fiability*
S	hard, crystalline, very pale	good	moderate	excellent	excellent
L	very hard, crystalline, pale	moderate	moderate	very good	very good
O, OP	very hard, colloidal, pale	excellent	excellent	excellent	medium
OM	hard, colloidal, pale	good	excellent	excellent	medium
Special	very hard, microcrystalline, dark	good	good	excellent	medium
F	hard, microcrystalline, very pale	medium	very good	excellent	medium
E	hard, crystalline, very pale	very good	medium[1]	very good	excellent
CR	very hard, crystalline, dark	medium	medium	very good	excellent
KPS	very hard, microcrystalline, pale	medium	good	excellent	excellent
KPE	similar to KPS, pale	—	—	excellent	excellent
KP	very hard, microcrystalline, pale	medium	good	good	good
BJ unbleached	soft, microcrystalline, pale	medium	very good	good	very good
BJ bleached	as BJ unbleached, but very pale	good	good	good	very good
N new	soft	—	—	very good	very good
Comparison	Carnauba Wax N.C.3 very hard, less pale	very good	good	excellent	good

[1] In combination with waxes of the O group: very good.

Table 7:22

Turbidity Temperatures of 10% Solutions of Hoechst Waxes

Hoechst Wax	*Trichlor-ethylene*	*Xylene*	*Gum Spirits of Turpentine*	*White Spirit*	*n-Butyl Acetate*	*n-Butanol*
S	30°C	40°C	40°C	50°C	55°C	75°C
L	35	40	45	45	50	70
O	30	30	50	55	65	—
OP	30	35	50	60	75	—
OM	25	25	15	55	60	80
Special	20	20	30	45	50	60
F	30	30	35	45	55	65
E	30	35	45	55	55	65
CR	20	25	30	40	40	60
KPS	30	30	35	55	45	65
KP	25	35	40	55	50	65
BJ unbleached	25	25	35	50	40	55
BJ bleached	20	30	40	50	45	60

Table 7:23

COMPATIBILITY OF HOECHST WAXES

Additive	*Wax S*	*Wax OP*	*Wax E*
Low Viscosity Ethyl Cellulose	+(120°C.)	+(120°C.)	+(120°C.)
Alkyd Resin	+	+	+
Chlorinated Paraffin 70 Solid	(+)	(+)	(+)
Gum Dammar	+	+	+
Rosin	+	+	+
Linseed Oil Stand Oil	+	+	+
Maleic Resin	+	+	+
Manila Copal	(–)	–	(–)
Methylcyclohexanone methyl adipate	+	+[1]	+
Pure Ozocerite or Amorphous, Microcrystall. Wax min. 68°C. (155°F) M.P.	+	+	+
Dibutyl Phthalate	+	+[1]	+
Polyethylene, M.W. Approx. 7000[2]	(+)	+	×
Low Viscosity Polymethacrylate	×	×	+
High Viscosity Polyvinylisobutyl Ether	+	×	+
Castor Oil	+	+	+
Shellac	(–)	(–)	(–)
Fully Refined Paraffin Wax 50/52°C	+	+	+
Tricresyl phosphate	+	+[1]	+

[1] Sweated out on cooling.

[2] Wax F is readily soluble in polyethylene.

Explanation of Signs

+ Completely dissolved in the melt; (+) Solution remains turbid; × Only partially dissolved and swelled; (—)Dissolved to only a small extent; — Insoluble.)

Compatibility

The compatibility of Hoechst Waxes with certain materials was determined as follows:

The melted wax was heated to 120°C, and its dissolving power determined. Table 7:22 lists some typical results. Wax S was employed as a wax acid, Wax E as an ester wax, and Wax OP as an ester containing wax soap.

The compatibilities refer to additions of 10–50% of the substances mentioned.

Table 7:24

SOLUBILITIES

(Temperatures in °C, at which 10% of the wax will dissolve in the solvent)

Hoechst Waxes	*Trichlor-ethylene*	*Xylene*	*Balsam turpentine oil*	*White Spirit*	*n-Butyl acetate*	*n-Butanol*
Acid waxes:						
S	35	40	40	50	50	65
L	35	40	45	50	50	70
LP	40	40	45	50	55	70
Ester waxes:						
E	30	35	40	55	50	65
F	30	30	35	45	55	65
KP	25	35	40	50	50	65
KPS	30	30	35	55	45	65
KSL	30	35	40	50	55	70
KSS	40	35	40	50	60	75
BJ	25	25	35	50	40	55
RT	40	40	35	30	60	75
Partially saponified ester waxes:						
OP	30	35	50	60	75	—
Special	20	20	30	45	50	60
O	30	30	50	55	65	—
OM	25	25	45	55	60	80
FL	35	30	40	45	55	—
Waxes based on polyethylene						
PA 520	60	65	65	70	*)	*)
PA 130	75	80	80	85	105	*)
PA 190	*)	85	85	90	*)	*)
PAD 521	55	60	60	65	80	*)
PAD 522	50	55	55	55	75	100
Waxes based on other materials						
R 21	60	65	65	70	90	105
V	0	5	5	10	30	100
C	75	85	90	100	100	80
GL 3	20	15	20	*)	*)	*)

*)=Not completely dissolved at the boiling point of the solvent.

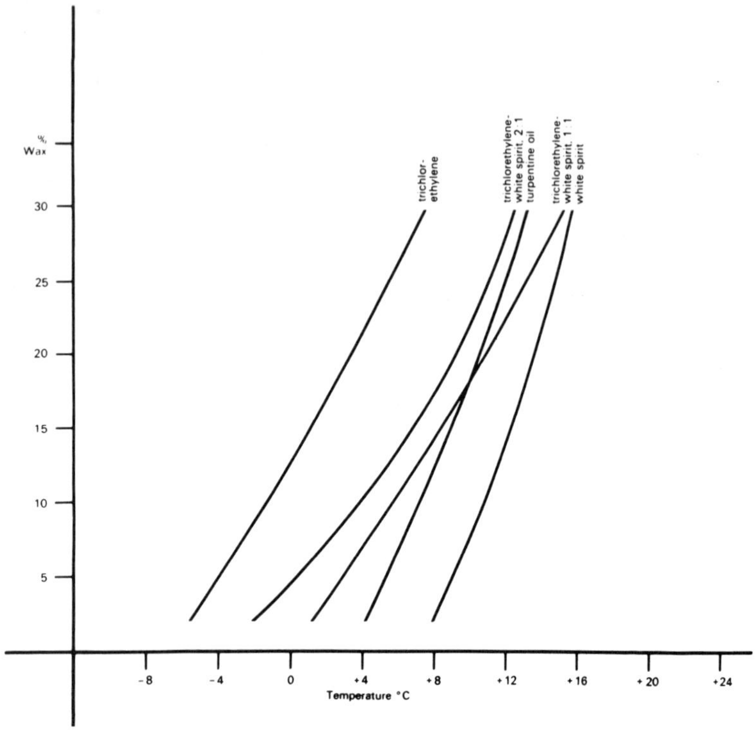

FIG. 7:6

CLOUD POINTS OF HOECHST WAX V SOLUTIONS OF DIFFERENT SOLIDS CONTENT IN VARIOUS SOLVENTS

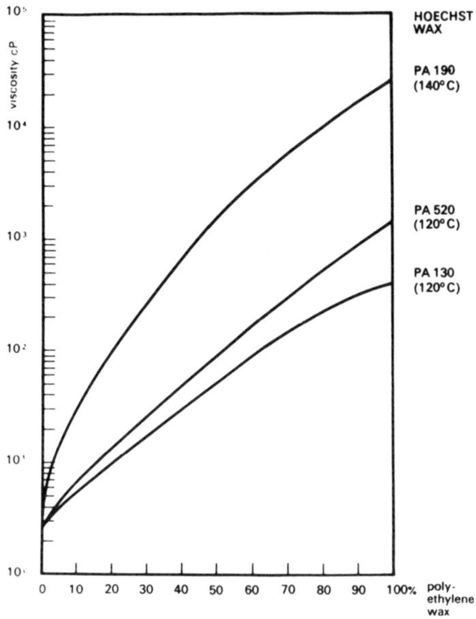

FIG. 7:7

MELT VISCOSITY OF MIXTURES OF HOECHST WAXES AND 52/54 SLAB PARAFFIN WAX

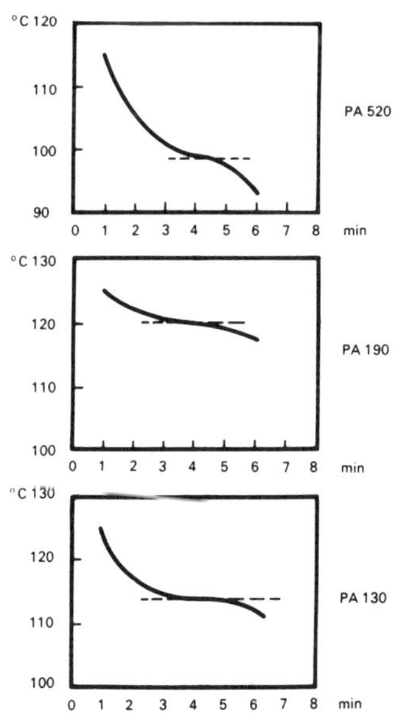

FIG. 7:8

COOLING CURVES WITH CRYSTALLIZATION REGRESSION POINTS OF HOECHST WAX

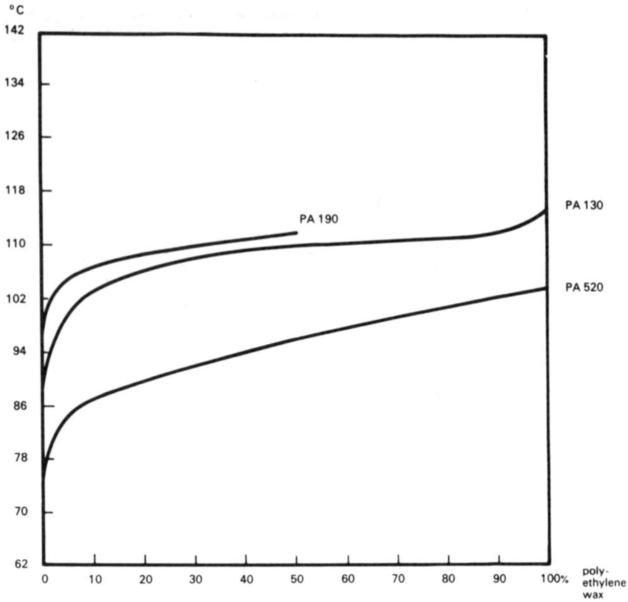

FIG. 7:9

CLOUD POINTS OF MIXTURES OF HOECHST WAX IN 52/54 SLAB PARAFFIN WAX

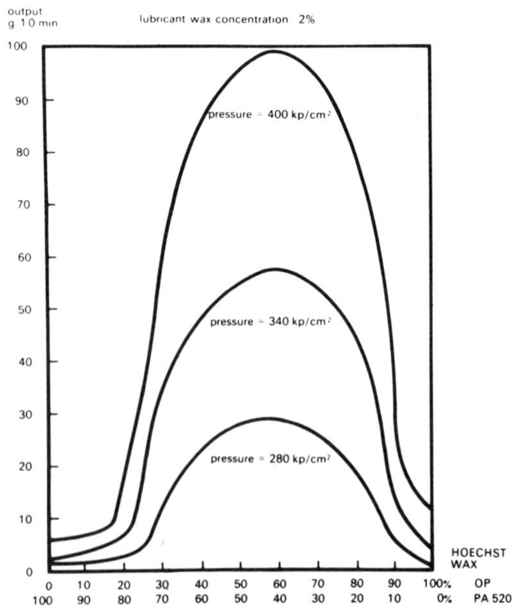

FIG. 7:10

LUBRICANT ACTION OF HOECHST WAX OP/PA 520 COMBINATIONS IN PVC AT THREE DIFFERENT PRESSURES

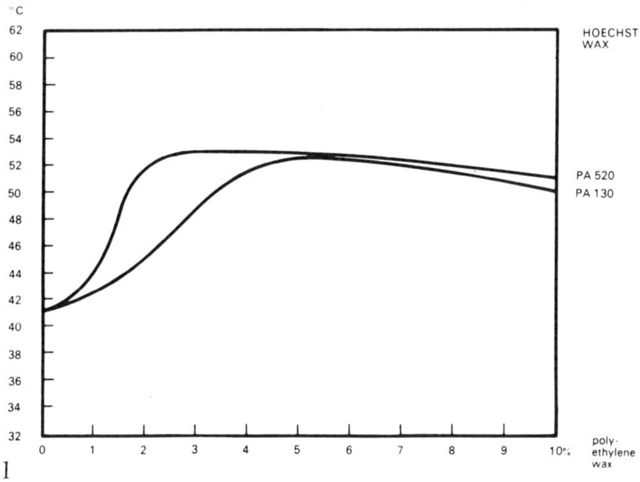

FIG. 7:11

BLOCKING POINTS OF MIXTURES OF HOECHST WAX AND 52/54 SLAB PARAFFIN WAX, MEASURED ON GREASEPROOF PAPER

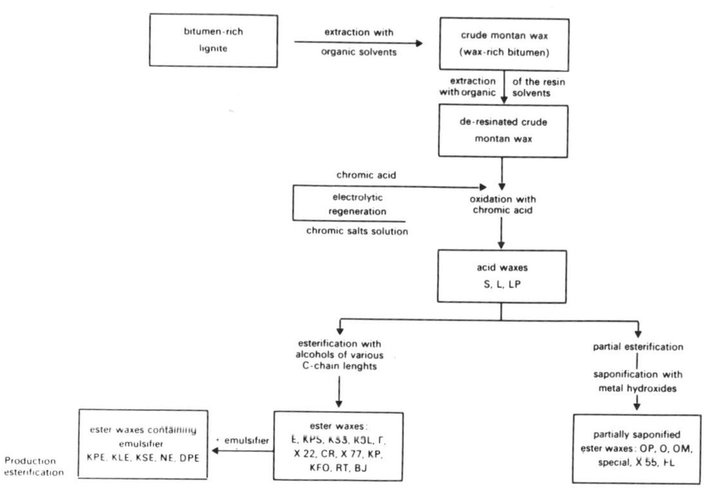

FIG. 7:12

SCHEMATIC REPRESENTATION OF THE MANUFACTURE (HOECHST) OF WAXES BASED ON MONTAN WAX

Table 7:25

LACER BRAND WAX "MAX"

Acid Value	26.6
Saponification Value	178.36
Drop Melting Point Ubbelohde	48°C
Penetration 100 g 5 sec 20°C	13
Congealing Point (rotating thermometer)	46°C

This is a modified Ester Wax in Powdered Form.

Solubility at 20°C: Soluble in all common organic solvents, 80% aqueous ethanol, and vegetable and mineral oil. Less than 4% in 70% alcohol.

Typical Chemical Properties

Iodine Value	72
Saponification Value	374
% Monoglyceride	0 to 2
Reichert-Meissl Value	147
Acid Value	Less than 4

PARACIN 9 Waxes

These waxes are tan, wax-like solids, based on hydroxystearates. **Paracin 9** is Propylene Glycol Monohydroxystearate.

Physical and Chemical Properties

Molecular Weight	360 (approx.)
Color, Gardner (melted)	2+
Acid Value	8
Sp. Gr. @ 25°C	0.983
Sapon. Value	158
Iodine Value	4
Hydroxyl Value	286
Volatile Matter, %	0.3
Melting Point, °C	55

Paracin 13 is a mixture of approximately 50% Glyceryl Monohydroxystearate, 40% Glyceryl Dihydroxystearate, and 10% Glyceryl Trihydroxystearate.

Physical and Chemical Properties

Color, Gardner (melted)	6
Acid Value	5.5
Viscosity, G.H. @ 25°C	Solid
Sp. Gr. @ 25°C	1.03
Sapon. Value	162.5
Iodine Value	4.5
Hydroxyl Value	320
Volatile Matter %	0.3
Melting Point, °C	65–67

Paracin 15 consists of Ethylene Glycol Monohydroxystearate with 15% Ethylene Glycol Dihydroxystearate.

Physical and Chemical Properties

Color, Gardner (melted)	3
Acid Value	5.0
Sp. Gr. @ 25°C	1.01
Sapon. Value	167.0
Iodine Value	3.0
Hydroxyl Value	263.0
Volatile Matter, %	0.4
Melting Point, °C	61–63

Paracin 32 is a saturated dihyroxy triglyceride.

Physical and Chemical Properties

Color, Gardner	2+
Acid Value	4
Sp. Gr. @ 25°C	0.954
Sapon. Value	180
Iodine Value	5
Hydroxyl Value	105
Volatile Matter, %	0.1
Melting Point, °C	64

Pentawaxes

Properties

Pentawax No.	*Softening Point °C*	*Flow Point °C*	*Sward Hardness*	*Compatibility*	*Soluble in*	*Insoluble in*	*Suggested Use*
145	60	81	18	X, O	O, W, E,	A	Emulsifier
177	70	73	22	X, O	O, H	W, A, E	Polishes
217	63	65	55	X, O	O, H	W, A, E	Carnauba substitute
218	54	63	10	X, O	O, H[1]	W	[1]Beeswax
227	38	80	4	X, O	O, H, E	W, A	Polishes
232	59	65	20	X, O	O, H, E	W, A	Polishes
235	53	54	28	X, O	O, H	W, A, E	Japan wax substitute

[1] Pentawax 218 is soluble hot in all classes of organic solvents. It has an odor similar to that of beeswax. It does not shrink on solidifying.

Soluble or *insoluble* refers to the following substances at room temperature:
O = oil, W = water, A = alcohols, E = esters, H = hydrocarbons
Compatibility refers to the following, when melted together:
X = other waxes, including montan, beeswax, carnauba, opalwax, ceresin, paraffin
O = vegetable and mineral oils

Chapter 8

HYDROGENATED OILS

Complete hydrogenation of cottonseed, coconut, peanut, soyabean, castor, fish, whale, or other vegetable, fish, or animal oils yields waxlike substances. The highest-melting commercial wax is produced from castor oil

Table 8:1

PROPERTIES OF HYDROGENATED OILS

I. PARTIALLY HYDROGENATED

Oil	*M.P.* °C	*Congelation Point* °C	*Butyrorefractometer Reading (40°C)*	*Acid No.*	*Saponification No.*	*Iodine No.*
Coconut	44.5	27.7	35.9	0.4	254.1	1.0
Sesame (Techn.)	62.1	45.3	38.4	4.7	188.9	25.4
Whale	45.1	33.9	49.1	1.2	192.3	45.2
Arachis	51.2	36.5	50.1	1.0	188.7	47.4
Sesame	47.8	33.4	51.5	0.5	190.6	54.8
Cottonseed	38.5	25.4	53.8	0.6	195.7	69.7

II. COMPLETELY HYDROGENATED

Hydrogenated Oil or Fat	*M.P.* °C	*Iodine No.*	*Saponification No.*	*Fatty Acid, M.P.,* °C
Tallow	62.0	0.1	197.7	64.0
Lard	64.0	1.0	196.8	62.0
Cacao Butter	63.5–64	0.0	193.9	65.5
Arachis	64–64.5	0.0	191.6	67.0
Cod Liver	65.0	1.2	186.2	59.0
Linseed	68.0	0.2	189.6	70.5
Sesame	68.5	0.7	190.6	69.5
Olive	70.0	0.2	190.9	71.0
Poppy	70.5	0.3	191.3	71.0
Almond	72.0	0.0	191.8	71.0

PROPERTIES OF SELECTED COMMERCIAL PRODUCTS
STRAIGHT CHAIN ALPHA OLEFINS

Chemical	*Predominant Chain Length*	*Calculated Molecular Weight*	*Approximate Composition %* C_8	C_{10}	C_{12}	C_{14}	C_{16}	C_{18}	C_{20}	C_{22}	*Unidentified Odd Chain*	*Mols α-Olefin per kg Min*	*Iodine Value*	*Bromine Number*	*Cloud Point* °C	°F	*Average Flash Point* °F *(Cleveland Open Cup)*	*Average Specific Gravity 25/25°C*	*Average Viscosity cps/25°C*
High Purity Alpha Olefins																			
Adecene A-12	C_{12}	169–172		2	90	8						5.27	145 (av)	93 (av)	−10	14	195	.757	
Adecene A-18	C_{14}	196–201		T	2	90	8					4.54	129 (av)	81 (av)	−7	19	240	.765	
Adecene A-52	C_{16}	224–230				4	90	6				3.94	108 (av)	68 (av)	3	37	270	.779	
Adecene A-62	C_{18}	247–263					4	90	6			3.63	102 (av)	64 (av)	17	62	310	.783	
Mixed Alpha Olefins																			
Adecene A-11	C_{12}	180–191	8	7	48	18	9	10				4.70	130 (av)	82 (av)	−6	22	185	.757	
Adecene A-54	C_{16}	232–240				5	60	30	5			3.75	110 (av)	69 (av)	4	40	295	.777	
Adecene A-51	C_{16}	225–235			7	16	40	30	7			3.80	100–115	65–70	2–10	35–50	265	.776	
Adecene A-71	C_{20}	280–309						15	50	35		3.09	88 (av)	55 (av)	28	83	330	.785	
Isomerized Olefins																			
Adecene 10-51	C_{16}	232–240			7	16	40	30	7				110–115	68–72	−18	0	270		
Adecene 10-71	C_{20}	280–309						15	50	35			90–96	56–60	0	32	335		

ADM ALIPHATIC STRAIGHT CHAIN HYDROCARBONS

Chemical	*Predominant Chain Length*	*Calculated Molecular Weight*	C_8	C_{10}	C_{12}	C_{14}	C_{16}	C_{18}	C_{20}	C_{22}	*Unidentified Odd Chain*	*Mols α-Olefin per kg Min*	*Iodine Value*	*Bromine Number*	°C	°F	*Flash Point* °F	*Specific Gravity*	*Viscosity cps/25°C*
Adakane 12	C_{12}	172		1	80	15	2	1			1				−13	8	205		1.7
Adakane 51	C_{16}	200			6	14	42	30	5		3				21	70	310		5.4
Adakane 71	C_{20}-C_{22}	287					3	15	42	40					32	90	335		

Adakane hydrocarbons are available in drum lots from pilot plant production through the ADM Development Department.
Compositions shown are based on gas chromatographic data. They are approximate and may vary ±10% of the values given depending upon the raw materials available.

Spermoils
(Hydrogenated Sperm Oils)

Chemical	FFA Max as Oleic %	Acid Value Max	I dine Value	Sap. Value	% Unsap.	Melting Point Open Tube °C	°F	Cloud Point °F Max	Pour Point °F Max	Color Gardner 1953 Max	Average Flash Point °F	Average Fire Point °F	Average Viscosity SSU/100°F	Specific Gravity 60°F	Approximate Composition % C_{10}	C_{12}	C_{14}	C_{16}	C_{18}	C_{20}	C_{22}	Unsaturated Content %
45° N.W. Spermoil	2	4	80–92	132–138	32–36			45	43	7	500–510	560–590	100–110	0.882–0.884	T	3	9 *T*	21 *14*	20 *20*	7 *2*	3	75 *60*
38° N.W. Spermoil	2	4	80–92	132–138	32–36			38	36	7	500–510	560–590	100–110	0.882–0.884	T	3	9 *T*	20 *12*	22 *20*	8 *2*	3	76 *60*
45° B.W. Spermoil	0.25	0.5	80–92	132–138	32–36			45	43	5	500–510	560–590	100–115	0.882–0.884	T	3	9 *T*	21 *14*	20 *20*	7 *2*	3	75 *60*
38° B.W. Spermoil	0.25	0.5	80–92	132–138	32–36			38	36	5	500–510	560–590	100–115	0.882–0.884	T	3	9 *T*	20 *12*	22 *20*	8 *2*	3	76 *60*
Spermafol 45	2	4	20–30	130–140	33–40	44–46	111–115			40Y/4R[1]	490	545		0.825	T	3	9 *T*	21 *14*	19 *21*	7 *2*	3	21 *37*
Spermafol 52	2	4	7 max	130–140	35–40	47–51	117–124			30Y/3R[1]	490	545		0.823	T	6	12 *2*	20 *20*	15 *15*	9 *T*		7 *7*
Cosmol 1000	1	2	6 max	125–135	41–46	46–50	115–122			25Y/2.5R[1]	490	550		0.842[2]	T	6	16 *2*	20 *29*	9 *13*	5		6 *6*
Liquid Cosmol	1	2	73 min	113–120	48–52			62	60	50Y/5R[1]	468	540	83	0.868	T	2	8 *2*	17 *28*	17 *20*	5	T	79 *60*
Polysperm 300	8.5	17	56 max	165–180	30–35			49 av	45 av	6		535	280–300	0.920–0.930								
ADM Fatty Acid G	45–60	90–120	78–90	130–140	30–35			45–60	50 av	12–15	385	425		0.875[3]	T	3	9 *T*	21 *14*	20 *20*	7 *2*	3	75 *60*

[1] 5¼″ Lovibond.
[2] 60/25°C.
[3] 25/25°C.

The italic numbers represent the fatty alcohol composition. Regular numbers represent fatty acids.

Characteristics of Hydrogenated Fatty Acids

Chemical	Predominant Chain Length	Titer °C	Iodine Value	Acid Value	Sap. Value	Color Lovibond 5¼" Max	Calculated Molecular Weight	Acetyl Value Min	Hydroxyl Value Min	Specific Gravity 100/25°C	C_8 & C_{10}	C_{12}	C_{14}	C_{16}	C_{18}	C_{20}	C_{22}	12-Hydroxy C_{18}	Unsaturated Content %
											Approximate Composition %								
Hydrofol Acid 630	40% C_{12}	31–37	1 max	245–255	246–256	10Y/1R[1]	220–228			0.838	10	40	17	13	20				
Hydrofol Acid 150	80% C_{18}	62–64	3 max	195–201	196–202	15Y/1.5R	279–286			0.839			T	20	80				
Distal 42	65% C_{18}	40–43	40–60	200–205	201–206	40Y/4R	273–280			0.892			5	30	65				50
Hydrofol Acid 405	65% C_{18}	56–60	6 max	200–205	201–206	25Y/2.5R	273–280			0.835			5	30	65				
Hydrofol Acid 410	65% C_{18}	58–61	1 max	201–206	202–207	10Y/1R	272–278			0.835			5	30	65				
Hydrofol Acid 420	53% C_{18}	54–56	1 max	204–209	205–210	10Y/1R	268–274			0.834			7	40	53				
Hydrofol Acid 46	38% C_{16}	45–48	20–30	198–203	198–204	30Y/3R	276–284			0.839			10	38	25	17	10		25
Hydrofol Acid 51	38% C_{16}	51–53	6 max	197–202	198–203	25Y/2.5R	277–284			0.832			10	38	25	17	10		
Snodotte Acid	30% C_{16}	54 min	3 max	193–198	194–199	10Y/1R	282–290			0.832			10	30	20	25	15		
Hydrofol Acid AB	55% C_{20}–C_{22}	60–63	5 max	178–185	179–186	25/Y2.5R	302–314			0.828			2	13	30	30	25		
Erucic Acid Hydrofol Acid 565	80% C_{22}	26–30	78–85	163–170	164–171	30Y/3.5R	329–343			0.837					8	12	80		98
Behenic Acid Hydrofol Acid 560	80% C_{22}	69–73	4 max	168–174	169–175	25Y/2.5R	322–333			0.828					8	12	80		
12-Hydroxystearic Acid Hydrofol Acid 200	88% 12-Hydroxy C_{18}	70–74	4 max	172–182	182–187	Light	300–309	133	147	0.880				2	10			88	

[1] Gardner 5 max after 1 hour @ 204°C.

Compositions shown are based on gas chromatographic data. They are approximate and may vary ±10% of the values given depending upon the raw materials available.

Characteristics of Hydrofol Glycerides

Chemical	Predominant Chain Length	Titer °C	FFA Max as Oleic %	Acid Value Max	Iodine Value	Sap. Value	Melting Point Open Tube °C	Melting Point Open Tube °F	Color Lovibond 5¼″ Max	Acetyl Value Min	Hydroxyl Value Min	Av. Flash Point °F	Av. Fire Point °F	Sp. Gravity 100/25°C	Approximate Composition C_{14}	C_{16}	C_{18}	C_{20}	C_{22}	12-hydroxyl C_{18}	Unsaturated Content %
Hydrofol Glyceride 42	34% C_{16}	41–44	3	6	32–45	188–195	44–47	111–117	40Y/4R					0.859	9	34	26	16	15		43
Hydrofol Glyceride 45	34% C_{16}	44–47	3	6	22–35	188–195	47–50	112–122	30Y/3R					0.859	9	34	26	16	15		28
Hydrofol Glyceride 50-51	35% C_{16}	50–53	3	6	6 max	188–195	53–57	127–135	25Y/2.5R					0.856	8	35	27	17	13		
Hydrofol Glyceride 150	92% C_{18}	63–66	2	4	3 max	188–193	65–68	149–154	25Y/2.5R					0.858		8	92				
Hydrofol Glyceride T-57-L	75% C_{18}	59–62	0.25	0.5	1 max	193–196	61–65	142–149	5Y/0.5R					0.847	1	24	75				
Hydrofol Glyceride T-57-N	70% C_{18}	57–61	0.5	1	1 max	193–198	59–64	138–147	10Y/1R					0.846	2	28	70				
Hydrofol Glyceride T-57	70% C_{18}	57–60	7	15	7 max	190–197	58–64	136–147	50Y/5R						2	28	70				
Hydrogenated Castor Oil Hydrofol Glyceride 200	88% 12-Hydroxy C_{18}	70–74	2	4	3 max	177–181	86–88	187–190	50Y/5R	138	155			0.899		2	10			88	
Heat Transfer Mediums																					
Hydrofol Tinfat 25		46–50	1	2	11 max	175–185	49–54	120–129				530	600	0.894							
Hydrofol Tinfat 50		49–53	1	2	12 max	187–194	49–55	120–131				585	655	0.854							

Table 8:2

PROPERTIES OF HYDROGENATED COCONUT AND COTTONSEED OILS

Hydrogenated Oil	*M.P. °F*	*Iodine No.*	*Saponification No.*	*Reichert-Meissl No.*	*Polenske No.*	*Refractive[1] Index*
Coconut	107.2	2.2	250	8.5	15.5	1.4486 (40°C)
Cottonseed	140.7	3.9	195	0.5	0.7	1.4468 (60°C)

Hydrogenerated vegetable oils have dieletric constants of 20–25[2]

[1] Langwill, *Mfg. Confectioner* **21**, 15 (1941).
[2] Valter, *J. Tech. Phys.* (U.S.S.R.) **10**, 1970 (1941).

Table 8:3

WHALE OIL AT DIFFERENT STAGES OF HYDROGENATION[1]

Oil Stages	*M.P. °C*	*Congelation Pt. °C*	*Acid No.*	*Saponification No.*	*Iodine No.*	*Mol. Equiv. of Fatty Acids*
Original Oil	Fluid	Fluid	9.50	192.2	144.8	287.7
Artificial Tallow	47.5	38.1	9.88	183.7	56.9	296.4
Artificial Stearin	54.3	47.3	7.80	187.7	11.7	297.0
Hydrogenated Whale Oil	41.9	31.9	5.30	190.9	57.8	282.9

[1] Sandelin, *J. Soc. Chem. Ind.* **33**, 1097 (1914).

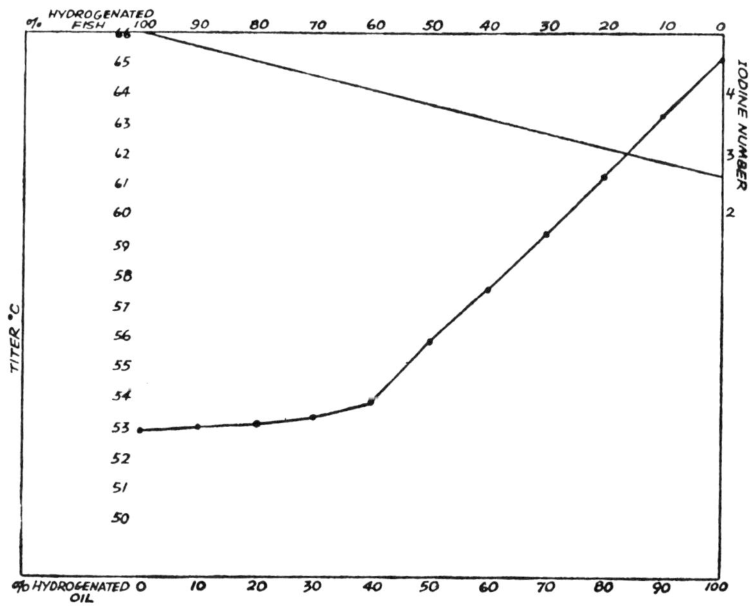

FIG. 8:1. TITER AND IODINE NUMBER OF HYDROGENATED SOYBEAN- AND FISH-OIL BLENDS.

Table 8:4

Properties of Hydrogenated Sardine Oil and Tallow

	45° Incompletely Hydro-generated Sardine Oil		53° Fully Hydro-genated Sardine Oil		Fully Hydro-genated Tallow	
	Glycerides	Acids	Glycerides	Acids	Glycerides	Acids
Titer	—	44–46	—	52–54	—	56–58
Melting Point (open tube)°C.	43	44.5–46.5	54	52.5–54.5	57	56.5–58.5
Iodine Number	27–34	28–35	2–4	2–4	2–6	2–6
Free Fatty Acids as Oleic	2–5%	95–97%	2–5%	95–97%	3–5%	96–98%
Acid Number	4–10	189–193	4–10	189–193	6–10	191–195
Saponification Number	185–189	193–199	185–189	193–199	190–193	199–201
Unsaponifiable	0.8–1.0%	0.8–1.0%	0.8–1.0%	0.8–1.0%	0.6–1.0%	0.6–1.0%
Average Molecular Weight	896	286	896	286	878	280
Specific Gravity 60°/15°	0.873	0.895	0.876	0.854	0.856	0.881
Flash Point °F.	530	405	530	385	515	405
Fire Point °F.	580	440	585	425	595	430
Unsaturated Content	37%	38%	3%	3%	4%	4%
Lovibond* Color	15–35 Y 2.5–5 R	10–15 Y 2–3 R	10–35 Y 2–5 R	10–15 Y 2–3 R	10–35 Y 2–5 R	10–15 Y 2–3 R
Thiocyanogen No.	No data. Should be same as iodine on 53° and hydrogenated tallow, only slightly lower than iodine on 45°.					

* Lovibond readings on glycerides, 5¼ in. column; fatty acids, 1 in. column.

Table 8:5

COMPOSITION OF HYDROGENATED SARDINE OIL AND TALLOW

	45° Incompletely Hydrogenated Sardine Oil %	53° Fully Hydrogenated Sardine Oil %	Fully Hydrogenated Tallow %
Saturated Acids:			
Myristic	7	7	4
Palmitic	26	30	30
Stearic	18	24	62
Arachidic (C_{20})	8	21	—
Behenic (C_{22})	3	14	—
Unsaturated Acids:			
Myristoleic	0.2	trace	trace
Palmitoleic	4	trace	0.5
Oleic	6	0.5	3.5
Eicosenoic (C_{20})	14	1.0	—
Docosenoic (C_{22})	13	2.0	—

Table 8:6

MELTING POINTS OF COMPLETELY HYDROGENATED OILS

Hydrogenated Oil	Melting Point, °C
Coconut	43–45
Palm Kernel	43–45
Herring; Cod	52–55
Whale	54–60
Palm	56–58
Cottonseed	62–63
Olive; Peanut	68–69
Soyabean; Linseed	69–71
Castor	86–90

Castorwax

Castorwax is a synthetic, waxlike compound obtained by the controlled hydrogenation of pure selected castor oil. The principal constituent is the glyceride of 12-hydroxystearic acid. There are also present minor quantities of mixed glycerides of this acid and dihydroxystearic and stearic acids.

Castorwax is a hard, brittle, high melting point product that is practically odorless (faint characteristic fatty or waxy odor) and tasteless. It is furnished in the form of uniform, free-flowing, white flakes. The color of solid Castorwax is cream to pearly by reflected light and bluish-white by transmitted light. When molten it is clear, transparent, very fluid, and colorless to straw-colored.

Solid Castorwax has an amorphous fracture, dull to waxy luster, and white streak. It is uniform in composition and free from extraneous matter.

Castorwax is relatively non-toxic; ingestion tests on mice indicate that no fatal effect will result up to the ability of a person to ingest the material. It can be considered as having a toxicity on the order of linseed or other vegetable oils.

Castorwax is extremely hard in comparison with most other waxes. Its penetration value (Penetrometer Needle No. 14 with 100 grams weight) is 0.2 mm as compared with a value of 0.1 for carnauba wax and 1.7 for beeswax.

Table 8:7

PROPERTIES OF CASTORWAX

I. PHYSICAL PROPERTIES	
Specific Gravity @ 20°C	0.99
Melting Point, °C	86 (187°F)
Ash Content, %	0.005
Dielectric Constant (0.1 KC) at 30°C	12
Flash Point, COC, °F	600
Fire Point, COC, °F	635
II. CHEMICAL PROPERTIES	
Acid Value	2
Iodine Value	4
Saponification Value	180

Compatibility

Castorwax is compatible with all insect waxes, natural vegetable waxes, and Montan wax. It is only moderately compatible with crystalline and microcrystalline paraffins and with ceresin. It mixes with ozocerite to varying degrees depending on the grade of ozocerite.

Castorwax is highly miscible with rosin, abietyl alcohol, hydro-abietyl alcohols, chlorinated aromatics, shellac, ethyl cellulose, cellulose acetate butyrate, and polybutyl and polyisobutyl methacrylates. It is moderately miscible with rosin esters, ester gums, and coumarone indene resins. It is highly compatible with all major natural and synthetic rubbers. Thus, Castorwax can be incorporated with GR-M, GR-A, GR-S and GR-1 elastomers in concentrations up to 60 parts PHR. It has limited compatibility with higher alcohols, such as cetyl alcohol.

Solubility

An outstanding property of Castorwax is its relative insolubility in most organic solvents at moderate temperatures. Castorwax will dissolve in a number of solvents and oils at elevated temperatures but these solutions on cooling form gels or paste-like masses. Castorwax is sparingly soluble in tetrahydrofurane and nitrobenzene. It is soluble up to 15% in chloroform.

Castorwax dissolves, with severe discoloration and decomposition, in concentrated sulphuric acid and is strongly attacked by concentrated nitric acid. Strong alkalies saponify Castorwax releasing glycerol.

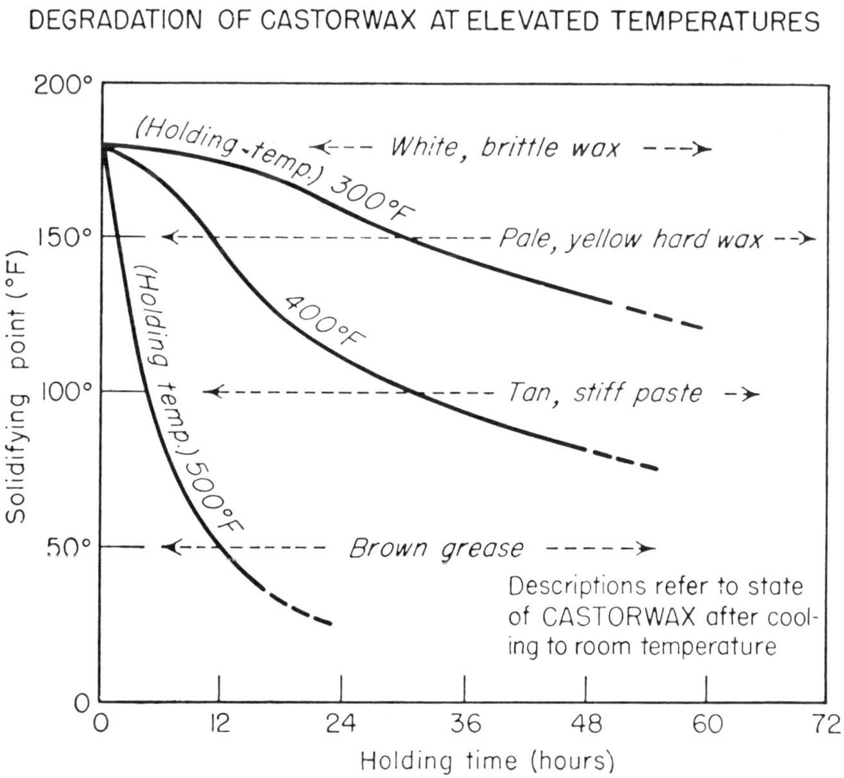

FIG. 8:2. DEGRADATION OF CASTORWAX AT ELEVATED TEMPERATURES.

Table 8:8

SOLUBILITY OF CASTORWAX
(grams per 100 ml of solvent)

Solvent	*Solubility at* 20°C	33°C
Acetic Acid, Glacial	0.4	1.2
Acetone	0.6	1.1
Butanol, Iso	0.3	0.9
Carbon Disulfide	1.3	1.9
Carbon Tetrachloride	1.4	1.5
Cellosolve	0.3	0.4
Cellosolve, Methyl	0.1	0.2
Chloroform	15.0	—
Chloroform 60/Butanol 40	15.0	—
Ethanol, Absolute	0.2	0.4
Ether	0.3	1.2
Ether, Petroleum	0.1	0.2
Methanol	0.2	0.2
Toluene	1.5	1.5
Xylene	1.2	1.2

Table 8:9

PROPERTIES OF CASTORWAXES

	Castorwax F-1	*Castorwax*
Melting Point, °F		
Capillary Method	180–183	186–188
Drop Point Method	185	186
Congeal Point, °F	160	164
Density (gm/cm³)		
32°F	0.975	1.034
75	0.971	1.021
100	0.969	1.015
150	0.964	1.000
200	0.901	0.897
250	0.881	0.884
300	0.862	0.865
Shrinkage, Molten Wax (%)		
Congeal point to 75°F	5.4	10.3
300°F to 75°F	11.2	15.3
Viscosity of Melt (poise)		
200°F	0.31	0.24
250	0.15	0.16
300	0.095	—
Penetration Hardness (0.1 mm)		
100g/77°F/5 sec.	11	2
Flexibility (1)	58	5
Surface	dull, white, translucent	white, translucent
Texture of Shavings	waxy	brittle

Heat Resistance

When Castorwax is subjected to high temperatures for prolonged periods of time it undergoes heat deterioration, so that on cooling, the original brittle wax is found degraded to a stiff paste and eventually, on further heating, to a brown grease. Figure 8:2 indicates the course of this breakdown in terms of solidification temperature *vs.* time of hold at different temperatures. An inspection of these plotted data reveals the controlling influence of temperature and indicates that if hot melt holds of several days or longer periods are contemplated, Castorwax hot melts should be maintained at temperatures considerably less than 300°F.

Table 8:10

Comparative Tests on Castorwaxes

	Flexibility	Grease Resistance	
	Flex Rating[1] 0 : Brittle 100:Very Flexible	TAPPI[2] Test T454 M-44	Grease[3] Transudation Test
CASTORWAX F-1	58	0.5 + hr.	Passes
CASTORWAX (unmodified)	5	0.5 + hr.	Passes
Stearic Acid	60	Fails	Fails
Microcrystalline Wax 190–195	60	Fails	Fails
Paraffin Wax 138–140	100	Fails	Fails

[1] T. C. Patton, "Physical Properties of Common Waxes"; *Soap & Sanitary Chemicals*, December 1957.
[2] TAPPI Turpentine Test for Determining Grease Resistance of Coated Paper.
[3] The Grease Transudation Test Procedure: Coat a one mil thick layer of the test wax on paper. Place a lump of lard containing 1% of an oil soluble dye on the waxed side of the paper and subject to a temperature of 40°C (104°F). No penetration of the dyed grease to the back face of the paper at the end of 18 hours is considered passing; otherwise the wax is rated as failing the test.

Emulsification

Smooth, stable anionic Castorwax emulsions can be prepared using triethanolamine stearate together with a non-ionic emulsifier.

Castorwax can also be emulsified with cationic emulsifying agents to give emulsions that are stable in acid solutions.

Special Products

Complementing the fully hydrogenated or regular Castorwax, there is also available a series of partially hydrogenated castor oils of intermediate melting points known as Castorwax MP70 and MP80. These waxes of lower melting point differ in composition from regular Castorwax by containing lower percentages of glyceryl trihydroxystearate. As the content of this saturated fatty acid glyceride is reduced, there is a corresponding lowering of the melting point and at the same time an increase in the iodine number as can be seen below.

These special MP Castorwaxes are slightly more compatible and soluble than Castorwax itself. Therefore they have utility in those applications where some sacrifice in melting point and hardness can be tolerated to achieve better compatibility and solubility, or where the lower melting point is actually desired.

PROPERTIES OF CASTORWAXES

	MP70	*MP80*
Melting Point, °C	70	80
Acid Number	2	2
Saponification Number	179	179
Iodine Number	48	38
Hydroxyl Number	168	168

Opalwax

Opalwax is produced from the hydrogenation of castor oil.

Properties

Opalwax is insoluble in solvents. None of the common solvents has been found to dissolve as much as 2% at temperatures up to 30°C. Toluene, carbon tetrachloride, and carbon disulfide are the most effective. Opalwax is dissolved, with discoloration, by concentrated sulfuric acid. It is disintegrated and partly dissolved by concentrated nitric acid.

It is only slightly compatible with paraffin wax, but more so with ceresin, ozocerite, montan, and carnauba waxes, and many natural and synthetic resins, such as rosin, ester gum, coumaroneindene, *n*-butyl methacrylate, ethyl cellulose, alkyd, urea-formaldehyde; and drying-oil types.

Molten Opalwax (100°C) is introduced with rapid agitation into about two and a half times its own weight of water, at 85°C, in which a small amount of a suitable emulsifying agent has been thoroughly mixed in advance, it will produce an emulsion that will remain stable when subsequently diluted and yield a continuous film on drying.

Uses

Opalwax is used for impregnating and coating papers, fiberboard, leather, cork, textiles, etc., to make them grease-, oil-, and waterproof for electrical insulation; as a lubricant; also in the manufacture of candles, rubber-coated fabrics, polishes and finishes, carbon paper inks, and cutting oils. The addition of 20% naphthalene tetrachloride, behenone, or stearone to hydrogenated castor oil (Opalwax) makes it approach carnauba wax in hardness, fracture, and solvent-retention.

Chapter 9

KETONES, AMINES, AMIDES

The symmetrical aliphatic ketones are water-white in the pure state.

High-molecular-weight aliphatic ketones are insoluble in water. They are fairly insoluble in most organic solvents at room temperature, but quite soluble above the melting point of the ketone. Because of their high molecular weight, they are inert chemically and do not react readily. They can be changed to reactive derivatives by halogenating.

Their solvent characteristics at elevated temperatures are excellent. The indicated fields of commercial utility are chemical synthesis and intermediates, waxes, polishes, moistureproofing agents and rust-preventing compositions, candle dipping, textile lubricating, and softening agents.

These ketones do not deteriorate in storage and may be held indefinitely. The chief impurities are *n*-hexadecane and palmitic acid.

Solubility

The aliphatic ketones are insoluble in water and have essentially no water absorption. They are only slightly soluble in organic solvents at room temperature, although at elevated temperatures they are soluble to considerable degrees in several solvents. Approximate solubilities in several solvents at various temperatures are summarized for each product. These data were obtained on ketones derived from relatively pure fatty acids. It is to be expected that those derived from commercial acids may show slightly higher solubilities than indicated.

16-Hentriacontanone

The average molecular weight of the commercial product is 454.7. It consists of 98% ketone divided in about the following proportions: 16-hentriacontanone 85 and 18-tritriacontanone 15.

The melting point of 16-hentriacontanone is 82.8°C and the iodine number 0. Commercial samples have a light, lemon-yellow color, the melting point is 76°C and the iodine number 2. The boiling range is 240–260°C at 0.5 mm. It is very heat stable and does not decompose readily.

STEARONE, LAURONE

Stearone and **Laurone** are symmetrical aliphatic ketones derived from their corresponding fatty acids. They are particularly inert and subject to only a few reactions under rigorous conditions. They begin decomposing at about 300°C.

Stearone is a symmetrical high molecular weight aliphatic ketone derived from commercial stearic acid; formula

$$CH_3(CH_2)_{16}\text{—}\overset{\overset{\displaystyle O}{\|}}{C}\text{—}(CH_2)_{16}CH_3$$

Table 9:1 PROPERTIES OF STEARONE

	Specifi-cations	*Typical Analysis*
Melting Point	80°C min.	85
Free Fatty Acid (as Stearic)	2% max.	1%
Iodine Value	9 max.	6
Color (Gard. 1933 Stds. at 90°C)	7 max.	5
Flash Point	265°C	
Fire Point	296°C	
Odor	Bland	Bland

Table 9:2 SOLUBILITY OF STEARONE

Solvent	*Grams of Stearone per 100 g. Solvent*			
	10°C	30°C	50°C	80°C
Cyclohexane	0.1	0.5	9.3	>100
Benzene	0.1	0.6	12.7	>100
Ethyl Acetate	—	—	0.1	>100
Butyl Acetate	—	0.2	0.9	>100
Acetone	—	0.1	0.4	—
Methanol	—	—	—	—
Ethanol, 95%	—	—	0.1	49
n-Butanol	—	—	0.1	>100
Naphtha	—	—	22.5	—
Turpentine	—	—	16.5	—
Cyclohexanone 50, MEK 50	—	—	4.7	—

Table 9:3

MELTING POINTS OF STEARONE PARAFFIN WAX (128–130°F) COMPOSITIONS

% *Stearone*	*Melting Point* (°C)
0.0	53.6
10.0	57.6
20.0	60.5
33.3	65.0
50.0	70.0
75.0	74.5
100.0	78.5

Laurone

Laurone is a symmetrical aliphatic ketone derived from commercial lauric acid; formula

$$CH_3(CH_2)_{10}-\overset{\displaystyle O}{\overset{\|}{C}}-(CH_2)_{10}CH_3$$

Table 9:4

PROPERTIES OF LAURONE

	Tentative Specifications	*Typical Analysis*
Melting Point	65°C min.	69
Free Fatty Acid (as Lauric)	3% max.	2%
Iodine Value	9 max.	6
Color (Gard. 1933 Stds. at 90°C)	7 max.	6
Odor	Bland	Bland

Table 9:5

SOLUBILITY OF LAURONE

Solvent	*Grams of Laurone per 100 g Solvent*			
	10°C	30°C	50°C	80°C
Cyclohexane	1.6	12.7	100	∞
Benzene	1.2	20.3	100	∞
Ethyl Acetate	0.4	2.2	45.4	—
Butyl Acetate	0.6	3.7	48.8	∞
Acetone	0.5	1.1	100	—
Methanol	0.1	0.5	1.7	—
Ethanol, 95%	0.2	0.6	6.0	—
n-Butanol	0.4	1.2	16.2	∞

Compatibility

The fact that these materials are unaffected by high temperatures, acids, alkalies, and other strong chemicals suggests a wide range of applications. They are compatible with such waxes as vegetable, microcrystalline, and paraffins. They are also compatible with triglycerides and fatty acids. In 20% to 30% concentrations, they form solid pastes at room temperature with many solvents:

Solid Solvent Pastes

75% Turpentine
25% Stearone

75% ASTM Naphtha
25% Stearone

80% Xylene
20% Stearone

Toxicity

Stearone is considered to be a non-toxic material. In laboratory tests, experimental animals have been fed Stearone in amounts as high as 10%

of the diet with no harmful effects. No indication of dermal toxicity has been found in tests on these animals. Data on the toxicity of Laurone are not available.

ALIPHATIC AMINES

Octadecyl–Hexadecyl Amine

Average Molecular Weight	267
Percentage of Amine by Titration	<98
Specific Monoamine Mixture:	
Octadecylamine	94
Hexadecylamine	6

PROPERTIES

Color	Water white to light yellow
Initial Boiling Point	340°C
90% Over	348°C
Final Boiling Point	350°C
Nonvolatile Matter	Trace
Melting Point	53°C
Iodine Number	<2
Moisture	None

Octadecylamine

PROPERTIES

Molecular Weight	269.5
Color	Water white
Boiling Point, 760 mm	348.8°C
Boiling Point, 2 mm	166.1°C
Vapor Pressure, 20°C	<0.1 (0.1 mm. at 115°)
Melting Point	55°C
Specific Gravity	About 0.844
Weight per Gallon	About 7 lb

ALIPHATIC AMIDES

High-molecular-weight aliphatic amides are readily hydrolyzed to the corresponding fatty acid by boiling in aqueous alkali or acid solutions. This reaction is quantitative over a period of a few hours. Care must be used in planning chemical reactions, to avoid conditions that will favor hydrolysis of the amide, to the exclusion of the desired reaction product.

These amides are unstable above 170°C in the absence of a catalyst, and at lower temperatures in its presence. This decomposition is simple, it does not change the length of the alkyl-chain and it produces equal molar proportions of the corresponding nitrile, fatty acid, and ammonia.

These amides readily react with organic halogens, to give the corresponding substituted amide, accompanied by the liberation of hydrogen halide.

One or both hydrogens, attached to the nitrogen atom in the amide group, can be substituted when heated with active metals, some organic acids, and metal oxides. This reaction can be widely used for the application of amides as chemical intermediates and in the synthesis of other products.

Properties

Pure aliphatic amides are water-white. Commercial samples have a very light lemon-yellow color. The melting point is 99°C and the iodine number 3. The specific gravity is about 0.8. The boiling point is 241–251°C at 12 mm; there is decomposition at this temperature.

High-molecular-weight aliphatic amides are insoluble in water and almost insoluble in the usual organic solvents, at room temperature. They are quite soluble in most organic solvents at elevated temperatures and are themselves excellent solvents for a wide variety of high-melting resinous substances at temperatures above their melting points. They are fairly reactive chemically. Thermal decomposition, which begins at temperatures above 175°C, converts half of the amide quantitatively into the corresponding nitrile, whereas the other half is converted into the corresponding free fatty acid. Strong acids and alkalies hydrolyze the amides to fatty acids. They can be condensed with formaldehyde to form substituted amides.

Amides do not deteriorate in storage and may be held indefinitely. The chief impurity is free stearic acid.

Uses

The aliphatic amides are used as solvents for organic dyes in manufacture of carbon paper. They are excellent neutral, nonionic textile-softening and sizing agents when applied from an aqueous dispersion containing ordinary emulsifying agents.

Because of their high melting points, amides are partially replacing carnauba, montan, and other waxes in commercial waxing and polishing. They are also used to some extent in waterproofing textiles. Because of their high melting points, they should be excellent for candle dipping and for any use where a soft, plastic, high-melting wax is desirable.

Octadecanamide–Hexadecanamide

The average molecular weight of this technically pure mixed product is 275. It consists of 95% amide in about the following proportions: octadecanamide 70, hexadecanamide 30.

Solubility of Amides[1]

	METHANOL			ETHANOL		
Amide	*10°C*	*30°C*	*64.7°C*	*10°C*	*30°C*	*78.5°C*
A	22.1	53.0	263	13.4	32.8	303
B	6.8	15.2	170	4.3	12.0	350
C	4.2	12.4	206	2.9	11.4	370
D	1.0	2.7	56	1.2	3.3	180
E	0.7	1.2	51	0.4	1.5	216
F	0.4	0.7	23	0.2	0.8	124

[1] A—Octanamide. B—Decanamide. C—Dodecanamide. D—Tetradecanamide. E—Hexadecanamide F—Octadecanamide.

All figures are gravimetric quantities per 100 g of solvent.

	ISOPROPANOL			n-BUTANOL		
Amide	*10°C*	*30°C*	*82.3°C*	*10°C*	*30°C*	*100°C*
A	15.7	26.3	288	17.1	25.8	1300
B	6.0	10.9	380	3.5	9.1	—
C	3.7	11.4	357	2.2	10.1	3550
D	1.2	3.2	208	0.8	3.8	1150
E	0.4	1.8	229	0.3	1.9	1050
F	0.2	1.0	172	0.1	1.1	590

	ETHYL ACETATE			BUTYL ACETATE		
Amide	*10°C*	*30°C*	*80.1°C*	*10°C*	*30°C*	*81.4°C*
A	2.2	5.7	115	2.6	6.0	740
B	0.5	2.2	123	0.9	2.9	—
C	0.5	2.5	133	0.4	3.0	2550
D	0.3	1.0	55	0.3	1.2	700
E	0.2	0.6	65	0.2	0.8	670
F	0.1	0.3	41	0.1	0.5	465

	TETRACHLOROMETHANE			NITROETHANE		
Amide	*10°C*	*30°C*	*76°C*	*10°C*	*30°C*	*100°C*
A	—	0.4	42.5	1.0	2.1	1600
B	—	0.2	59.0	0.8	1.1	—
C	—	0.4	72.0	0.8	1.1	5100
D	—	<0.1	27.6	0.3	0.5	1500
E	—	0.1	46.5	0.2	0.3	1075
F	—	<0.1	28.4	0.1	0.2	61

	BENZENE			CYCLOHEXANE		
Amide	*10°C*	*30°C*	*80.1°C*	*10°C*	*30°C*	*81.4°C*
A	0.5	0.6	153	—	0.1	53
B	0.6	0.8	195	—	0.4	150
C	0.6	1.0	238	—	0.5	163
D	0.3	0.4	95	—	0.2	61
E	0.3	0.4	145	—	0.3	122
F	0.3	0.4	89	—	0.3	88

	ACETONE			2-BUTANONE		
Amide	*10°C*	*30°C*	*56.5°C*	*10°C*	*30°C*	*79.6°C*
A	3.4	7.8	48.4	3.4	8.3	163
B	2.0	3.8	23.9	2.2	3.8	189
C	1.3	3.4	30.5	1.2	3.8	194
D	0.4	1.0	6.4	0.6	1.2	100
E	0.3	0.8	6.9	0.4	1.0	118
F	0.2	0.5	3.7	0.3	0.7	79

Octadecane Nitrile

PROPERTIES

Molecular Weight 265.4	
Color	water white
Boiling Point, 760 mm	357.4°C
Boiling Point, 2 mm	174.6°C
Refractive Index (Solid at 20°C)	1.4405^{40}
Vapor Pressure, 20°C	<0.1 mm (0.1 mm at 121°)
Melting Point	41°C
Specific Gravity, 41°C	0.8147
Weight Per Gallon	about 7 lb

Octadecane–Hexadecane Nitrile

Average Molecular Weight	263
Nitrile, %	99
Specific Nitrile Mixture:	
Octadecane Nitrile	94
Hexadecane Nitrile	6

PROPERTIES

Color	white to light yellow
Distillation:	
Initial Boiling Point	354°C
95% Over	355°C
Final Boiling Point	357°C
Nonvolatile Matter	Trace
Melting Point	39°C
Iodine Number	<3
Moisture	none

PROPERTIES OF SOME COMMERCIAL PRODUCTS

Adogens

Properties

Type of Adogen	*Amide Content % (min.)*	*FFA % (max.)*	*Iodine Value*	*M.P. (closed tube)°C*	*Moisture (max.)*	*Color Gardner53 (approx.)*
72 Oleyl Amide	90	4	88–95	68–72	0.5	7
57 Erucyl Amide	90	5	80–85	75–80	0.5	10
42 Stearyl Amide	90	4	0.5 max.	100–104	0.5	8
40 Hydrogenated Tallow Amide	90	5	3 max.	98–102	0.5	8
60 Coco Amide	90	4	8–11	85–89	0.5	6
1 Arachidyl Behenyl Amide	90	4	3 max.	97–102	0.5	10

Type of Adogen	*Physical Form*	*Predominant Chain Length*	*Molecular Weight*
72 Oleyl Amide	solid	90% C_{18} (oleyl)	278–283
57 Erucyl Amide	solid	80% C_{22} (erucyl)	328–342
42 Stearyl Amide	beads	94% C_{18}	282–286
40 Hydrogenated Tallow Amide	beads	65% C_{18}	271–277
60 Coco Amide	solid	48% C_{12}	204–210
1 Arachidyl-Behenyl Amide	beads	55% C_{20}–C_{22}	300–314

Approximate Composition (%)

Type of Adogen	C_8	C_{10}	C_{12}	C_{14}	C_{16}	C_{18} *Stearyl*	C_{18} *Oleyl*	C_{20}	C_{22}
72 Oleyl Amide							90		
57 Erucyl Amide							8	12[1]	80[1]
42 Stearyl Amide					8	92			
40 Hydrogenated Tallow Amide				5	30	65			
60 Coco Amide	8	7	48	18	9	5	5		
1 Arachidyl-Behenyl Amide				2	13	30		30	25

[1] Approximately 98% Unsaturated.
Compositions shown are based on gas chromatography. They are approximate and may vary ±10% of the values given depending upon the raw materials available.

Solubility

Fatty amides are insoluble in water, and generally only slighly soluble in organic solvents at room temperature. However, they increase in solubility with a rise in temperature at or near the melting point of the amide. Melted amides are compatible with ketones, esters, alcohols, turpentines, natural fats, petroleum oils, and fatty acids.

Representative ADM fatty amides have the following solubilities in common solvents:

Solvent	*Type of Amide* AB	Stearyl	Hydrogenated Tallow	Coco	Oleyl	Erucyl
Isopropanol	I	I	I	I	S	S
Ethanol	I	I	I	I	Sl S	Sl S
Acetone	I	I	I	I	Sl S	Sl S
Ethyl acetate	I	I	I	I	Sl S	Sl S
Cyclohexane	I	I	I	I	Sl S	Sl S
Acetonitriles	I	I	I	I	Sl S	Sl S

S—Soluble, Sl S—Slightly Soluble, I—Insoluble.

ARMOUR WAXES

Armids

COMPOSITION AND CONSTANTS

Compound	*Carbon Chain*	*Armid 18*	*Armid HT*	*Armid C*	*Armid O*
Hexanamide	6				
Octanamide	8			8	
Decanamide	10			7	
Dodecanamide	12			49	
Tetradecanamide	14			17	
Hexadecanamide	16	3	22	9	
Octadecanamide	18	93	75	2	6
9-Octadecanamide	18	4	3	6	91
9.12- Octadecadienamide	18			2	3
Minimum Melting Point °C		99	98	85	68
Maximum Unreacted Fatty Acid %		5	5	4	3.5
Maximum Iodine Value		2	5	10	90
Min. Color, Gardner		10	10	10	7
Physical Form		**	*** *	*	** *

* Flake. **Solid. *** Powder.

Armid HT

Armid HT is a fairly hard, wax-like material. It is light tan in color and has a compact crystal structure.

CHEMICAL AND PHYSICAL PROPERTIES

	Typical Analysis
Free Fatty Acid	5.0%
Iodine Value	5.0
Color—Gardner	10
Solidification Point (ASTM D-87-42) °C	98 to 103
Odor	Bland
Specific Gravity (approx.)	0.960
Moisture	0.5%
Flash Point °C (approx.)	225
Fire Point °C (approx.)	250

APPROXIMATE COMPOSITION

Hexadecanamide (Palmitamide)	22%
Octadecanamide (Stearamide)	75%
9-Octadecenamide (Oleamide)	3%

SOLUBILITY OF ARMID HT IN 100 ML SOLVENT

Temperature °C	*10*	*30*	*65*	*76*	*80*	*100*
Methanol	0.475	0.825	30			
Ethanol	0.25	0.975			147	
Isopropanol	0.25	1.20			184	
n-Butanol	0.15	1.30				706
Ethyl Acetate	0.125	0.375			47	
Butyl Acetate	0.125	0.575			517	
Tetrachloromethane	—	0.100		32		
Nitroethane	1.25	0.225				314
Benzene	0.300	0.400			103	
Cyclohexane	—	0.300			97	
Acetone	0.225	0.575	4.5			
2-Butanone	0.325	0.770			89	

MELTING POINTS[1] OF ARMID HT—ROSIN COMPOUNDS

% Rosin	*% Armid HT*	*Melting Point[1] °C*
0	100	99
10	90	96
20	80	93
30	70	88
40	60	84
50	50	80
60	40	75
70	30	70
80	20	61
90	10	54
100	0	78

[1] Drop point.

Armid O

Armid O is a soft wax-like material that becomes ductile and smooth when worked between the fingers. It is light amber in color.

CHEMICAL AND PHYSICAL PROPERTIES

	Typical Analysis
Free Fatty Acid	3.5%
Iodine Value	80 to 90
Color—Gardner	7
Solidification Point °C (ASTM D-87–42)	68
Odor	Bland
Specific Gravity (approx.)	0.90
Flash Point °C (approx.)	207
Fire Point °C (approx.)	228
Moisture %	0.5

APPROXIMATE COMPOSITION

Octadecanamide (Stearamide)	6.0%
9-Octadecenamide (Oleamide)	91.0%
9,12-Octadecadienamide (Linoleamide)	3.0%

COMPATIBILITY OF ARMID HT

Wax or Oil	*10%* Compatibility @ 25°C	*Armid* Cloud Point °C	*HT* Soft. Point °C	*50%* Compatibility @ 25°C	*Armid* Cloud Point °C	*HT* Soft. Point °C	*90%* Compatibility @ 25°C	*Armid* Cloud Point °C	*HT* Soft. Point °C	*No Armid* Cloud Point °C	*HT* Soft Point °C
Paraffin Wax 51°C A.M.P.	C	90	50	C	93	93	C	97	98.5	51	51
Carnauba	C	87	83	LC	94	93	C	97	95	83	85
Castor Oil	LC	74	*	LC	95	94.5	C	98	99	—	—
Rosin W. W.	C	Clear	57	C	80	76	C	96	95	—	74
Stearic Acid (Neo-Fat 18)	C	63.5	67	C	76	79	C	96	94	66	69
Oleic Acid (Neo-Fat 90–14)	LC	46	—	C	78	80	C	96	95	—	—
Ceresin	C	94	67	LC	94	95	C	93	97	64	66
Ozocerite	C	93	80	LC	95	94.5	C	97	99	67	66
Spermaceti	C	84	46	LC	93	93	C	97	98	45	46
Beeswax	C	77	64	LC	94	96	C	97	98	63	52
Mineral Oil	I	88	—	I	92	93	C	96	97	—	—
Ethyl Cellulose* 7 cps. N-Type	C	—	—	C	—	—	C	98	98	—	—
Armid HT										98	100

* Hercules Powder Company, I—Incompatible, C—Compatible, LC—Limited Compatibility.

Solubility was determined by adding Armid O to the more common solvents and heating to effect solution. The solutions were cooled to 25°C and checked for clarity. The figures in the chart below represent the approximate grams of Armid O that can be dissolved in 100 mls of solvent and still remain in solution on cooling to 25°C.

SOLUBILITY OF ARMID O

Water	Insoluble	
Isopropanol (95%)	59 grams	
Ethanol (95%)	30 grams	
Acetone	15 grams	
Petroleum ether	Insoluble	
Ethyl Acetate	11 grams	
VM & P Naphtha	1.5 grams	
Mineral Oil (White)	9% dispersible	(solidifies at higher concentrations)
Methyl isobutyl ketone	11 grams	
Trichloroethylene	15 grams	

Solubility in the more common plasticizers was determined by weighing various amounts of Armid O and plasticizer, gently warming to effect solution, and then cooling to 25°C. Samples were allowed to age for 48 hours and then checked for clarity. The figures given represent the approximate amounts that will stay in solution at the times indicated.

SOLUBILITY IN PLASTICIZERS

	After 2 hours	*After 24 hours*	*After 48 hours*
Tricresyl phosphate	5%	1%	1%
Triethyl phosphate	5	5	5
Dibutyl sebacate	5	5	1.5
Dioctyl sebacate	5	5	3
Dimethyl phthalate	5	5	1
Dibutyl phthalate	5	5	1.5
Dioctyl phthalate	5	5	1.0

Compatibility

Compatibility was determined by first dissolving the resin or wax in a suitable solvent. Weighed amounts of Armid O were then added to the mixture and warmed to effect solution. After cooling to room temperature, the solutions were cast on tin-mercury panels using a Boston-Bradley applicator adjusted for 20 to 30 mils wet film thickness, and allowed to air-dry. The resulting free-films were checked for clarity, flexibility, color, tackiness and toughness.

Uses

Armid O and Armid HT are being utilized as "anti-block" or "slip" agents in the production of plastic films and wax-coated papers that are ultimately used as food packaging materials. The Food and Drug Administration has sanctioned the use of both oleic acid amide and stearic acid amide as release agents on plastic food wraps. Evidence demonstrates that these agents are split in the body and digested as normal fatty acids.

Properties of Films Modified with Armid O

Additive	1% Armid O					5% Armid O				
	Clarity	Flex.	Color	Tack.	Tough.	Clarity	Flex.	Color	Tack.	Tough.
Ethyl Cellulose[1] 7 cps. N-type	N.C.	N.C.	N.C.	None	N.C.	N.C.	N.C.	N.C.	None	Better
138–140 AMP Paraffin Wax	N.C.	N.C.	N.C.	None	—	N.C.	N.C.	Sl.D.	None	—
Methyl Methacrylate	N.C.	N.C.	N.C.	None	N.C.	Sl.H.	N.C.	N.C.	None	N.C.
Poly Vinyl Chloride-Acetate (Vinylite Resin VYHH)[2]	N.C.	N.C.	N.C.	None	N.C.	H	N.C.	N.C.	None	N.C.
Poly vinyl Acetate (Vinylite resin AYAT)[2]	N.C.	N.C.	N.C.	Tacky	N.C.	H	N.C.	N.C.	Sl-Tacky	N.C.
Polyethylene PM-1[2]	N.C.	N.C.	N.C.	None	N.C.	H	N.C.	N.C.	None	N.C.
Cellulose Acetate[1] LL-1	N.C.	N.C.	N.C.	None	N.C.	H	N.C.	N.C.	None	N.C.
Carnauba Wax # 1 Grade	—	—	N.C.	Sl.	—	—	—	Sl.D.	Sl.	—

[1] Hercules Powder Company; [2] Bakelite Company, Div. of Carbide and Carbon.
N.C.—No Change, Sl.—Slight, D—Discoloration, H—Hazy.

Protective Films: Armid O can be dissolved in warm solvents, cooled to room temperature, and used as a dip or spray coating for such metals as brass and copper alloys and parts plated with precious or semi-precious metals. The Armid O coating affords protection against attack of fungi and of chemical fumes such as hydrogen sulfide. Surfaces coated with Armid O show improved resistance to salt water or marine conditions.

Dye Solubilizing: Armids O and HT are effective solubilizers on dyes in wax formulations with either oil-soluble or alcohol-soluble dyestuffs. Armid O is the more effective for straight solubilizing but Armid HT is recommended if a higher melting point is desirable. A few of the dyes that benefit from the use of Armid O include:

Oil Soluble: Methyl Violet Base (du Pont)
Victoria Blue B Base (du Pont)
Alcohol Soluble: Crystal Violet Extra Pure APN (du Pont)
Methyl Violet S. Concentrate (du Pont)
Victoria Blue B Concentrate (du Pont)

This application is of proven value in the manufacture of carbon paper coatings. For example, Methyl Violet Base can be dissolved in concentrations up to 80% by weight in Armid O. An oil solution of this dyestuff containing 5% of Armid O (basis 100 parts 138–140°F AMP wax) is fluid and pourable. Best effects with the Armids will be obtained in predominantly hydrocarbon formulations such as pencil and typewriter carbon coatings.

A few examples of the efficiency of Armid O when combined with 100 grams of 138°–140°F AMP paraffin wax are indicated in the following table.

Dye	*Grams of Dye*	*Grams of Armid O*
Crystal Violet (duPont) Extra Pure APN	2.0	8.0
Methyl Violet S Concentrate	2.0	8.0
Methyl Violet Base	8.0	2.0

Polyolefin Plastics: Armid O, and Armid HT are being used as slip and anti-block agents in polyethylene. As little as 0.05% of Armid O gives "slip" to the polyethylene. Armid-modified polyethylene films may be used in today's high speed packaging operations with no stoppages due to "blocking" . . . Armids act as internal mold release agents for both injection and extrusion molding. Molding equipment cycles faster, operators spend less time cleaning the molds, and between-cycle release sprays are eliminated.

The Armids are compatible with polyethylene at temperatures above the melting points of the Armid and polyethylene. It is believed that, because of the limited compatibility with polyethylene at room temperature, the Armid migrates to the surface of the polyethylene where it functions as an anti-block or slip agent.

Wax Papers: Armid O is used to advantage in paraffin wax paper coatings. In the wet waxing of paper it is desirable to include modifiers that will, in small percentages, improve the penetration of the wax, flexibility, translucency, and the "slip" or "seal" properties. It is especially desirable when increased transparency and "slip" are desired.

Armid O acts as a mutual solvent for the primary ingredients of quality wax coatings, which are paraffin wax and ethyl cellulose. The latter, although not used in all formulation on wax paper coating, is often added in small amounts (1% of Armid O based on total weight of coating) to give better flexibility and film strength, and to improve heat-stability.

The wetting property of Armid O improves the transparency of the wax coating by allowing better penetration of the wax into the paper fibers—also a better bond of the coating to the paper. At the same time, the "slip" or "feel" of the wax coating will be enhanced.

Armowax

Armowax (methylene bis-stearamide) is a hard, brittle, high-melting wax produced by the reaction of Stearamide and paraformaldehyde in the presence of a suitable acid catalyst. It possesses a compact, amorphous structure and will, upon impact, give a conchoidal fracture. It is cream colored with a very faint, pleasant odor. Formula

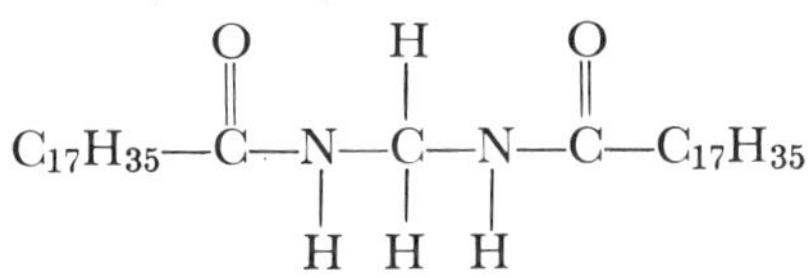

Compatibility

Armowax has some interesting compatibility characteristics. For example, when mixed to the extent of 10% in paraffin wax, it gives internal crystallization on slow cooling, but at 50% the resulting compound is perfectly amorphous. Certain other wax-like materials show this same effect with Armowax.

Chemical and Physical Properties

	Typical Analysis
Free Fatty Acids	5.0%
Color–Gardner	16
Ash	0.03%
Specific Heat	0.532 @ 60°C
Flash Point	250°C
Fire Point	258°C
Molecular Weight	579
Form	Lump or Powder
Solidification Point	130°C (ASTM D-87-42)
Moisture %	0.5

Solubility

Armowax is not soluble to any appreciable extent at room temperature in the more common solvents. This property is advantageous, in that dispersions or gels can often be made that find use in wax polishes, and the like. It will dissolve at elevated temperatures and it can, therefore, be incorporated into wax mixtures. The following chart will serve as a guide.

COMPATIBILITIES IN SYNTHETIC AND NATURAL WAXES

	% Armowax @ 25°C			*Cloud Points °C*			*Cloud Point of Resins, Waxes, Oils, °C*
Resin, Oil, Wax	10	50	90	10	50	90	
Paraffin wax 51°	I	C	C	121	126	131	51
Carnauba wax	C	C	C	199	126	133	83
Castor Oil	LC	C	C	110	128	133	
Rosin W W	LC	LC	C	—	113	131	
Stearic Acid	C	LC	C	91	114	132	66
Oleic Acid	LC	LC	LC	91	113	129	
Ceresin	C	C	C	120	126	132	64
Ozocerite	C	C	C	120	126	128	67
Ethyl Cellulose	C	C	C	—	—	133	
Spermaceti	LC	C	LC	112	126	132	45
Beeswax	C	C	C	111	122	133	63
Mineral Oil	LC	C	C	115	120	131	—

COMPATIBILITY IN PARAFFIN

% Armowax	*Cloud Point*	*Compatibility @ 25°C*
0	51	—
2	108	I
5	113	I
10	121	I
25	123	I
50	126	C
75	130	C
90	131	C
100	134	—

I—Incompatible, LC—Limited Compatibility, C—Compatible.

TEMPERATURE OF CRYSTALLIZATION FROM SOLUTIONS

Solvent	*Concentration of Armowax in Solution* 10%	20%	30%
Turpentine	86°C	97°C	102°C
Pine Oil	80	91	98
Xylene	91	97	100
V M & P Naphtha	101	105	108
Mineral Oil	115	118	121
Butanol	90	95	98
Methyl Cellosolve	104	108	109
Butyl Cellosolve	106	105	107
Methyl Isobutyl Ketone	98	104	108
Benzene	Insoluble @ 80°C		
Ethyl Acetate	Insoluble @ 74		
Ethanol (3-A)	Insoluble @ 79		
Ethylene Dichloride	Insoluble @ 84		
Carbon Tetrachloride	Insoluble @ 75		
Glycerine	Insoluble @ 140		

CRODAMIDE 'E'

Erucamide

Neutral surface active amide prepared from erucic acid. Cream colored flaked waxy solid. Insoluble in water, low solubility in organic solvents at ambient temperatures. Substantially stable to heat, light and air. Resistant to hydrolysis by dilute acids and alkalis.

	Specification	*Typical Analysis*
Acid Value	10 max	6.9
Melting Point	75/80°C	77.5°C
Iodine Value	75/85	77.0
Color 5% w/v in 5¼" cell	15Y:3.OR max	5.0Y:1.1R
Amide Purity	90% min.	91%

Uses:

Decreases friction of polythene film by 50% at 0.02% concentration.

F.D.A. approved for use in food packaging materials as slip agent (F.D.A. Regulation No. 121.2509 & 2514)

Improves ink adherence to polythene films and of resins to glass fibers.

Mould release agent and processing aid for rubber goods.

Anti-blocking agent for pressure sensitive tapes.

Foam stabilizer in sulphonate-type detergents.

Corrosion inhibitor.

CRODAMIDES L

Lauramides

These amides are available in several qualities. Lauramide L.90 is based upon 90% Lauric Acid and is available straight run or refined and in bead form.

Lauramide L.50 is based upon distilled coconut fatty acids and is also available in the same forms. All these qualities are low odor white waxy materials.

The materials are insoluble in water and most solvents at room temperature, although slightly more soluble than higher carbon amides. Lauramide is soluble to around 10% in butanol, ethanol, isopropanol and methanol at 30°C.

Specifications

	D.8320 Lauramide L.90	D.8330 Lauramide L.90.R (Refined)	D.8325 Lauramide L.50	D.8335 Lauramide L.50.R (Refined)
Melting Point °C min.	90	90	85	85
Titre °C	95	95	83	85
Iodine Value	0.5	0.5	12	12
Acid Value max.	10	1.0	10	2
Gardner Color max.	7	1.0	10	2
Amide Purity min %.	92	94	92	92
Moisture max.	0.5	0.5%	0.5	0.5

Uses

Lauramide acts as a foam booster and stabilizer in detergents of the sulphonate type. It has also been employed as a gelling agent in solvent-soaps.

Lauramides have also been proposed as thickening agents for inflammable-type paint removers in which they retard solvent evaporation and allow for better penetration of solvents into the paint films.

The product is F.D.A. approved (Subpart F) as a defoaming agent in the manufacture of paper and paper board; and also for use as a release agent in the production of polymer films. It can be used as a co-solvent for making blends of polymers or hydrocarbons and resins. It is compatible with some urea and melamine formaldehyde resins.

It is also of value as an anti-blocking agent in vinyl copolymer solution coatings and improves the 'slip' properties of stoved resin film surfaces.

CRODAMIDE 'O' & 'OR'

Oleamides

Neutral surface active amides from oleic acid. Waxy material of low odor. Insoluble in water, low solubility in organic solvents at ambient temperatures. Substantially stable to heat, light and air. Resistant to hydrolysis by dilute acids and alkalies.

Specification	*O*	*OR (Refined)*
Acid Value	10 max	3 max
Melting Point	66/72°C	72/76°C
Iodine Value	85/95	80/92
Color	20Y:5R max (5% w/v 5¼″ cell)	5 max (Gardner)
Amide purity	90% min	94% min
Moisture	0.5% max	0.5% max

Uses

Decreases friction of polythene films by 50% @ ca 0.05%. F.D.A. approved for use in food packaging materials as slip agent, (F.D.A.) Regulation 121,2509,2514.

Improves ink adherence to polythene films and of resins to glass fibers.

Anti-static agent.

Extrusion and moulding aid for plastics.

Mould release agent and processing aid for rubber goods.

Improves solubility of dyestuffs in printing inks.

Anti-blocking agent for pressure sensitive tapes.

Foam stabilizer for sulphonate-type detergents.

Blending agent for paraffin waxes and polyamide resins.

Corrosion inhibitor.

Intermediary for textile waterproofing, anti-foams and synthetic waxes.

Viscosity reducer in hot melt compositions.

Provides anti-corrosive film for metals.

Additive in fusible color coatings for ceramics helping to give sharp clean designs.

CRODAMIDE 'S' AND 'SR'

Stearamides

Neutral surface active amides prepared from stearic acid. White waxy material of low odor. Substantially stable to heat, light and air. Resistant to hydrolysis by dilute acids and alkalis. Insoluble in water, low solubility in organic solvents at ambient temperatures.

Specification	*S*	*SR (Refined)*
Acid value	10 max	2.0 max
Melting point	96/101°C	97/102°C
Iodine Value	2 max	2 max
Color	5.0Y:1.5R	2.0 max
	(5% w/v 5¼" cell)	(Gardner)
Amide Purity	90% min	94% min

Uses

F.D.A. approved for use in food packaging materials as anti-blocking agent.

Improves ink adherence to polythene films, and of resins to glass fibers.

Mould release agent and processing aid for rubber goods.

Aids in preventing sulphur bloom and ozone attack in rubber.

Non-blooming mould release agent for many resin systems.

Foam stabilizer in sulphonate-type detergents.

Corrosion inhibitor.

Additive for increasing melting point of other waxes.

Textile oil auxiliary emulsifier used in conjunction with soap.

Viscosity reducer in hot melt compositions.

Anti-tack agent in plastisols and vinyl solution coatings.

Improves anti-blocking properties of hydro-carbon waxes for paper coatings.

Gelling agent for solvent soaps.

Bestows no-tack properties in resins and waxes used in water proofing.

In cosmetic formulations Stearamide can act as a useful auxiliary emulsifier for o/w emulsions or as a weak emulsifier for w/o emulsions.

Kemamides

PRODUCT	Description	Average Molecular Weight	% FFA Max.	Iodine Value	Capillary M. P. °C	Color Gardner Max.	% Moisture Max.
Kemamide B	Behenamide	312	2.0	0.5-2.0	98-108	5	0.25
Kemamide S	Stearamide	278	2.0	1.0-3.0	98-108	5	0.25
Kemamide U	Oleamide	275	2.0	72-90	68-78	5	0.25
Kemamide E	Erucamide	335	2.0	70-80	76-86	5	0.25
Kemamide 1288	12-Hydroxy Stearamide	285	2.5	1.0-5.0	105-112	8	0.25

Kemamide Bisamides

Product	Acid Value	Amine Value	Color Gardner	Approx. Melting Pt.
Kemamide W 45	10 Max.	10 Max.	3 Max.	145°C
Kemamide W 40	10 Max.	10 Max.	3 Max.	140°C
Kemamide W 40-OH	10 Max.	10 Max.	8 Max.	140°C
Kemamide W 35	10 Max.	10 Max.	6 Max.	140°C
Kemamide W 20	10 Max.	10 Max.	6 Max.	120°C
Kemamide W 10	10 Max.	10 Max.	6 Max.	110°C

Kemstrene Waxes

Typical Analysis	*% Amide*	*Average Molecular Weight*	*% FFA*	*Iodine Value*	*Capillary M.P.-C °C*	*Color 1″ Lovibond*	*% Moisture*
Kemstrene Amide S (Stearamide)	93	278	2.0	1.5	98–104	8Y–1.8R	0.1
Kemstrene Amide U (Oleamide)	95	275	1.3	91.3	67–74	35Y–5.0R	0.2
Kemstrene Amide SU-25 (Olea/Stearamide)	94	275	1.9	56.5	80–87	6Y–1.7R	0.1
Kemstrene Amide B (Behenamide)	92	312	1.8	1.0	100–103	3Y–1.0R	0.1
Kemstrene Amide P (Palmitamide)	94	254	2.0	1.2	98–102	5Y–1.5R	0.1
Kemstrene Amide UR (Oleamide—Refined)	95	279	2.0	90.0	75–79	15Y–3.0R	0.1

Glyco Amide Waxes

	PRODUCT	CERAMID	GLYCOLUBE VL	GLYCONOL
TYPICAL PROPERTIES	Color & Form	Yellow, hard	Tan, fairly hard	Amber, soft, tacky
	Melting Point, °C.	78 - 81	110 - 117	100 - 108
	Flash Point, °C.	245	271	215
	Specific Gravity @ 20°C.	0.98	0.97	
SOLUBILITIES	Water	I	I	I
	Alcohol	SH	I	S
	Toluol	SH	SH	S
	Naphtha	SH	SH	G
	Mineral Oil	SH, GC	SH	G
	Vegetable Oil	SH, GC	SH	G

Glyconol

Melting Point, °C	100–108
Color (Gardner, melt)	12 max.
Clarity (melt)	clear
Penetrometer Hardness (cm @ 200g/min.)	1.73

Solubility:

Low to moderate solubility in most solvents and oils at ambient temperatures. High solubility in all solvents at their boiling points, also in oils at 100°C. Most hot solutions gel on cooling.

	Solubility, g/100g.	*At Solv.*	
Solvent	*Ambient*	*b.p.*	*Temp. °C*
Water	I	I	100
Ethyl alcohol	4.8	S	78
MEK	6.8	S	79
Isopropyl alcohol	13.5	S	82
Trichlorethane	9.3	S	113
Toluene	40.5	S	110
Mineral spirits	gel	S	154
Soybean oil	gel	S	100
Mineral oil	gel	S	100

I=<0.5 g. solute per 100g. solvent
S=>100g. solute per 100g. solvent

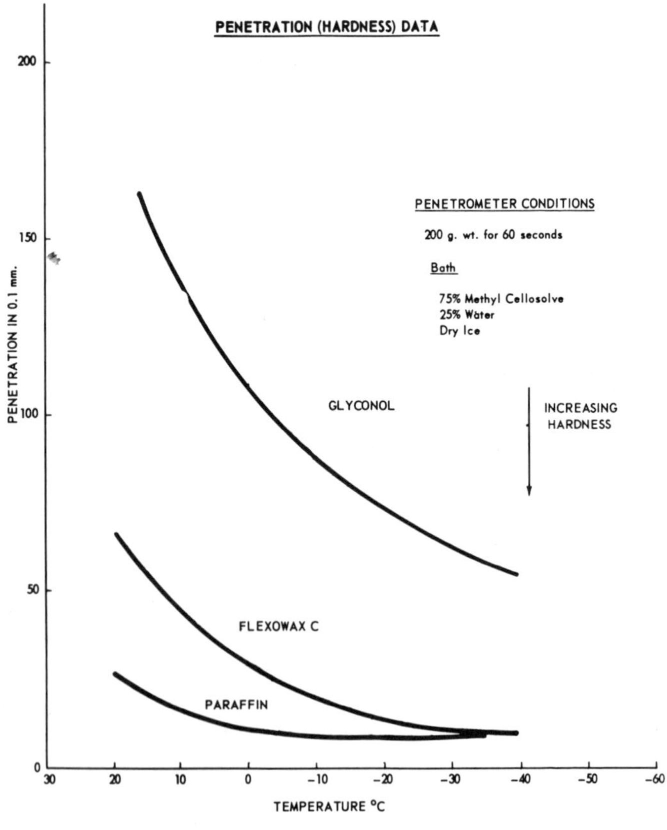

FIG. 9:1

CINN. MILACRON WAXES

Cinn. Milacron 225 Wax is a hard, fatty amide wax.

Typical Properties,

Melting Range	223–228°F
Acid Value	12 Max
Smoke Point, C. O. C.	350°F
Flash Point, C. O. C.	525°F
Fire Point, C. O. C.	600°F
Color	Tan
Form	Finely Beaded

Solubility:

225 Wax is soluble in hot naphtha, hot toluene, hot benzene, hot methanol and most hot organic solvents. It is insoluble in water and alcohol. 225 Wax is miscible at elevated temperatures with natural and synthetic waxes, asphalts, tars, and similar compounds.

Cinn. Milacron 240 Wax is a hard, high-melting synthetic wax.

Typical Properties

Melting Point Range	235–245°F
Acid Value	10 Maximum
Smoke Point, C. O. C.	335°F
Flash Point, C. O. C.	515°F
Fire Point, C. O. C.	590°F
Color	Tan
Form	Finely Beaded

Solubility

Milacron240 Wax is soluble in hot naphtha, hot toluene, and most hot organic solvents; insoluble in water and alcohol. Miscible at elevated temperatures with natural and synthetic waxes, asphalts, tars, resins, and similar compounds.

Cinn. Milacron 267 Wax is a moderately hard, high-melting synthetic wax that exhibits low shrinkage upon cooling.

Typical Properties

Melting Range	265–270°F
Color	Tan
Form	Finely Beaded
Smoke Point, C. O. C.	330°F
Flash Point, C. O. C.	510°F
Fire Point, C. O. C.	570°F
Shrinkage upon cooling	6%

Solubility

Soluble in hot solvents such as ethanol, VM & P naphtha, benzene, xylene, 2-ethoxyethanol, dimethyl formamide, methyl isobutyl ketone, heptane, octene, and A-pinene.

Cinn. Milacron 272 Wax is a moderately hard, high-melting synthetic wax.

TYPICAL PROPERTIES

Melting Range	269–275°F
Color	Yellow-Tan
Form	Finely Beaded
Smoke Point, C. O. C.	335°F
Flash Point, C.O.C.	515°F
Fire Point, C. O. C.	600°F
Shrinkage upon cooling	13%

Solubility

Soluble in hot solvents such as ethanol, VM & P naphtha, benzene, xylene, 2-ethoxyethanol, methyl ethyl ketone, methyl isobutyl ketone, heptane, octene, and A-pinene.

Cinn. Milacron 275 Wax is a dark tan, high-melting synthetic wax.

TYPICAL PROPERTIES

Melting Range	275–280°F
Acid Value	12 maximum
Flash Point, C. O. C.	540°F
Fire Point, C. O. C.	575°F
Color	Dark tan
Form	Finely Beaded or Solid in Drums

Solubility

Milacron 275 Wax is soluble in hot naphtha, hot toluene, and most hot organic solvents; insoluble in water and alcohol. The 275 Wax is miscible at elevated temperatures with natural and synthetic waxes, resins, asphalts, tars, and similar compounds.

Cinn. Milacron 280 Wax (N, N′-ethylenebis-stearamide) is a hard, light colored, high-melting synthetic wax.

TYPICAL PROPERTIES

Melting Range	280–290°F
Acid Value	10 Maximum
Smoke Point, C. O. C.	325°F
Flash Point, C. O. C.	530°F
Fire Point, C. O. C.	595°F
Color	Off White
Form	Finely Beaded

Solubility

Milacron 280 Wax is soluble in hot naphtha, hot toluene, and most hot organic solvents; insoluble in water and alcohol. The 280 Wax is miscible at elevated temperatures with natural and synthetic waxes, asphalts, tars, and similar compounds.

Electrical Properties

Dielectric Strength	430 volts/mil
Insulation Resistance @ 25°C	68,000 megohms
Oxidation Potential	450 millivolts
Power Factor @ 1000 cycles	0.028

MICROTOMIC 280 WAX

Cinn. Milacron Microtomic 280 Wax (N, N'-ethylenebis stearamide) is a hard, light colored, high melting synthetic wax of extremely small particle size.

TYPICAL PROPERTIES

Particle Size, % thru 100 mesh U.S. Standard Sieve	100
Melting Range, °F	280–290
Color	Off White
Chemical Name	N, N'–Ethylenebis Stearamide
Acid value	10 Maximum
Smoke Point, °F, C. O. C.	325
Flash Point, °F, C. O. C.	530
Fire Point, °F, C. O. C.	595

Solubility

Microtomic 280 Wax is soluble in hot naphtha, hot toluene, and most hot organic solvents; insoluble in alcohol and water. Microtomic 280 Wax is miscible at elevated temperatures with natural and synthetic waxes, asphalts, tars, and similar materials.

Electrical Properties

Dielectric Strength	430 volts/mil
Insulation Resistance @ 25°C,	68,000 megohms
Oxidation Potential	450 millivolts
Power Factor @ 1000 Cycles	0.028

Cinn. Milacron 290 Wax is a hard, bisamide wax.

TYPICAL PROPERTIES

Melting Point Range	288–298°F
Acid Value	10 maximum
Flash Point, C. O. C.	550°F
Fire Point, C. O. C.	590°F
Color	Light Tan
Form	Finely Beaded

Solubility

Milacron 290 Wax is soluble in hot naphtha, hot toluene, and most hot organic solvents; insoluble in water and alcohol. Miscible at elevated temperatures with natural and synthetic waxes, asphalts, tars, resins, and similar compounds.

Cinn. Milacron 315 Wax is a moderately hard, high-melting synthetic wax.

TYPICAL PROPERTIES

Melting Range	313–320°F
Color	Tan
Form	Finely Beaded
Smoke Point, C. O. C.	340°F
Flash Point, C. O. C.	420°F
Fire Point, C. O. C.	500°F

Solubility

Soluble in hot solvents such as ethanol, VM & P naphtha, xylene, 2-ethoxyethanol, methyl ethyl ketone, and A-pinene.

Cinn. Milacron 325 Wax is a moderately hard, high-melting synthetic wax.

TYPICAL PROPERTIES

Melting Range	324–329°F
Color	Tan
Form	Finely Beaded
Smoke Point, C. O. C.	360°F
Flash Point, C. O. C.	450°F
Fire Point, C. O. C.	525°F

Solubility

Soluble in hot solvents such as ethanol, xylene, 2-ethoxyethanol, methyl ethyl ketone, A-pinene, dimethyl formamide, benzene, kerosene, and VM & P naphtha.

Cinn. Milacron 350 Wax is a hard, high-melting synthetic wax.

TYPICAL PROPERTIES

Solidification Point	350°F
Color	Dark Tan
Form	Finely Beaded
Smoke Point, C. O. C.	395°F
Flash Point, C. O. C.	555°F
Fire Point, C. O. C.	650°F

Solubility

Soluble in hot 1, 3-butylene glycol, and hot dimethyl formamide. Partially soluble in hot 2-ethoxyethanol.

Cinn. Milacron 350-A Wax is a hard, translucent, high-melting synthetic wax. In sheets 350-A Wax exhibits plastic tendencies permitting slight deformation before breaking.

TYPICAL PROPERTIES

Solidification Point	350°F
Color	Yellow–Tan
Form	Finely Beaded
Smoke Point, C. O. C.	358°F
Flash Point, C. O. C.	380°F
Fire Point, C. O. C.	380°F

Solubility

Soluble in hot 1, 3-butylene glycol, and hot dimethyl formamide; partially soluble in hot 2-ethoxyethanol.

Cinn. Milacron 400 Wax is a hard, exceptionally high-melting synthetic wax.

TYPICAL PROPERTIES

Solidification Point	400°F
Color	Beige–Tan
Form	Finely Beaded
Smoke Point, C. O. C.	400°F
Flash Point, C. O. C.	520°F
Fire Point, C. O. C.	630°F

Solubility

Soluble in hot 1, 3-butylene glycol. Slightly soluble in hot naphtha, hot xylene, hot A-pinene, and hot 2-ethoxyethanol.

SOLUBILITY OF (CINN. MILACRON) WAXES

SOLVENTS	267	272	315	325	405	440	450
Ethyl Alcohol	PS[a]	S[b]	S	S	I[c]	I	I
1,3-Butylene Glycol	S	S	S	S	S	S	S
Acetone	I	I	I	I	I	I	I
Benzene	S	S	S	PS	PS	PS	I
Heptane	S	S	PS	PS	I	I	I
Dimethylformamide	S	S	S	S	S	S	PS
Carbon Tetrachloride	S	S	PS	PS	I	I	I

[a]PS = Less than 10% solution at boiling point of solvent

[b]S = 10% solution by weight

[c]I = Insoluble even at boiling point of solvent

[1]Where it is not possible to make a solution, up to 10% of these waxes can be suspended in a solvent by use of a ball mill or colloid mill.

BAKER WAXES

Flexricin and Paricin Waxes

The following waxes are hydroxylated amides of castor oil and hydrogenated castor oil, selected to provide a wide range of melting points. Colors are tan to light brown.

PHYSICAL AND CHEMICAL PROPERTIES

Condition	*Flexricin 115* Unsaturated	*Flexricin 185* Unsaturated	*Paricin 220* Saturated	*Paricin 285* Saturated
Melting Point, °F	120	185	220	280
Acid Value	3	7	2	8
Iodine Value	80	86	4	4
Hydroxyl Value	285	173	305	178
Volatile, %	0.2	0.3	0.4	0.7
Sp. Gr. @ 25°C	0.995	1.017	1.049	1.020
	SOLUBILITY, %			
in: Water	Insoluble	Insoluble	Insoluble	Insoluble
Ethyl Alcohol	19.0	17	15	6
Mineral Spirits	Insoluble	Insoluble	Insoluble	Insoluble
Toluol	0.5	0.3	0.6	0.8
Cellosolve	17	0.5	7.7	1.3
Butyl Acetate	20	0.2	1.4	2.5
Methyl Ethyl Ketone	21	0.5	1.6	4.5

Suggested Applications

Anti-static agents (particularly Flexricin 115 and Paricin 220)
Carbon paper
Ethyl cellulose compounds, hot melts, etc. (particularly Paricin 285)
Greases, soda-based and gasoline-resistant (Paricin 285)
Lubricants for plastics and metals

Glokem Waxes

Two types are available. One is a long-chain fatty amide, the other is a long-chain fatty ester amide of Glokem Waxes. The chain lengths of the substituents are varied to produce grades having modified properties.

PROPERTIES

Glokem Wax	*Color*	*Titre °C*	*Acid Value*	*Saponification Value*
D2S	Cream	139–143	2–6	<6
D2P.	Cream	144–148	2–6	<6
D2B.	Cream	136–140	5–10	<10
D2L.	Cream	150–154	5–10	<10
D2M.	Light Brown	123–127	5–15	5–15
DMS.	Yellow	126–130	5–15	5–15
DNIS.	Yellow	103–107	5–15	5–15
DN2S.	Yellow	94–98	5–15	5–15
E2S.	Off-White	74–78	20–25	105–120
E2M.	Light Brown	84–88	22–27	80–95
E2B.	Off-White	89–93	15–20	90–110
ES/SM.	Yellow	71–75	18–23	105–120
ESU.	Off-White	84–88	<5	<5
AS.	Fawn	77–81	5–20	5–20
XS.	Fawn	58–62	5–20	5–20
PSZ.	Grey	190–200	5–15	5–15
P2S.	Grey	170–180	5–15	5–15
O.	Yellow Brown Semi Liquid	60–65[1]	8–13	10–25

[1] Rather indefinite

The melting points of these waxes can generally be taken as being from 3 to 4° higher than the quoted titres. The acid and saponification values given above are those obtained by application of standard methods and hence saponification values approaching zero are recorded for some waxes which, under extremely rigorous conditions, could, in fact, be saponified. They are soluble to some extent in most hot solvents and on cooling crystallization or gelation is normal.

Dispersions of Glokem Waxes

Stable, aqueous dispersions of the higher melting point Glokem Waxes may be prepared by using finely divided wax (to pass 300 mesh) and certain quaternary ammonium compounds as dispersing agents. In general, 2% of quaternary ammonium compound on the total weight is required and the recommended procedure is to dissolve the quaternary ammonium compound in a portion of the water, add to the finely powdered Glokem Wax and stir to a paste or cream. The remainder of the water may be added to this concentrate gradually and with stirring.

The quaternary pyridinium compounds such as Morpan CPB are particularly effective, but more economical dispersions can be prepared using Morpan CW.

A typical formulation is:

Glokem D2S (300 mesh)	10%
Morpan CW.	2%
Water to	100%

EFFECT OF GLOKEM WAXES ON POLYETHYLENE

(Average values)

Glokem Wax Additive	*Coeff. of Friction (Ratios).*	*Minimum Heat Sealing Time*
Control (No Additive).	1.0	1.0
E2S.	0.55	1.25
ESU.	0.77	2.0
D2S.	0.82	3.5

These figures are ratios clearly demonstrating that the coefficient of friction is materially reduced, particularly with Glokem E2S, and, in addition, this wax in particular does not significantly alter the sealing time. Other factors, such as film tensile strength and elongation etc. are virtually unaffected by the wax additives.

Glokem Wax Emulsion

Cationic emulsions can be obtained by using a C_{12} to C_{18} amine plus excess acetic acid as emulsifying agent, e.g.

Glokem D2S	10%
Oleylamine	2%
Glacial Acetic Acid	0.8%
Water	100%

The first three ingredients are heated to 140°C and poured slowly into boiling water with vigorous stirring until uniform or homogenizing after preliminary emulsification.

Waxes of the E type are usually easy to emulsify with standard emulsifying agents exemplified by non-ionic surfactants, Texofors A10, A1 or B1, etc. Suitable anionic emulsifying agents are triethanolamine stearate and sulphated alcohols although a wide range of materials of this type will give satisfactory results.

ANTIBLOCKING EFFECT ON POLYVINYL CHLORIDE FILM

Glokem E2S Additive	*Cohesive Force*
0.0	2.8 lb/in
0.5	0.7
1.0	0.3
2.0	0.0

The above additions do not affect tensile strength, and they slightly increase the modulus of rigidity.

Acrawax[1]

Acrawax is a reaction product of hydrogenated castor oil and monoethanolamine.

[1] Glyco Chemicals Inc.

PROPERTIES OF ACRAWAX

Color and Form	Hard, tan wax
Luster	High
Melting Point	94–97°C
Flash Point	230°C (open cup)
Specific Gravity, 24°C	1.04

Soluble in: Hot alcohol, hot toluene (forms gel on cooling), hot butyl acetate, hot turpentine; insoluble in water and mineral spirits.

Used as: Antitack agent for vinyl resin films; for impregnating leather packings; in water- and oil-resistant greases.

Compatibility

9 parts Acrawax to 1 part of Congo, Acrawax B, Shellac and Candelilla Wax; 4 parts Acrawax to 1 part of Gum Manila Loba, Diolin, Ester Gum, and Candelilla Wax; 1 part Acrawax to 1 part of Albacer, Aldo 33, Ester Gum, Rosin, Shellac, Stearic Acid, Acrawax B, and Acrawax C.

Acrawax B is a reaction product of stearic acid and monoethanolamine.

PROPERTIES

Color and Form	Hard, light-brown wax
Luster	Good
Melting Point	83–86°C
Flash Point	235°C (open Cup)
Acid Number	<2
Specific Gravity, 25°C	0.97

Soluble in: Hot alcohol, hot toluol, hot butyl acetate, hot mineral spirits (forms gel on cooling), and mineral oil. Insoluble in water.

The properties of Acrawax B are very similar to those of Acrawax except that Acrawax B is compatible with paraffin wax and forms gels with mineral spirits and kerosene.

Compatibility

9 parts Acrawax B to 1 part of: Acrawax, Ester Gum, Gum Dammar, Cumar, and Shellac; 3 parts Acrawax B to 1 part of Flexo Wax C and Paraffin Wax; 1 part Acrawax B to 1 part of Albacer, Aldo 33, Ester Gum, Flexo Wax C, Candelilla Wax, Carnauba Wax, Beeswax, Paraffin Wax, Reso Wax B, and Rosin.

Acrawax C

PROPERTIES

Color and Form { Lumps / Powder	Hard, tan wax
Luster	High
Melting Point	140–143°C
Flash Point	285°C (open cup)
Specific Gravity @ 25°C	0.97
Moisture Absorption @ 96°F for 48 hours	0.3 g moisture per 50 g wax
Oxidation Potential	400–500 millivolts
Dielectric Strength	428 volts per mil
Insulation Resistance at 25°C	68,000 megohms
Power Factor at 1,000 Cycles	0.0277

Compatibility

1 part Acrawax C to 1 part of Candelilla Wax, Carnauba Wax, Flexo Wax C, Glyco Wax S932, Paraffin Wax, Pentalyn, Rosin, etc.

It is compatible with tricresyl phosphate which can be used with it as an efficient plasticizer.

Solubility

Acrawax C is insoluble in boiling water and low-boiling organic solvents, such as alcohol, acetone, benzene, and carbon tetrachloride. It is soluble in hot high-boiling solvents, such as kerosene, toluene, Cellosolve, turpentine, etc., at and above its melting point (140–143°C). On cooling, the wax precipitates; from a turpentine solution a gel is formed.

Uses: As antiblocking or antitack agent in heat-sealing adhesives, plastics, synthetic, and natural rubbers; antistatic agent; for raising the softening point of asphalts; in dental waxes; in electrical insulating coatings and cable fillers; as a lubricant in wire drawing, molded metal parts, metal coatings; plastic mold release; stop-off wax in electroplating; moistureproofing and antimar agent in paints, varnishes, and lacquers; moistureproofing and nonblock agent for textiles; printing inks and bake finishes.

Acrawax C can be incorporated by three methods into lacquer sealers and top coats:

1. Acrawax C beads are dissolved in hot toluene and the solution added, with rapid agitation, to the nitrocellulose wetted with cold toluene.
2. Acrawax C atomized is added to an alcoholic solution of nitrocellulose and mixed thoroughly.
3. Acrawax C powdered is dispersed in the lacquer by millgrinding technique.

HALOCARBON WAXES

Halocarbon waxes are saturated low-molecular-weight polymers of chlorotrifluoroethylene and have the general formula: $—(CF_2–CFCl)_n—$. They are made by a controlled polymerization technique and are then stabilized in another step so that the terminal groups are completely halogenated and inert. The product is then separated into various fractions by vacuum distillation.

Properties Halocarbon Waxes

Type and Serial Number	*Distill. Range at 0.5 mm Hg °F*	*Initial Boiling Point, 1 Atm. °F*[1]	*Melting Pt.*[2] *or Pour Pt.*[3] *°F*	*Viscosity in Centistokes at 100°F*	*Density at 210°F*	*Appearance at 70°F*
Wax 6–00	Residue Abv. 410	Abv. 500	140[2]	Solid	1.89	White
Wax 8–00	Residue Abv. 410	Abv. 500	178[2]	Solid	1.89	White
Wax 12–00	Residue Abv. 410	Abv. 500	248[2]	Solid	—	White
Wax 15–00	Residue Abv. 410	Abv. 500	292[2]	Solid	—	White

[1] Vapor Phase.
[2] A.S.T.M. D127-30.
[3] A.S.T.M. D97-47.

The single *most* outstanding property of the Halocarbon waxes is chemical inertness. This property, together with other outstanding properties such as high density, lubricity, and high dielectric strength, provides the economic justification for the use of Halocarbon products in a wide variety of applications.

In tests established in specifications for anti-seize compounds for oxygen service, including oxidation and explosion tests, these halocarbons have shown:

1) No spontaneous ignition in oxygen at 2000 psig at temperatures up to 440°C (824°F).
2) No explosion in an atmosphere of pure oxygen under an impact of 114 foot pounds.
3) No drop in oxygen pressure in the oxygen bomb test (Method 3453.1, Federal Specification VV-L-791e).
4) No change in color when oxygen is bubbled through a Halocarbon oil for 24 hours at 160°F and at 300°F.

Electrical Properties

The electrical properties of Halocarbon polymers will vary slightly with the molecular weight. The following data are typical:

Halocarbon Wax, Series 8–00

Frequency	*Dielectric Constant*	*Power Factor*
1 Kilocycle	3.01	9.1×10^{-3}
10 Kilocycles	2.94	15×10^{-3}
100 Kilocycles	2.81	27×10^{-3}
63 Megacycles	2.50	43×10^{-3}
Surface Resistivity (ambient atmosphere)		4×10^{13} ohms/cm^2
Surface Resistivity (after 28 days at 165°F, 90% humidity)		1×10^{11} ohms/cm^2

Toxicity

At temperatures below 550°F, in fifteen years of experience with these low polymers, there has been no report of acute or chronic poisoning, dermatitis, or other adverse physiological effects from Halocarbon waxes.

At temperatures above 550°F, these polymers are subject to thermal cracking, with rapid breakdown occurring above 620°F. This breakdown may occur from heating in a vessel or operating equipment, or from the vapors coming in contact with an open flame or passing through a lighted cigarette. Breakdown products in air are toxic, acidic substances which produce harmful physiological effects from prolonged exposure. When these waxes are heated above 400°F for long periods in open vessels, it is recommended that adequate ventilation be provided.

Solubility

All halocarbon materials are insoluble in aqueous solutions, both acid and alkaline. Most of these materials are only partially soluble in methyl and ethyl alcohol.

Halocarbon materials will dissolve the halogens (except fluorine) and will dissolve volatile anhydrous inorganic salts (such as titanium tetrachloride).

The following substances are insoluble in all Halocarbon Waxes:

Acetamide, agar, aqueous solutions, anthracene, casein, cellulose acetate, ethylene glycol, gelatin, glycerol, nitric acid, nitrocellulose, phenol, salicylic acid, shellac, sugar, sulphuric acid, sulfonal, tartaric acid, thiourea, urea, water (see *Electrical Properties*).

Effects on Materials of Construction

Effects on Elastomers and Plastics

Halocarbon waxes have no adverse effect on polyvinyl alcohol, polysulfide elastomers, Buna-N (butadiene/acrylonitrile), Hycar, Neoprene GN rubbers, or Teflon and Viton polymers.

Effect on Metals

The halocarbon materials wet metallic surfaces readily and form lubricating films similar to the more common lubricants.

Halocarbon waxes are non-corrosive toward metals at temperatures up to 350°F, with the exception of copper and some of its alloys which will discolor at temperatures over 120°F. Prior testing should be done on all metals for applications above 350°F and on copper for applications above 120°F.

Halocarbon materials are not recommended for application to aluminum thread. The extremely high localized temperatures of minute seizures and/or welding points of aluminum have been known to cause a chemical reaction between Halocarbon Waxes and aluminum, with a resulting detonation.

VYDAX® 78U

Fluorotelomer Powder

VYDAX 78U fluorotelomer powder is a white, *wax-like,* low molecular weight polymer of tetrafluoroethylene. This micro-sized powder exhibits two outstanding characteristics: (1) low coefficient of friction and (2) excellent release or anti-stick properties.

Typical Properties:

Density, 77 F (25C), g/cc	2.16
Softening Point (ASTM E-28-58T)	510 F (265C)
Melting Point (crystalline)	572 F (300C)
Molecular Weight	3,700 approx.
Particle Size	60% below 5 microns
	100% below 30 microns

Ross Waxes

Ross Wax 160[1]

This is a tan, medium hard, high melting point synthetic (amide) wax with the following properties:

Melting Point	157–159°C
Acid Value	5–10
Flash Point	310–312°C
Iodine Value	7.5
Specific Gravity	1.0232
Dielectric Constant (1000 cycles)	3.0
Power Factor	0.00977

Ross Wax 160 is soluble in hot naphtha, xylene, isopropanol, trichlorethylene, and Carbitol, but is not soluble in solvents at ordinary temperatures. It is insoluble in water.

It is compatible with hydrocarbon waxes, ester type waxes, vegetable waxes, hydrogenated castor oil, fats, and rosin.

When added to other waxes and esters Ross Wax 160 raises their melting (dropping) points (ASTM Method—D127-49) in accordance with the following table:

	*50/50**	*75/25**	*90/10**
Carnauba Wax (82.5–86°C)	148–150°C	141–143°C	131–134°C
Castor Wax (86–88°C)	148–150°C	141–143°C	130–131°C
Hydrogenated Tallow (59–64°C)	149–151°C	143–146°C	136–138°C
Paraffin Wax (56–57°C)	147–149°C	145–147°C	142–144°C
Micro Wax (88°C)	152–154°C	151–153°C	149–153°C

* The second figure is the percentage of Ross Wax 160.

[1] Frank B. Ross Co.

Table 9:6

RISE IN DROP MELTING POINT BY ADDITION OF ROSS WAX 160

Base		*10% Ross Wax 160*	*M.P. without Wax 160*
90%	Naphthalene	100.1°C	80.2°C
,,	Beeswax	130.9	65.0
,,	Degras (Sulfur free)	119.4	
,,	Ester Gum #8	126.7	89–96
,,	Double Pressed Stearic Acid	119.6	55.5
,,	Phenol	70.4	41.0
,,	Oleo Stearine	132.5	50.0
,,	Coconut Oil	128.4	21–24
,,	Petrolatum	145.0	50.0–60.0
,,	Carbowax 6000	149.5	60.0–63.0
		20% Ross Wax 160	
80%	Carbowax 6000	156.5°C	
		25% Ross Wax 160	
75%	Carbowax 6000	157.5°C	

PENETRATION TEST COMPARISON (ASTM METHOD — D5-52) AT 100 GRAMS.

Paraffin Wax	12–14
Microcrystalline Wax	4–7
Ross Wax 160	3.5
Carnauba Wax	0.5
Candelilla Wax	1.0

Ross Wax 160 is stable under molten conditions, although the color darkens when it is kept in a melted condition under agitation for long periods. The melting point is not adversely affected.

Santowaxes

The Santowaxes are solid hydrocarbons ranging in melting point from a little above room temperature to as high as 210°C (410°F).

Properties

Boiling 10% sulfuric acid or caustic soda has no detectable effect on Santowaxes.

Santowax R is not a true wax but it exhibits wax-like properties. At room temperature it is a light buff, crumbly solid as shipped. Chemically, Santowax R is a mixture of *ortho-*, *meta-*, and *para*-terphenyls. These high-melting hydrocarbons show unusual stability towards heat, acids, and alkalis.

Manila Resin (Loba B)	C		I	I
Ester Gum	C		I	I
N Wood Rosin (Hercules)	C	C	I	I
Piccolite Resin, 85°C	C		I	I
Staybellite Resin	B		I	I

Santowax O

Other Components	*% Santowax O in Mixture*				
	10	25	50	75	90
Paraffin, 120/125°F	C		C		C
Barnsdall Wax, 160/165°F	C		C		C
Beeswax	C	C	C		I
Opal Wax	C		I		I
Stearic Acid	I	I	I		I
N Wood Rosin	C	C	I	I	

Santowax P

Other Components	*% Santowax P in Mixture*				
	0.5	1–2	3	5	10–75
Paraffin Wax, 120/125°F	I				
Paradene No. 1				C	I
Paradene No. 2		C	I		
N Wood Rosin					I

C—Compatible; I—Incompatible; B—Almost Compatible.

Physical Properties of the Santowaxes

Santowax	O	M	P
Color, NPA	< 1.8	2.2–4.2	0–1.25
Density, g/cc	1.14	1.164	1.236
Melting Point	Santowaxes do not have sharp melting points		
Begin to Soften	35°C. (95°F.)	75°C. (167°F.)	200°C. (392°F.)
Completely Liquid	50°C. (122°F.)	85°C. (185°F.)	215°C. (419°F.)
First Crystals	—	< 90°C. (< 194°F.)	209–213°C. (408–415°F.)
Hold Point	52–54°C. (125.6–129.2°F.)	83–85°C. (181–185°F.)	
Distillation Range ASTM D-20 (corrected)	330–345°C. (626–653°F.)	370–378°C. (698–712°F.)	381–388°C. (718–730°F.)
Flash Point	171°C. (340°F.)	207°C. (405°F.)	207°C. (405°F.)
Flame Point	183°C. (390°F.)	229°C. (445°F.)	238°C. (460°F.)
Viscosity at 210°F. (Saybolt Universal Seconds)	40.6	39.3	28.8 (225°C.)

Electrical Properties of Santowaxes

	Dielectric Constant	Resistivity—ohm/cm³ 100°C.	Resistivity—ohm/cm³ 250°C.
Santowax O	2.54	$8{,}200 \times 10^9$	—
Santowax M	2.62	$2{,}600 \times 10^9$	—
Santowax P	—	—	30×10^9

Compatibility of Santowaxes

Santowax M

Other Component	% Santowax M in Mixture 10	25	50	75	90
Paraffin Wax, 120/125°F.	incompatible with 2% or more Santowax M				
Barnsdall Wax, 160/165°F.		C	B	I	I
Ceresin	I		I		I
Carnauba Wax	C		C	C	I
Candelilla Wax	C	C	I		I
Montan Wax	I		I		I
Beeswax	I		I		I
Japan Wax	C	I	I		
Opal Wax	C	C	I		I
IG Wax O	I		I		I
Stroba Wax	I		I		I
Acrawax	C		C		B
Stearic Acid	I		I		I
Shellac	C		I		I

Table 9:7

Comparative Hardness of Some Waxes

(Smaller figure denotes greater hardness)

Wax	*Average penetration in cm × 10^{-2} in 60 seconds at 25°C under load indicated. ASTM D5-25.* *Load* 50 g	150 g	250 g
Santowax M	0	0	2.3
Santowax P	0	0	3.3
Carnauba Wax	0	2.3	4.7
Candelilla Wax	0	3.3	4.7
Santowax O	1.3	4.0	10.7
50% Paraffin : 50% Santowax M	3.7	12.3	18.0
75% Paraffin : 25% Santowax M	7.3	18.0	25.3
Ceresin Wax	9.3	15.7	26.0
95% Paraffin : 5% Santowax M	10.7	21.0	29.7
98% Paraffin : 2% Santowax M	11.7	26.0	32.7
Paraffin Wax	14.3	23.3	31.3
Santowax R, aged 24 Hours	28.7	—	—
Santowax R, nonaged	60.0	—	—

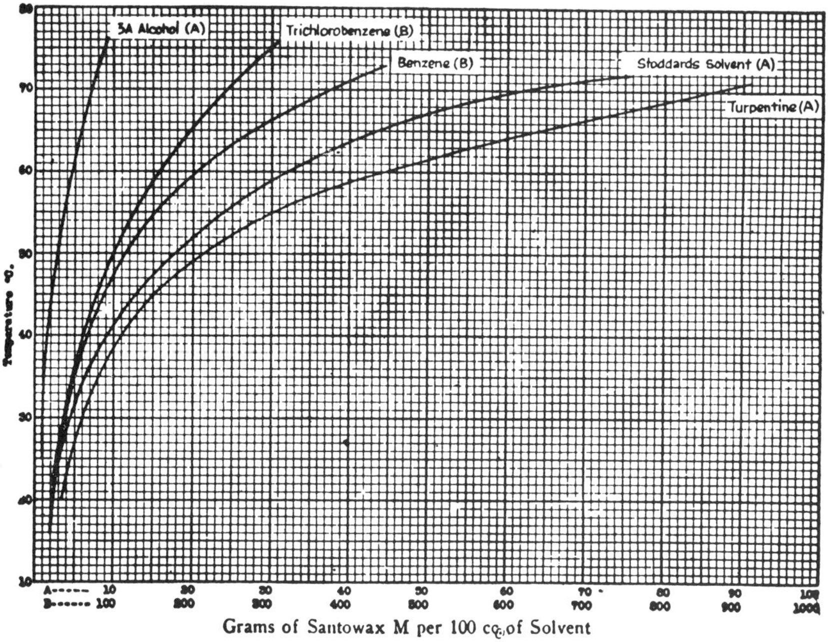

FIG. 9:2. SOLUBILITY OF SANTOWAX M

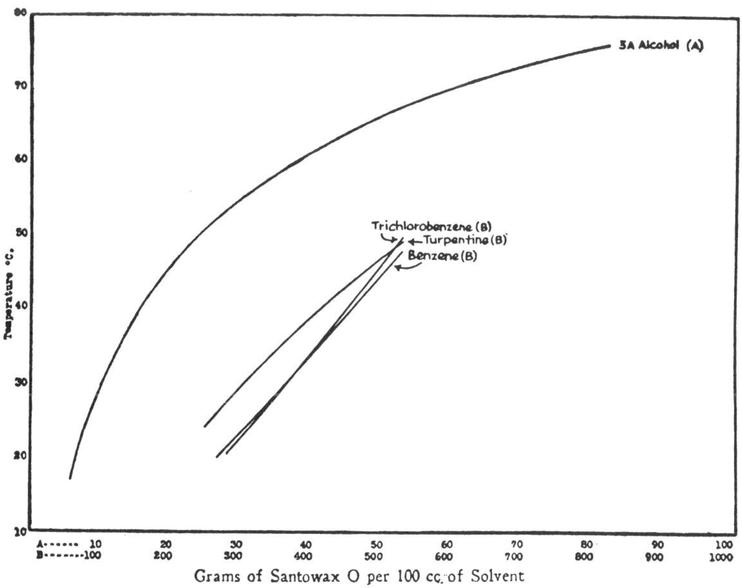

FIG. 9:3. SOLUBILITY OF SANTOWAX O

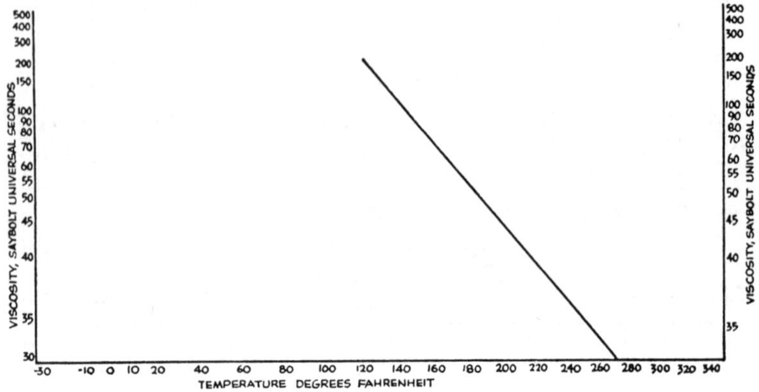

Fig. 9:4. Solubility of Santowax P

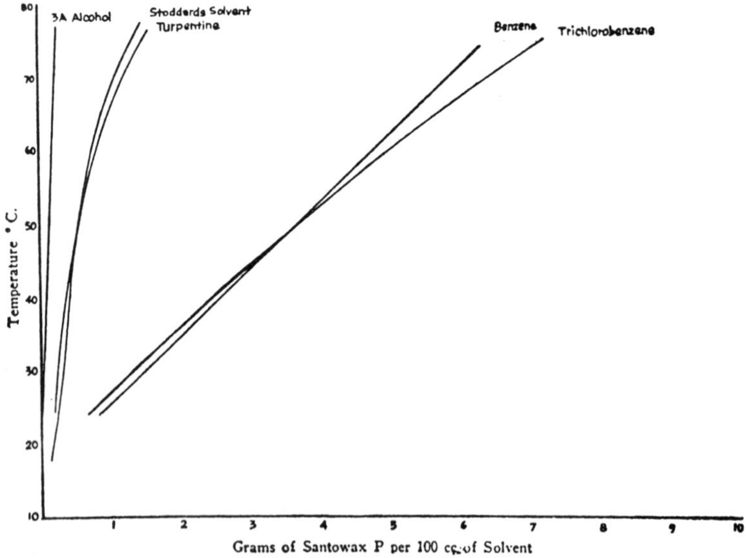

Fig. 9:5. Viscosity of Santowax O

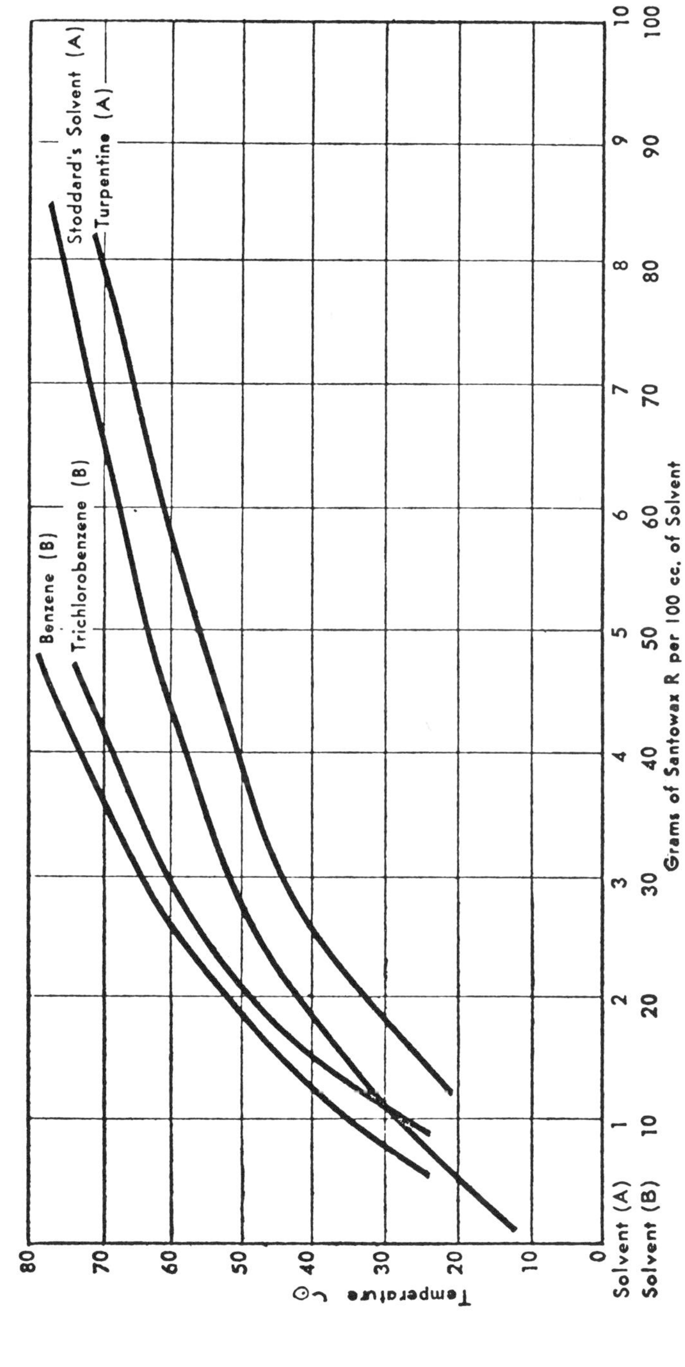

FIG. 9:6. SOLUBILITY OF SANTOWAX R IN FOUR COMMON SOLVENTS

Santowax R

Physical Properties

Density at 25°C—g./cc	1.133
Color (molten)[1]—NPA	3.5–4.5 (amber)
Melting Point	
Santowax R does not have a sharp melting point.	
Begins to soften at	60°C (140°F)
Completely liquid	145°C (239°F)
Solidification Temperatures:	
Upper hold point[1]	137–145°C (278–293°F)
Lower hold point[1]	48–63°C (118–145°F)
Distillation Range ASTM D-20 modified	364–418°C (687–784°F)
Flash point	191°C (375°F)
Flame point	238°C (460°F)
Vapor Pressure—mm of Mercury	

T °C	*mm Hg*	*T °C*	*mm Hg*	*T °C*	*mm Hg*
150	1.3	230	26	310	220
160	2.0	240	36	320	270
170	3.0	250	48	330	340
180	4.6	260	63	340	410
190	6.8	270	83	350	500
200	9.8	280	110	360	610
210	14.0	290	140	370	730
220	19.0	300	180	380	870

[1] Monsanto Specifications.

Electrical Properties

Dielectric Constant		2.58
Volume Resistivity—ohm-cm	100°C	$140{,}000 \times 10^9$
	135°C	$35{,}000 \times 10^9$
	155°C	550×10^9

SOLUBILITY

Solvent	*G Santowax R per 100 ml of Solvent* 25°C	75°C
3A Alcohol	0.008	0.6
Turpentine	1.2	9.0
Fuel Oil, API 36°, 38 S.U.S. at 100°F	1.66	8.4
Benzene	6.0	42.0
Trichlorobenzene	9.0	45.0
Stoddard's Solvent	0.7	8.0
Nitromethane	0.2	7.5
Nitroethane	2.0	15.0
1-Nitropropane	2.0	20.0
2-Nitropropane	2.0	20.0

(See also Figs. 1 and 2.)

Dissolution Temperatures for 20% Solutions of Santowax R in Various Solvents

Solvent	*Minimum Temp. for complete solution (20 gm of Santowax R in 80 gm of solvent* °C	°F
Tetrachloroethane	53	127.4
Dioxane	58	136.4
*Ortho*dichlorobenzene	58	136.4
Benzoyl chloride	60	140.0
Monochlorobenzene	60	140.0
Trichlorobenzene (mixed isomers)	60	140.0
Benzyl chloride	62	143.6
Benzene	64	147.2
Trichloroethylene	64	147.2
Pyridine	66	150.8
Chloronaphthalene	72	161.6
Furfural	72	161.6
Xylene	72	161.6
Bromonaphthalene	73	163.4
Diphenyl oxide	76	168.8

Compatibility with Resins

The resin compatibility of Santowax R has not been fully determined. However, this involves only a simple experimental procedure based on melting together Santowax R and the resin being studied. On cooling, the mixture is observed for separation or crystallization. After determining the room temperature compatibility, the blend can be placed in cold storage and observed further for separation. A series of mixtures should be studied.

Comparative Hardness of Waxes and Mixtures
(Smaller figure denotes greater hardness)

Wax Tested	*Average Penetration in cm× 10^{-2} in in 60 seconds at 25°C under load indicated. ASTM D-5-52* 50g	150g	250g
meta-Terphenyl (*Santowax M*)	0	0	2.3
para-Terphenyl (*Santowax P*)	0	0	3.3
Carnauba Wax	0	2.3	4.7
Candelilla Wax	0	3.3	4.7
ortho-Terphenyl (*Santowax O*)	1.3	4.0	10.7
50% Paraffin and 50% *meta*-Terphenyl	3.7	12.3	18.0
75% Paraffin and 25% *meta*-Terphenyl	7.3	18.0	25.3
Ceresine Wax	9.3	15.7	26.0
95% Paraffin and 5% *meta*-Terphenyl	10.7	21.0	29.7
98% Paraffin and 2% *meta*-Terphenyl	11.7	26.0	32.7
Paraffin Wax	14.3	23.3	31.3
Montan Wax	15.0	28.7	37.3
Santowax R: Aged 24 hours	28.7	—	—
Unaged	60.0	—	—

Penetration Tests (ASTM D-5-52) on Mixtures of Santowax R and Soft Paraffin

Composition of Mixture		*Approximate Temperature For First Crystals, °C*	*Apparent Segregation*	*60-Sec. Penetration in cm × 10^{-2} at 25°C*		
Percent Santowax R	Percent Paraffin			50 g Load	150 g Load	250 g Load
100	0	143	Yes	[1]38	80	161
				[2] 8	21	28
75	25	140	No	11	27	43
50	50	135	No	17	42	56
25	75	120	No	26	58	109
0	100	52	No	16	42	53

[1] Soft Portion. [2] Hard Portion.

Coefficients of Thermal Expansion of Santowax R

Temp. Range	*Coefficient of Cubical Expansion*	*Coefficient of Linear Expansion*
	Vol/unit vol/°C	Length/unit length/°C
25–60°C	8.3×10^{-4}	2.8×10^{-4}
60–77°C	11.5×10^{-4}	3.8×10^{-4}
25–77°C	9.4×10^{-4}	3.1×10^{-4}

Limited studies indicate that Santowax R is relatively incompatible with high-molecular-weight (21,000 or more) polyethylene, but is compatible with low-molecular-weight (4,000 or less) material. In the latter case a 50-50 mixture was miscible at 140°C and short term observation showed no separation.

Santowax R has some compatibility with polyvinyl chloride resin. Small amounts greatly improve the light stability of formulations which already use a standard stabilizer system. For example, the following data show how a typical PVC formulation behaved in the weatherometer using a commercial stabilizer system with and without Santowax R.

Formulations Tested:

	Additive:	
A	*Opalon* 300 PVC Resin	100 parts
	Dioctyl Phthalate	25 parts
	Santicizer 160	25 parts
	G-60 (Rohm & Haas)	3 parts
	Mark XI (Argus Chem.)	2 parts
	Mark XX (Argus Chem.)	1 part
B	Same formulation plus of *Santowax R.*	2 parts

Both formulations produced clear plastics. These yellowed slightly in the Weatherometer and color thereafter remained essentially constant until degradation occurred. This was indicated by the development of rusty "speckles" in the plastic. Degradation was rapid thereafter.

Results of Weatherometer Tests

Formulation "A" withstood 750 hours in the Weatherometer before speckling. Formulation "B" withstood 1300 hours before speckling. There was equal improvement with the more expensive Santowax M (*meta*-isomer) and not as much improvement with Santowax O, the *ortho*-isomer.

Monsanto has not tested other resins but will be glad to assist with such projects.

Toxicity

Santowax R has a very low order of toxicity (MLD for rats is greater than 10 grams per kilo of body weight). So far as is known, no particular hazard attends the use of this product in industrial formulations. The product is not to be taken internally or maintained in contact with the skin, eyes or other body parts.

Casawax $R_1CONHCH_2CH_2NHCOR_2$

The following products are available in the standard CASAWAX range:

Casawax	R_1R_2
C103	Oleyl
A93/W	Stearyl
A93/G	Oleyl/Stearyl
A93/B	Oleyl/Stearyl
A24	Behenyl
C29	Behenyl (Modified)
C30	Behenyl (Modified)

Additionally available is CASAWAX W10, a mixed ester/amide wax of formula R CONH CH_2CH_2 OOC R where R is a stearyl radical.

Casawax	C103	A93/W	A93/G	A93/B	A24	C29	C30	W10
Melting point (Ring and Ball)	115-120°C	140-145°C	135-140°C	130-135°C	140-145°C	200-210°C	220-230°C	90-95°C
Acid value	<10							
Ash content	NEGLIGIBLE							
Hydrocarbon content	NEGLIGIBLE							
Penetration 25°C/100 gm/5 secs	30-40	5-10	5-15	5-15	5-10	0-5	0-5	0-5
Specific Gravity	0·95	0·95	0·95	0·96	0·99	0·99	1·00	0·97
Flash point	>280°C	>280°C	>280°C	>280°C	>300°C	>300°C	>300°C	>200°C

Chapter 10

CHLORINATED WAXES

The synthetic waxes produced by the chlorination of naphthalene are both crystalline and amorphous. They occur in translucent, black, light, and various colors.

Important properties of the chloronaphthalene waxes are that they are good dielectrics, are free of moisture, and impervious to moisture. They do not support combustion, they are inert to acids and alkalies, are neutral, and do not corrode metals. They have sharp melting points, and melt to liquids of very low viscosity. The more highly chlorinated products are less volatile.

The chloronaphthalene waxes are soluble in benzene, toluene, carbon tetrachloride; generally insoluble in acids and alkalies.

They should be used with caution as they have adverse physiological and dermatological effects on many individuals.

Halowax 1001

Properties

Halowax 1001 is a mixture of tri- and tetrachloronaphthalenes. It is a white crystalline waxlike solid with 50 per cent chlorine content.

At 30°C Halowax 1001 is a crystalline solid. In the range of 90 to 95°C it melts sharply to a very mobile liquid that has a flat viscosity characteristic. At the melting point, the vapor pressure is less than one mm. The initial boiling point of the mixture is 310°C. There is no decomposition up to or during boiling so that this product vaporizes completely without residue.

At temperatures from 30°C to minus 60°C, there is no physical change in Halowax 1001.

Halowax 1001 is soluble in aromatic and chlorinated solvents and petroleum naphthas. Many organic materials such as alkyd, polystyrene, phenolic, and vinyl resins, natural and synthetic rubber as well as many natural and petroleum waxes are soluble in chloronaphthalene 1001. As many of the materials soluble in 1001 are insoluble in each other, its mutual solvent action is the basis for a number of important applications

Table 10:1

COMPARATIVE PROPERTIES OF CHLORONAPHTHALENES

	Number of Halowax		*1001*	*1099*	*1013*	*1014*	*1051*
1	Composition		Tri- + Tetra-Chlor	Tri- + Tetra-Chlor	Tetra- + Penta-Chlor	Penta- + Hexa-Chlor	Octa-Chlor
2	Chlorine Content, % (Approximate)		50	52	56	62	70
3	Specific Gravity	@ 25°C	1.58	1.59	1.67	1.78	2.00
		@ 60°C					
4	Initial B. Pt.	@ 30 mm	200°C	205°C	222°C	242°C	310°C
		@ 100 mm	234°C	241°C	258°C	278°C	—
		@ 760 mm	308°C	315°C	328°C	344°C	—
5	Distillation Range						
6	Softening Point (M.P.), °C (Approx.)		93	102	120	137	185
7	Flash Point, °C		141	141	180	180	None to 430
8	Fire Point, °C		None to Boiling	None to Boiling	None to Boiling	None to Boiling	None to Boiling
9	Specific Heat g cal/g/°c		0.22 @ 15° 0.65 @ 100°			0.19 @ 15° 0.48 @ 100°	
10	Color		White to Pale Yellow	White to Pale Yellow	Light Straw	Light Yellow	Light Yellow
11	Acidity, Maximum (Mg. KOH/g)		0.05	0.05	0.05	0.05	0.10
12	Viscosity, S. U. Sec. (Approx.)		30 @ 130°C	31 @ 130°C	33 @ 130°C	35 @ 150°C	
13	Volatility: 200 g., 9.5 sq. in., 10 days @ Room Temp. gm/sq in/hr @ 105°C		0.010		0.005	0.001	

			25°C	*100°C*	*25°C*	*100°C*	*25°C*	*130°C*	*25°C*	*150°C*	*25°C*
14	Dielectric Constant	@ 60 Cycles/Sec		4.10		4.10	4.82	3.80	4.42	3.78	
		@ 1000 Cycles/Sec	5–6	4.10	5–6	4.10	4.82	3.80	4.42	3.70	3.7
15	Power Factor	@ 60 Cycles/Sec		0.37		0.37	0.0017	0.451	0.0009	0.991	
		@ 1000 Cycles/Sec	0.002	0.05	0.002	0.05	0.0003	0.336	0.0002	0.437	0.0011
16	Resistivity megohm cm		Over 1×10^8	1×10^5	Over 1×10^8	1×10^5	Over 1×10^8	1.45×10^5	Over 1×10^8	1.17×10^4	Over 1×10^8

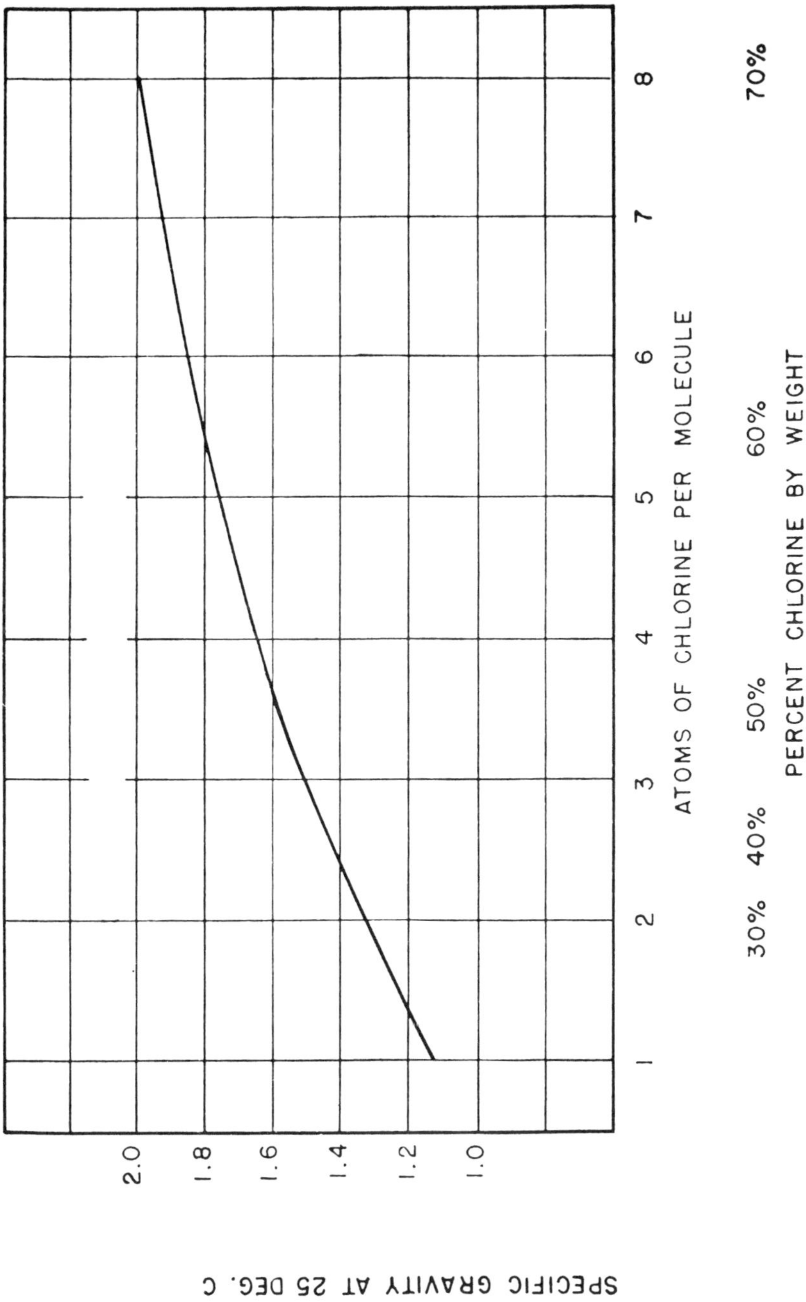

FIG. 10:1. SPECIFIC GRAVITY VS. CHLORINE CONTENT.

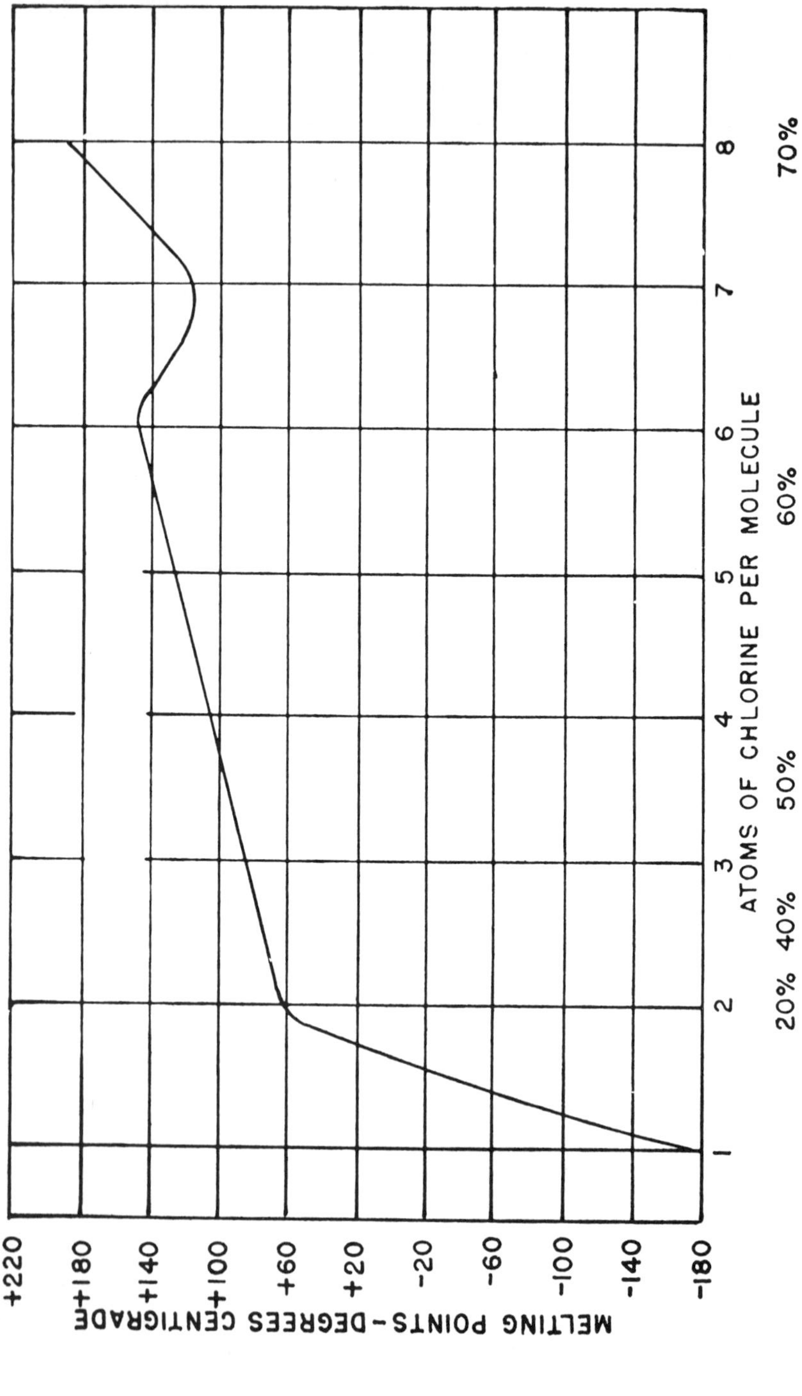

Fig. 10:2. Melting Point vs. Chlorine Content.

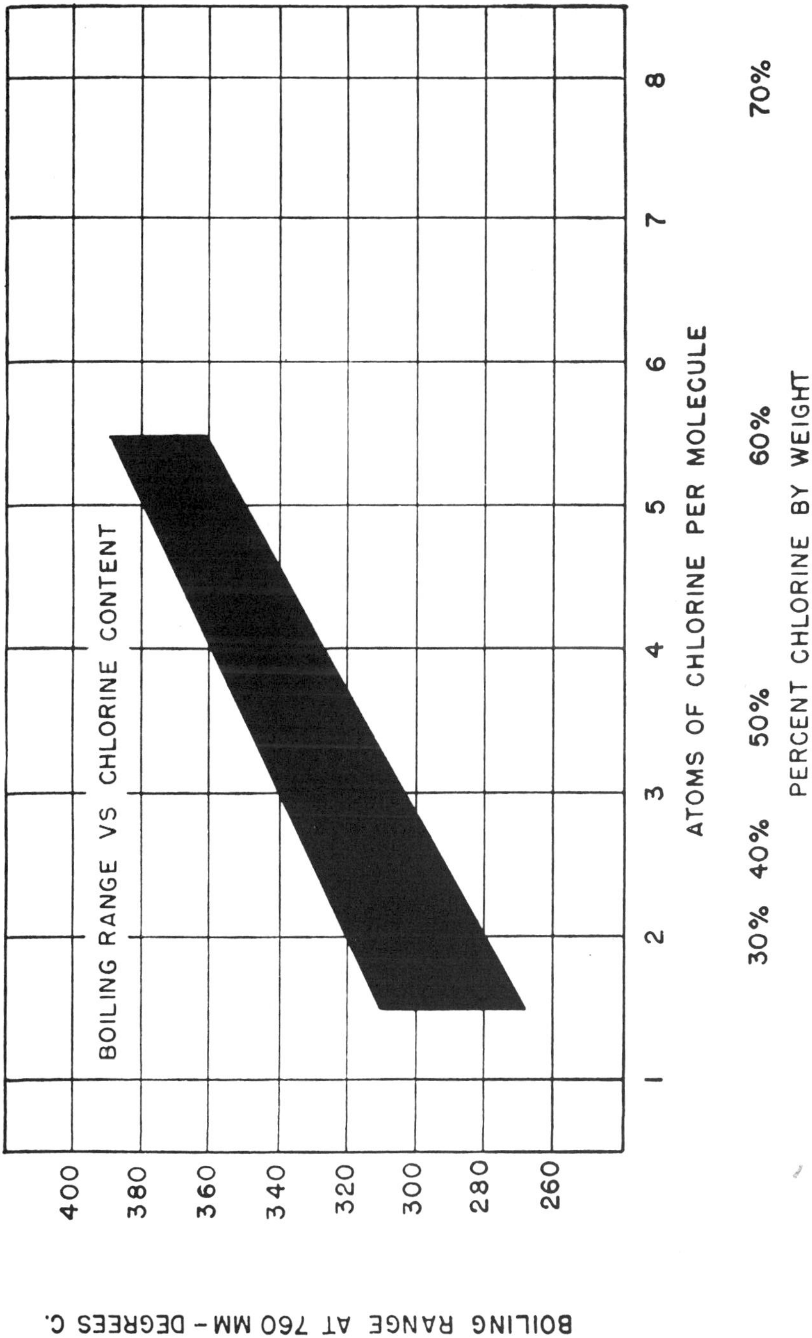

FIG. 10:3. BOILING RANGE VS. CHLORINE CONTENT.

Halowax 1001 is manufactured to take maximum advantage of the high dielectric constant (K) possible in the chlorinated naphthalenes. Depending on technique of measurement, the 1000 cycle K is 5.5 at 30°C and 4.1 at 100°C.

Other important electrical properties obtained are low DC conductivity and AC losses. The losses, as measured by 1000 cycle power factor, are below 0.1 per cent at 30°C. The direct current resistivity of chloronaphthalene 1001 is of the order of 1×10^9 megohm cm at 30°C and 1×10^5 megohm cm at 100°C.

As is typical of polar materials, Halowax 1001 goes through a power factor maximum and a change in dielectric constant with variations in temperature and in frequency of applied voltage. The 1000 cycle dieletric constant changes to approximately 3.0 at −30°C.

As a final step in the manufacture of 1001, it is treated with an adsorbent before shipment from the Halowax plant, to provide a high quality dielectric. However, processing in users' plants will add impurities which degrade the electrical characteristics. These can be readily restored by the user through filtration with the use of an activated filter aid.

The unique combination of high dielectric constant and low losses coupled with the low viscosity of the liquid has made Halowax 1001 an important factor in the electronics components industry as a capacitor impregnant that yields more microfarads per unit of volume than competitive products. Its use in the same industry as a binder for electrical grade ceramics is a result of the strength which it imparts to the preform and its complete volatilization without residue in firing.

The paper coating, precision casting, and chemical processing industries have found other combinations of properties of 1001 suitable to their uses.

Solubility

Halowax 1001 is generally soluble in the following (or related) chemicals:

Carbon tetrachloride, ethyl ether, methyl ethyl ketone, petroleum naphtha, toluene. (First three are limited in some concentrations.)

Compatibility

Halowax 1001 is generally compatible with the following materials (or related products):

Amorphous wax 190/195, alkyd resins, dioctyl phthalate, phenolic resins, polystyrene (e.g. Koppers Dylene 3), synthetic rubber, tricresyl phosphate.

Halowax 1099

Halowax 1099 is a mixture of trichloro- and tetrachloronaphthalenes, and is similar to Halowax 1001 in composition and application.

This product was designed specifically for use as a capacitor impregnant (dielectric) where higher operating temperatures are encountered.

Prolonged service at elevated temperature is provided by virtue of the relatively high melting point of Halowax 1099.

Halowax 1099 is generally compatible with the same materials as Halowax 1001.

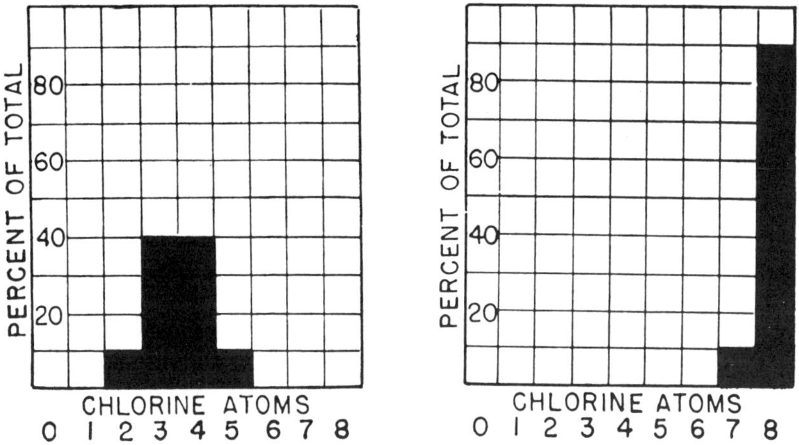

(Above: Left, 1001; Right, 1051).

1013 **1014**

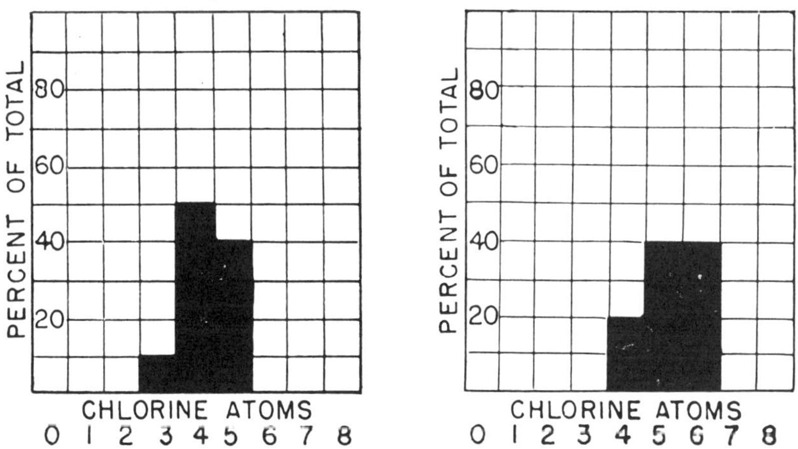

Fig. 10:4. Amount of Chlorine in Halowaxes.

Halowax 1013 and 1014

Halowax 1013 and 1014 are hard amorphous waxes with high melting points. Halowax 1013 is essentially a mixture of tetra- and pentachloronaphthalenes having a chlorine content of 56 per cent. Halowax 1014 is a mixture of penta- and hexachloronaphthalenes having a chlorine content of 62 per cent.

These waxes have sharp melting points and become low viscosity liquids as these temperatures are reached. They retain their mechanical strength to temperatures near their melting points. 1013 and 1014 are stable to temperatures beyond their distillation ranges which are above 320°C. Both materials are light in color.

These materials have good electrical insulating properties. They are flame-retardant. They have excellent resistance to moisture-vapor transmission. They will not hydrolyze in the presence of water. These waxes are resistant to acids and alkalies except very strong oxidizing agents. They are also resistant to fungi.

1013 and 1014 are generally compatible with various resins, plasticizers, waxes, rubber, and synthetic elastomers. However, the percentage of 1013 and 1014 that can be compounded with these materials is limited in many cases.

These waxes are generally soluble in chlorinated solvents, aromatic solvents, and petroleum naphthas. They are not soluble in water and alcohols. Their solubility is limited in ketones, acetates, and esters.

Current applications include electrical insulating compounds and stop-off compounds, as well as additives in automobile and industrial gear oils and cutting oils.

Solubility

Halowax 1013 and 1014 are generally soluble in the following (or related) chemicals:

Benzene, carbon tetrachloride (limited in some concentrations.), petroleum naphtha, toluene, xylene.

Compatibility

Halowax 1013 and 1014 are generally compatible with the following materials (or related products):

Amorphous wax, ethyl cellulose, phenolic resins, polystyrene (e.g. Koppers Dylene 3), synthetic rubber, vinyl resin.

Halowax 1051

This product, essentially octachloronaphthalene, is a hard resinous solid with a very high melting point, about 185°C. It has the highest chlorine content possible in these compounds, about 70 per cent. It has extremely low volatility. It is stable at temperatures up to its melting point, but decomposes rapidly at elevated temperatures with a rapid evolution of chlorine containing gases. This combination of properties makes Halowax 1051 exceptionally flame resistant.

The material has very limited solubility in resins, plasticizers, and solvents. It can be pulverized readily and employed as an organic filler by milling or grinding; or it can be dispersed by intensive mixing as in a Banbury mixer.

Toxicity

At temperatures below 550°F in fifteen years of experience with these low polymers, there has been no report of acute or chronic poisoning, dermatitis, or other adverse physiological effects from Halocarbon waxes.

At temperatures above 550°F, these polymers are subject to thermal cracking, with rapid breakdown occurring above 620°F. This breakdown may occur from heating in a vessel or operating equipment, or from the vapors coming in contact with an open flame or passing through a lighted cigarette. Breakdown products in air are toxic, acidic substances which produce harmful physiological effects from prolonged exposure. When these waxes are heated above 400°F for long periods in open vessels, it is recommended that adequate ventilation be provided.

Halowax 0077 is a refined chlorinated terphenyl having an approximate chlorine content of 60%. It is a wax-like thermoplastic resin with some stability to heat.

Halowax 0077 is non-corrosive to metals and is not hydrolyzed by water, alkali, or acids. Owing to its high chlorine content, low volatility, and wide range of compatibility the product is recommended as a compounding agent to provide fire retarding properties to plastic and resin materials. Because of its good dielectric constant, Halowax 0077 is particularly desirable as a flame-retardant component in the formulation of clear wire saturants.

Property	*Testing Method*	*Unit*	*Required Value*
Chlorine Content	612	per cent	60 (nominal)
Specific Gravity at 25°C	604	—	1.62–1.69
Softening Point	E-28	° Centigrade	100–110
Acid Number	616	mg of KOH/g	0.10 (max.)

Halowax 0091 is a crude chlorinated Terphenyl having an appoximate chlorine content of 60 per cent. It is used as an ingredient in making black wire saturant compounds for asbestos wire to impart flame resistance at high temperatures.

Property	*Testing Method*	*Unit*	*Required Value*
Chlorine Content	WC-37-B-1	per cent	60 (nominal)
Specific Gravity at 25°C	WC-1-D-4	—	1.58–1.66
Softening Point	WC-71-A	Centigrade	95–100
Acidity	WC-208-D	—	Neutral to slightly alkaline
Benzene Insoluble Matter Content	WC-15-V	per cent	0.38 max.

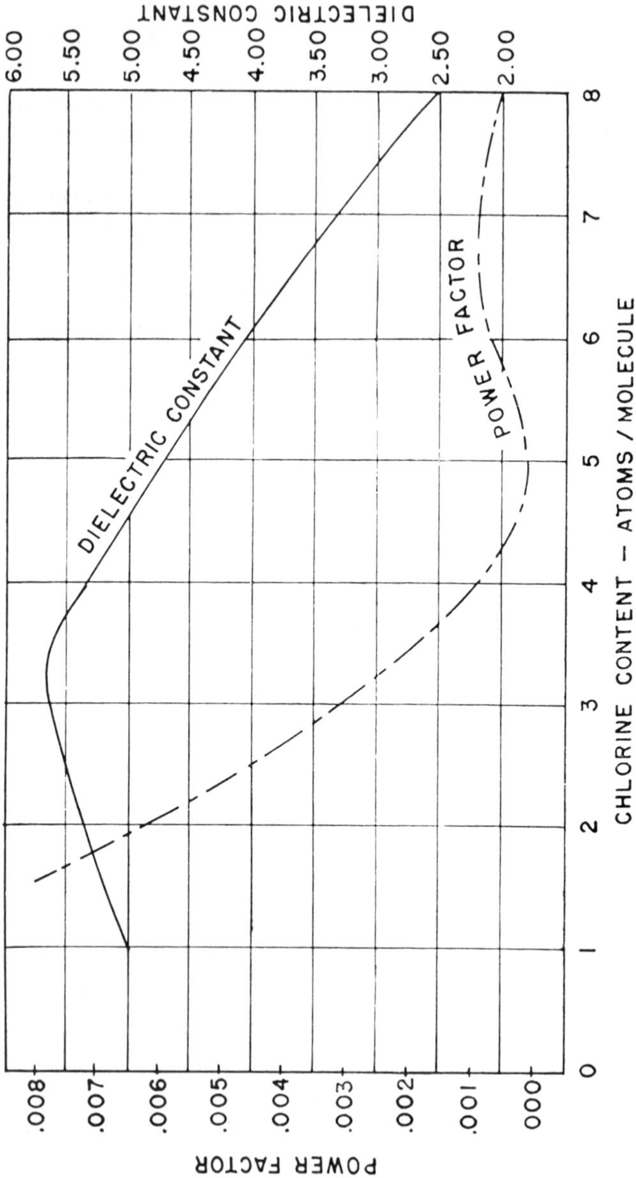

FIG. 10:5. DIELECTRIC CONSTANT AND POWER FACTOR VS. CHLORINE CONTENT MEASURED AT 30°C AND 1,000 CPS.

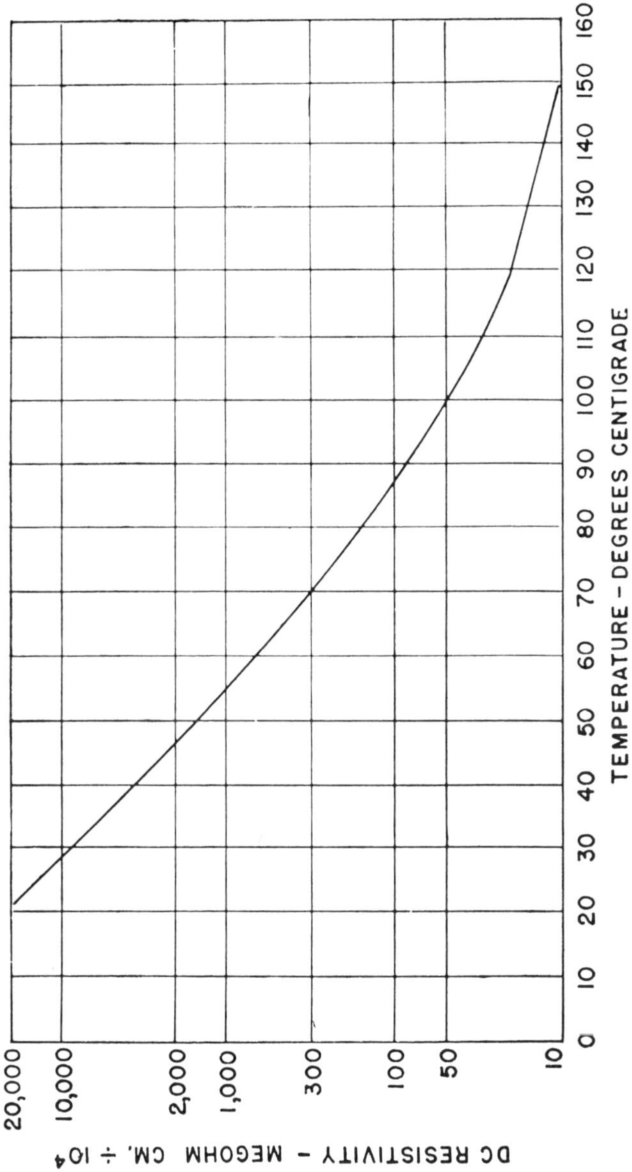

Fig. 10:6. Direct Current Resistivity vs. Temperature Halowax 1001.

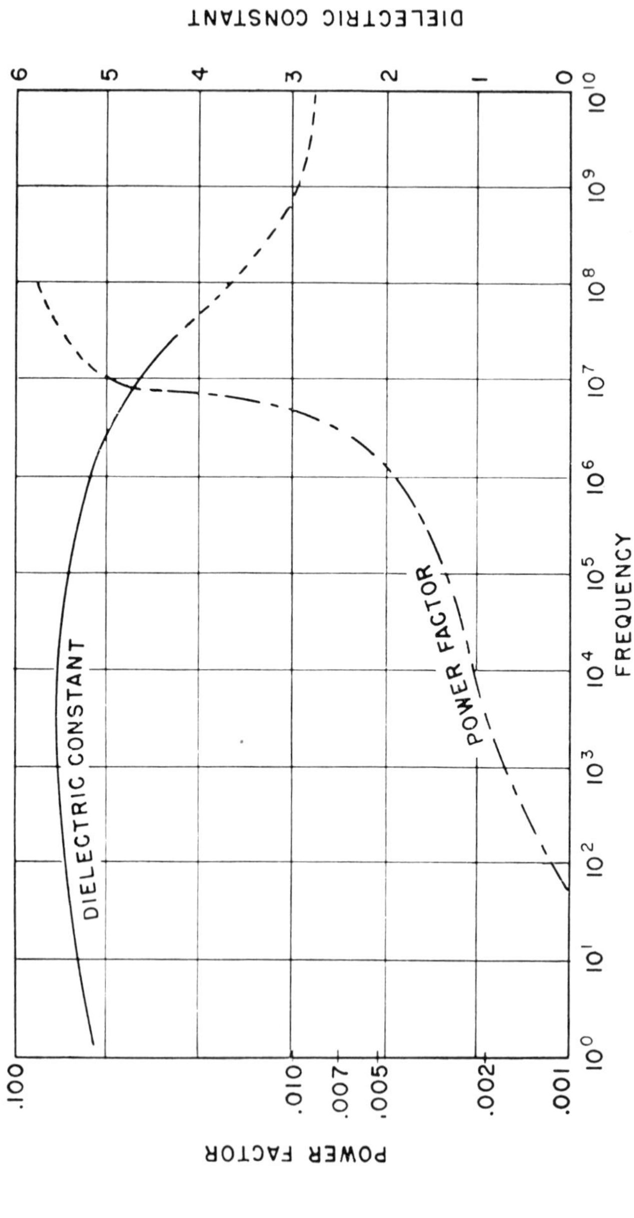

Fig. 10:7. Dielectric Constant and Power Factor vs. Frequency at 30°C, Halowax 1001.

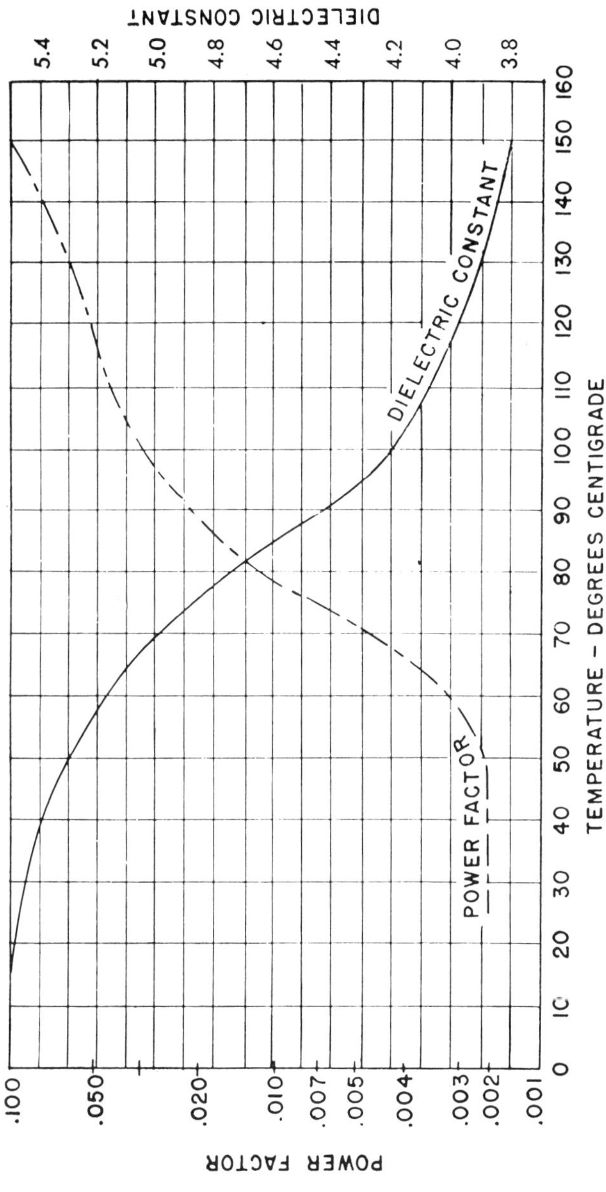

FIG. 10:8. DIELECTRIC CONSTANT AND POWER FACTOR VS. TEMPERATURE AT 1,000 CPS HALOWAX 1001.

Testing Methods

In the preceding tables, the properties described are measured by current issues of the testing methods identified. The prefix "E" identifies standard ASTM test methods. All other numbers are standard Koppers Co., Inc. methods which are available upon request.

In general, the properties of Halowax chloronaphthalenes as presented in this chapter are based on laboratory tests employing established Halowax test procedures. These procedures follow the ASTM methods as closely as possible.

Listed in the following table are the equivalent Halowax test procedures and ASTM methods where they coincide. Where there is no ASTM equivalent, as indicated by an (*), the Halowax test procedure has been designed to measure the inherent characteristics of the chloronaphthalenes.

Property	*Halowax Procedure*	*ASTM*
Chlorine Content (Thompson—Oakdale Method)	612	*
Specific Gravity: Solids	604	*
Solutions	603	*
Boiling Point	605C	*
Distillation Range	none	D86-56
Softening Point (Melting Point)	none	E28-58T
Flash and Fire Points (Cleveland Open Cup)	none	D92-57
Acidity	616	*
Viscosity	609	*
Dielectric Constant	none	D150-59T
Power Factor	none	D150-59T
Resistivity	613	*

Seekay Waxes[1]

Seekay waxes are British chlorinated naphthalenes made in different grades; color, yellow to black; melting point, 65–125°C. They are insoluble in water and in most hydrocarbon solvents.

Grades Available			
A Quality	*R Quality*	*Rc Quality*	*Setting Point °C*
A68	R68	—	67–73
A93	R93	R93c	90–95
A700*	—	—	89–93
—	R700*	—	83–88
A123	R123	R123c	120–125

A quality—Black; may contain a small amount of inorganic matter. R quality—Pale yellow to yellow; produced by refining A quality. Rc quality—Pale yellow to yellow; similar to R quality but conforms to special electrical specifications.

* Blends of A93 and R93 respectively with small amounts of plasticisers.

[1] Imperial Chemical Industries, Ltd.

PHYSICAL PROPERTIES OF SEEKAY R93

		Approx.
Specific gravity at 20°/20°C		1.58–1.60
Boiling range at 15 mm Hg	°C	170–230
Vapour pressure at 25°C	mm Hg	0.003
Vapour pressure at 100°C	mm Hg	0.4
Volatility. Loss from surface during first 5 hours at 100°C	$g/cm^2/h$	0.0004
Viscosity at 118°C	cS	2
Specific heat at 18°C	cal/g°C	0.2
	Btu/lb°F	0.2
Latent heat of fusion	cal/g	20
	Btu/lb	36
Odor		slight

Compatibility

Seekay waxes are compatible with bitumens, most natural waxes, polystyrene, alkyd and phenolic resins, natural and synthetic rubber.

Toxicity

Poisoning from chlorinated naphthalene may take two forms: damage to the skin, producing acne, particularly of the face; and toxic jaundice, produced by either an acute or subacute necrosis of the liver, or by acute yellow atrophy of the liver. A number of cases of dermatitis and one of acute atrophy of the liver among workers who had been exposed to the fumes or dust of chlorinated naphthalene were reported. If cases of poisoning are to be prevented, the wax must not be overheated, and the fumes and dust must be properly evacuated by exhaust ventilation.[1]

Uses

Dielectric impregnant for electrical condensers

The Rc grades of Seekay Wax have particular value as impregnants for the paper of electrical condensers because of their relatively high dielectric constant.

Their high melting point is an advantage where condensers are required for use under tropical conditions.

Some of the electrical properties of Seekay Wax grades R93c and R123c, when used as condenser impregnants, are given in the preceding table.

Electric cable coating

Seekay Wax is used as a nonflammable and water-resistant impregnant for electric cable wrappings. Underground cable is made resistant to attack by fungi and termites. The plasticized grades A700 and R700 are particularly suitable since they allow the cable greater flexibility. Seekay Wax can be pigmented for quick identification, suitable dyestuffs and pigments being available from I.C.I. Dyestuffs Division.

---

[1] E. Collier, Lancet, **72** (1943).

In electroplating

Seekay Wax is suitable for stopping-off in electroplating because it has good resistance to acids and alkalis, and the high melting point of the recommended grade (R123) permits use of it even in hard chrome baths operating at relatively high temperatures. The Seekay Wax is recoverable.

Wood preservation

Seekay Wax protects timber against attack by insects (including termite) and fungi. Grade R93 is added to fluids containing BHC, DDT, or Dieldrin, used for eradicating wood beetles (e.g. common furniture beetle, death-watch beetle, etc.), to increase the period of protection against possible subsequent reinfestation.

Chlorowax

Chlorowax is a trade mark applied to several liquid and resinous products which are derived from the carefully controlled chlorination of paraffin waxes and liquids. As a group, they are noted for non-flammability and general chemical inertness. They are resistant to the attack of dilute mineral acids and dilute alkalies and do not readily oxidize.

These products are insoluble in water, the lower alcohols and glycols, and they are distinguished by an unusually wide range of solubility in and compatibility with most organic solvents, resins, and plastics.

Chlorowax can be processed at temperatures up to 325°F for 30 to 45 minutes, depending on temperature and processing conditions. If higher temperatures or longer processing times are required, Chlorowax 70-S, a heat stabilized version of Chlorowax 70, is available. Stabilizers of the type normally used for processing polyvinyl chloride resins may also be employed.

Formulating for Fire Retardancy

In most cases, Chlorowax is used to replace, wholly or partially, conventional resins and plasticizers to become part of a finished formulation which provides the desired degree of flame retardancy for the intended application. Fire retarding characteristics of a given formulation wll generally improve with greater usage of Chlorowax and, conversely with lesser usage of flammable components in the organic binder of the formulation. The organic base or polymeric portion of typical formulations relying solely on Chlorowax for flame retardancy will contain 25% to over 90% Chlorowax.

When heat from a flame is encountered, Chlorowax provides flame retardancy by liberating a nonflammable gas.

Chlorowax is often used in combination with other non-chlorinated materials, such as those listed below, to obtain a greater flame retardant effect than is possible by the use of single additives. The greater efficiency of a combination may allow the use of a lower total percentage of flame retardant additives, thereby helping to maintain desired physical properties of the prime plastic, rubber or polymer.

PROPERTIES OF CHLOROWAX

	Chlorine Content Percent	*Ball and Ring Softening Point °C*	*Specific Gravity 25°C/25°C*
Chlorowax 70	71	102	1.66
Chlorowax 70 HMP	74	130	1.68
Chlorowax 70 LMP	70	87	1.64
Chlorowax 70 S	69	102	1.63

Chlorowax LV

Color (Gardner Scale 1933)	2
Viscosity	
Poises 25°C (Brookfield Viscometer)	7
Gardner Holdt (Bubble tubes) 25°C	U
Saybolt Universal Seconds 210°F	80
Saybolt Universal Seconds 100°F	1050
Viscosity Index	68
Specific Gravity 25°C/25°C	1.125
Weight (pounds per gallon)	9.4
Bulking Value (gallons per pound)	0.106
Evaporation rate	
(grams per square cm per hour) @ 100°C	0.00008
Refractive Index	1.501
Pour Point (degrees C)	20
Heat Stability	0.250
Chlorine Content (percent)	39
Molecular Weight	460

Chlorowax 40

Color (Gardner Scale 1933)	2
Viscosity	
Poises 25°C (Brookfield Viscometer)	31
Gardner Holdt (Bubble tubes) 25°C	Z-Z1
Saybolt Universal Seconds 210°F	150
Saybolt Universal Seconds 100°F	4000
Viscosity Index	43
Specific Gravity 25°C/25°C	1.17
Weight (pounds per gallon)	9.85
Bulking Value (gallons per pound)	.102
Evaporation rate	
(grams per square cm per hour) @ 100°C	0.00006
Refractive Index	1.505
Pour Point (degrees C)	0
Heat Stability	0.40
Chlorine Content (percent)	43
Molecular Weight	560
Surface Tension dynes/cm @ 25°C	39

Chlorowax 50

Color (Gardner Scale 1933)	2
Viscosity	
Poises 25°C (Brookfield Viscometer)	125
Gardner Holdt (Bubble tubes) 25°C	Z5-Z6
Saybolt Universal Seconds 210°F	265
Saybolt Universal Seconds 100°F	1500
Viscosity Index	22
Specific Gravity 25°C/25°C	1.23
Weight (pounds per gallon)	10.25
Bulking Value (gallons per pound)	0.098
Evaporation Rate	
(grams per square cm per hour) @ 100°C	0.00004
Refractive Index	1.516
Pour Point (degrees C)	10
Heat Stability	0.40
Chlorine Content (percent)	48
Molecular Weight	660

Chlorowax 500C

Resinous

Chlorowax 70

Color (Gardner Color Scale)*	3 maximum
Melting Point °C (Ball and Ring)	102
Specific Gravity—solid	1.65
Density-apparent-pounds per gallon (powder)	7.45
Density-actual-pounds per gallon	13.75
Bulking Value—gallons per pound	0.073
Refractive Index (25°C)	1.535
Particle Size	100% through 20 mesh
Solubility in Water	Insoluble
Odor	None
Dielectric Constant—10^3 cycles	2.89
—10^6 cycles	2.70
Power Factor—10^3 cycles	0.530%
—10^6 cycles	0.065%
Volume Resistivity ohm-cm	3.2×10^{14}
Dielectric strength, volts per mil, 60 cycles	374
Molecular Weight	1,060

Also available:

CHLOROWAX 70 LMP Melting Point—87°C
CHLOROWAX 70 HMP Melting Point—125°C

Chlorowax 70-LP

Color (Gardner Color Scale)*	6 maximum
Melting Point °C (Ball and Ring)	102
Specific Gravity—solid	1.65
Density-apparent-pounds per gallon (powder)	7.45
Density-actual-pounds per gallon	13.75
Bulking Value—gallons per pound	0.073
Refractive Index (25°C)	1.535
Particle Size	100% through 20 mesh
Solubility in Water	Insoluble
Odor	None

* 60% Solution in Toluol

Solubility of Chlorowax 70

CHLOROWAX 70 is soluble in a wide variety of organic solvents, including aliphatic and aromatic hydrocarbons, hydrogenated naphthas, chlorinated aliphatic and aromatic hydro-carbons, ketones, esters and drying oils.

Solvent	*Practical Upper Limit Percent Chlorowax 70 by Weight at Room Temperature*
Acetone	50
Amyl Acetate	50
Benzol	50
Butanol, Normal	Insoluble
Carbon Tetrachloride	50
Chinawood Oil	50
Dioxane	50
Ethanol	Insoluble
Ethyl Acetate	50
Ethylene Dichloride	50
Glycerine	Insoluble
Isopropanol	Insoluble
Linseed Oil	50
Methanol	Insoluble
Methyl Ethyl Ketone	70
Mineral Spirits	50
Monochlorobenzene	50
Orthodichlorobenzene	50
Perchlorethylene	50
Perilla Oil	50
Solvesso 100	70
Soya Bean Oil	50
Toluol	70
Turpentine	50
Xylol	70

Solutions of CHLOROWAX 70 are made most readily by milling CHLOROWAX 70 and solvent in a ball mill, or by adding CHLOROWAX 70 to the solvent with good agitation. Slight warming decreases dissolving time.

The resinous grades have a slight insoluble portion in most solvent systems. Practically complete solubility is possible in all types of oil or resin intermediates.

Compatibility of Chlorowax

MATERIALS	RATIO OF CHLOROWAX TO MATERIALS 1:1	1:2	1:4	METHOD
Acrawax	C	C	C	hot melt
Alkyd Resins				
Phthalic-Drying Oil Modified	C	C	C	solution
Phthalic-Non-Drying Oil Modified	C	C	C	solution
Resin Modified	C	C	C	solution
Styrene Modified	C	C	C	solution
Rosin Modified	C	C	C	solution
Amberol ST-137	C	C	C	hot melt
Aroclors	C	C	C	solution
Admex 710	C	C	C	solution
Asphalt, Petroleum	C	C	C	hot melt
Benzoflex 2-45	C	C	C	solution
Benzoflex 9-88	C	C	C	solution
Butyl Oleate	C	C	C	hot melt
Carbowax[3]	C	C	C	hot melt
Cellulose Acetate	C	C	C	hot melt
Cellulose Acetate Butyrate	C	C	C	solution
Chlorinated Rubber	C	C	C	solution
Coumarone-Indene Resins	C	C	C	solution
Dibenzyl Sebacate	C	C	C	solution
Dibutyl Phthalate	C	C	C	solution
Dicapryl Phthalate	C	C	C	solution
Di-iso-butyl Adipate	C	C	C	solution
Dioctyl Adipate	C	C	C	solution
Dioctyl Phthalate	C	C	C	solution
Di-iso-octyl Phthalate	C	C	C	solution
Dioctyl Sebacate	C	C	C	solution
Drapex 3.2	C	C	C	solution
Epoxy Resins	C	C	C	solution
Flexol CC-55	C	C	C	solution
Flexol TOF	C	C	C	solution
Flexol 3 GO	C	C	C	solution

Heat Stability of Chlorowax

The excellent heat stability of CHLOROWAX insures maximum freedom from decomposition at normal processing temperatures. Each production lot is thoroughly tested to maintain the uniformly good heat stability associated with this chlorinated paraffin.

In the test for heat stability, a sample is maintained at 347°F (175°C) for 4 hours. The liberated hydrogen chloride is collected and quantitatively measured. All grades of Diamond CHLOROWAX liberate less than 0.5% hydrogen chloride by this severe test.

If temperatures in excess of 250°F are encountered during processing of compounds containing CHLOROWAX, 1 to 3% of stabilizer should be added to retard decomposition. The normal barium-cadmium, tin, lead, and epoxy stabilizers used for polyvinyl chloride resins are also quite effective for CHLOROWAX.

Compatibility of Chlorowax

MATERIALS	RATIO OF CHLOROWAX TO MATERIALS			
	1:1	1:2	1:4	METHOD
Flexol 4 GO	C	C	C	solution
Maleic Resins	C	C	C	solution
Methyl Methacrvlate	C	C	C	solution
Monoplex S-71	C	C	C	solution
Neville Resin R Series	C	C	C	solution
Petroleum Resins	C	C	C	solution
Paraplex G-60	C	C	C	solution
Paraplex G-62	C	C	C	solution
Phenolic Resin, Non-heat Hardening	C	C	C	solution
Pliolite Resins	C	C	C	solution
Polydichlorstyrene	C	C	C	solution
Polyester Resins	C	C	C	solution
Polyethylene Resins	C	C	C	solution
Polyisobutylene-Vistanex	I	I	I	solution
Polystyrene Resins	C	C	C	solution
Polyvinyl Chloride Resins	C	C	C	solution
Rosin	C	C	C	solution
Rosin Ester Resins	C	C	C	solution
Rubber				
Natural	C	C	C	solution
Synthetic Nitrile	C	C	C	solution
Synthetic GR-S	C	C	C	solution
Synthetic Neoprene	C	C	C	solution
Synthetic Butyl	C	C	C	solution
Santicizer B-16	C	C	C	solution
Santicizer 141	C	C	C	solution
Santicizer 160	C	C	C	solution
Terpene Resins	C	C	C	solution
Tetrahydrofurfural oleate	C	C	C	solution
Tricresyl Phosphate	C	C	C	solution
Triphenyl Phosphate	C	C	C	solution
Urea Formaldehyde Resins	C	C	C	solution
Waxes				
Mineral	C	C	C	hot melt
Natural	C	C	C	hot melt
Paraffin	C	C	C	hot melt

Specific Gravity Solutions of Chlorowax

Solvent	*Concentration of Chlorowax 70*	*Specific Gravity at 25°C*
Mineral Spirits	45%	1.01
	50%	1.05
	55%	1.09
	60%	1.14
Methyl Ethyl Ketone	45%	1.06
	55%	1.09
	65%	1.23
	75%	1.33
Turpentine	45%	1.08
	50%	1.12
	55%	1.16
	60%	1.19
Toluol	45%	1.10
	55%	1.16
	65%	1.24
	75%	1.36
Xylol	45%	1.11
	55%	1.17
	65%	1.26
	75%	1.37

Viscosity of Chlorowax 70 Solutions (Centiposes)

	15°C	20°C	30°C	40°C	50°C
30% CHLOROWAX 70 in Mineral Spirits	10	10	10	10	10
45% CHLOROWAX 70 in Mineral Spirits	50	40	30	25	20
55% CHLOROWAX 70 in Mineral Spirits	160	125	60	35	25
60 CHLOROWAX 70 in Mineral Spirits	385	250	120	60	30
45% CHLOROWAX 70 in Turpentine	100	70	45	30	25
50% CHLOROWAX 70 in Turpentine	200	130	70	50	40
55% CHLOROWAX 70 in Turpentine	850	575	325	165	70
60% CHLOROWAX 70 in Turpentine	......	......	680	415	200
60% CHLOROWAX 70 in Toluol	35	25	18	12	10
65% CHLOROWAX 70 in Toluol	165	125	70	45	30
70% CHLOROWAX 70 in Toluol	450	300	150	85	50
75% CHLOROWAX 70 in Toluol	......	......	700	385	185
55% CHLOROWAX 70 in Xylol	25	20	15	12	9
60% CHLOROWAX 70 in Xylol	45	35	25	18	14
65% CHLOROWAX 70 in Xylol	150	120	65	40	35
70% CHLOROWAX 70 in Xylol	750	475	230	120	75
75% CHLOROWAX 70 in Xylol	......	......	1000	535	250
45% CHLOROWAX in Methyl Ethyl Ketone	7	7	7	6	5
60% CHLOROWAX in Methyl Ethyl Ketone	20	18	15	12	10
70% CHLOROWAX in Methyl Ethyl Ketone	125	90	60	40	25
75% CHLOROWAX in Methyl Ethyl Ketone	......	470	265	145	85
30% CHLOROWAX in Carbon Tetrachloride	18	14	12	10	8
45% CHLOROWAX in Carbon Tetrachloride	275	180	90	50	30
50% CHLOROWAX in Carbon Tetrachloride	850	535	235	125	70
55% CHLOROWAX in Carbon Tetrachloride	......	......	1000	465	170

Polyethylene Wax

BARECO Polywax 2000. A unique new low molecular weight homopolymer of ethylene.

Properties of BARECO Polywax 2000

Molecular Weight, Vapor Pressure	2,000
Molecular Weight Distribution, Mw/Mn.	1.3
Density, grams/cc	0.96
Melting Point, °F., ASTM D-127	257
Viscosity, Saybolt, SUS, @ 300° F., ASTM D-88	196
Viscosity, Brookfield, cps, @ 300° F., ASTM D-2669	34
Needle Penetration, @ 77° F., ASTM D-1321	0.5
Needle Penetration, @ 160° F., ASTM D-1321	3.5
Color, Saybolt, ASTM D-156	16
Acid Number, mg KOH/g of wax, ASTM D-664	0
Saponification No., mg KOH/g of wax, ASTM D-94	0
Coefficient of Friction, ASTM D-2534	0.187
Odor, ASTM D-1833	0

Chapter 11

SYNTHETIC HYDROCARBON WAXES

FISCHER-TROPSCH (F-T) WAXES

Franz Fischer[1] and Hans Tropsch in Germany, passed carbon monoxide and hydrogen over alkalized iron turnings at a pressure of 100 atmospheres and a temperature of 400°C. They thus produced a mixture of liquid hydrocarbons with some oxygenated compounds. As finally developed commercially by the Ruhrchemie A.G. in Germany in 1935–40, the process consisted of compressing "synthesis gas" (two volumes of hydrogen to one of carbon monoxide) to about seven atmospheres and passing it through a granular bed of cobalt catalyst at 185–205°C. The major products were wax, hydrocarbon oil, water, and gaseous hydrocarbons such as methane.

Post-war research has continued on the process in both the United States and Germany. In the latter country, present economics have emphasized the production of waxes as a major product, which is easily possible by an adjustment of process conditions. Aside from German developments, a new Fischer-Tropsch plant using coal as raw material has been erected in South Africa. The wax plant unit produces two main types of hydrocarbon wax: a paraffinic type petroleum wax, of low melting point; and a hard wax of higher melting point (105°C).

Hardening Fischer-Tropsch Wax

Hardening is increased and melting point is raised by adding $< 0.3\%$ cumarone resin (softening point $< 70°C$) and heating the mixture.to 150C°.

FT-200 and FT-300 Wax[2]

The FT-Waxes are purely synthetic polymers of carbon monoxide and hydrogen which can best be described chemically as mineral waxes. They are snow-white in color, as hard as the hardest vegetable waxes, and capable of taking an extremely high gloss. Being long-chain aliphatic hydrocarbons, with relatively short side-chains, they have a short softening

---

[1] Dura Commodities Corp.

[2] German Patent: 1,093,085, Nov. 17, 1960.

range. They are absolutely free of residual oil and sulphur. When melted, the FT-Waxes are fluid and water-clear. Their dielectric constant, electrical resistance, and electrical breakdown strength are superior to those of any petroleum waxes. One of the characteristics of the FT-Waxes is their definite tendency to form microcrystals, which is evident from their fracture. When FT-Waxes are blended with ordinary petroleum paraffin, the hardness, and especially the melting point, of the blend is considerably higher than theoretical values would indicate.

SPECIFICATIONS

	FT-95	*FT-100H*	*FT-150*	*FT-200*	*FT-300*
	Physical Properties				
Molecular Weight	500	600	700	600	730
Melting Point °F	212–215	212–221	215–223	212–219	221–225
Congealing Point °F	194–198	194–203	197–205	194–202	205–212
Penetration @ 77°F	5–10	4–8	1–3	3–5	below 1
Acid Value	0	0	0	0	0
Saponification value	*0.5	*0.1	*0.5	*0.5	0
Hydroxyl value	*1	*1	*1	*1	0
Iodine value	0.1	0.1	0.1	0.1	0.1
Viscosity in cps @ 248°F	11	11	12	11	12
Ball Pressure Hardness, Kg/cm^2	90/97	—	305/265	190/150	355/340
Percent Ash	*0.01	*0.01	*0.01	*0.01	*0.01
Oil Content	*1%	*1%	*1%	0	0

* less than

Solubilities and Compatibilities:

FT-Waxes are compatible with all mineral waxes, almost all natural waxes and many resins. In blends, the high solidification point of the FT-Waxes will predominate. At elevated temperatures, the FT-Waxes are soluble in fatty and hydracarbon solvents, but oxygenated solvents have less solubility for the FT-Waxes. Solubility decreases in proportion to the percentage of oxygen in the solvent so that methyl and ethyl-alcohol have hardly any capacity for dissolving them. FT-Waxes are not water-soluble, and, being non-polar, have no solubility for soaps.

DUROXON WAXES

The Duroxon Waxes, first made available to industry in 1954, are the products of research into special applications of the Fischer-Tropsch hydrocarbon synthesis.

At the large-scale plant of Krupp Kohlechemie G.m.b.H., West Germany, the basic output of the reactor, called *gatsch*, is first fractionated to separate the liquid components (Aliphatine and Kogasine) from the solid components (crude mineral wax and paraffin). The further treatment of crude mineral wax, solvent extraction and bleaching, produces the hydrocarbon products known in the trade as F-T[r] Wax. The latter is chemically neutral, very hard and high-melting, snow-white in color, and free from sulphur and residual oil. F-T Wax has gloss-producing properties equalling those of carnauba wax.

By subjecting the F-T Wax to various chemical treatments, particularly oxygenation and esterification, its properties can be adjusted to practically all requirements of the hard-wax industry. The products resulting from such chemical modification of F-T Wax are known as Duroxon[r] Waxes.

Table 11:1

SPECIFICATIONS OF DUROXON WAXES

Type of Wax	*Specification*	*Typical Test*	*Specification*	*Typical Test*
EMULSIFIABLE DUROXONS	J-324[3]		C-60A	
Solidification Point[1] °F	183/190	188.5	190/198	191
Acid Value	5/15	10	20/30	23
Saponification Value	25/40	35	50/65	62
Penetration[2]	2/4	2.5	2/4	2.5
Flash Point °F	—	505	—	466
Fire Point °F	—	577	—	556
Color	Pale Straw		Light Yellow	
EMULSIFIABLE DUROXONS	H-111		H-120	
Solidification Point[1] °F	181/189	185	190/194	192
Acid Value	10/20	13.7	25/35	31.5
Saponification Value	40/60	55	60/70	65
Penetration[2]	2/4	2.8	2/4	2.8
Flash Point °F	—	477	—	485
Fire Point °F	—	595	—	585
Color	Pale Straw		Light Amber	
EMULSIFIABLE DUROXONS	B–60		B–120	
Solidification Point[1] °F	187/194		187/194	
Acid Value	20/30		40/60	
Saponification Value	50/70		90/110	
Iodine Number	5/7		5/7	
Penetration[2]	10/13		20/24	
Color	Pale Yellow		Pale Yellow	
DUROXONS for SOLVENT SYSTEMS	E-302		E-321	
Solidification Point[1] °F	187/191	190	181/187	182
Acid Value	5/10	7	5/15	11
Saponification Value	35/50	39	40/55	49
Penetration[2]	1/3	1.8	1/4	2.5
Flash Point	—	504	—	462
Fire Point	—	594	—	574
Color	Pale Straw		Pale Tan	

[1] Rotating Thermometer.
[2] 100 g/5 sec/77°F.
[3] The data for J-400 are identical except penetration limits 1/3, typical value 1.6.

Duroxon Waxes—Group B

Characteristic Properties	Duroxon	
	B-60	B-120
Solidification Point, rotating thermometer (ASTM D938-49) °F	187–194	187–194
Specific Gravity 20°C/4°C	0.96	0.98
Color	Light Yellow	Light Yellow
Acid Value	20–30	40–60
Saponification Value	50–70	90–110

The synthetic Fischer-Tropsch waxes, Duroxon B-60 and B-120, consist of oxygenated compounds of high molecular weight, chiefly carboxylic and waxy acids and esters. Laboratory analysis has indicated the presence of wax acids such as cerotic, montanic, and melissic and their esters. The B waxes differ in acid and saponification values as well as hardness. When emulsified, they closely resemble beeswax in many respects, especially in their reaction with alkaline materials. In the melt they are miscible and compatible with a wide range of waxes and resins. They are soluble hot in many organic solvents and cold in some.

Uses

The most important property of Duroxon B-60 and B-120 is the ease with which they can be saponified with aqueous solutions of inorganic or organic alkaline substances to give outstanding dispersions.

The saponified waxes are excellent dispersing agents; thus they permit the incorporation of paraffin, high molecular weight synthetic Fischer-Tropsch wax and microcrystalline wax in appreciable proportions. Such emulsions are perfectly smooth and capable of keeping inorganic or organic pigments effectively suspended. They permit the incorporation of solvent and thus, the manufacture of mixed base polishes or creams.

Duroxons of the B group also serve as additives in the manufacture of lubricating greases for the purpose of raising their dropping point and improving the consistency.

Duroxon Waxes—Group C

The C group of Duroxon waxes differs from the B group mainly with respect to hardness; other properties are similar. Duroxon C-60A finds considerable use in the manufacture of partially saponified water-emulsified paste waxes, cream waxes, as a dispersant for nigrosine in carbon paper and shoe polishes, and impregnant for textiles and paper.

Characteristic Properties	Duroxon	
	C-60	C-60A
Solidification Point, rotating thermometer (ASTM D938-49) °F	190–198	190–198
Specific Gravity 20°C/4°C	0.97	0.97
Color	Pale Yellow	Pale Yellow
Penetration (100 g/5 sec/77°F)	3–7	2–4
Acid Value	20–30	20–30
Saponification Value	50–65	50–65

Duroxon

	C-20	C-460
Melting Pt., °F	219	214
Congeal. Pt., °F	196	196
Acid Value	8–11	25–30
Saponification V	18–26	55–65
Penetration	1–2	1–3
Color	Snow-White to sl. Yellowish	
Form	—— Pellets ——	

Duroxon Waxes—Group E

Duroxon E-302 is a very hard wax, completely opaque, and with a conchoidal fracture. Its most characteristic feature is its ability to act as a homogenizing agent and protective colloid in wax–solvent preparations. In this respect it behaves somewhat similar to Ozocerite; it is, however, much harder and exhibits good glossy effects.

Duroxon E-321 likewise a very hard wax, completely opaque and with a conchoidal fracture, is capable of forming stable fluid dispersions in mineral spirits and similar organic solvents. Films obtained from solvent dispersions of Duroxon E-321 dry without chalking. The use of this wax makes it possible to produce liquid polishes of high concentration which neither set at low temperatures nor settle appreciably in summer. Liquid floor finishes of good quality are made, which can be sprayed without any difficulty, even though they may contain 10% or even 20% of total wax.

Both Duroxon E-302 and E-321 are soluble hot in most organic solvents.

Characteristic Properties	Duroxon	
	E-302	E-321
Solidification Point, rotating thermometer (ASTM D938-49) °F	187–191	181–187
Specific Gravity 20°C/4°C	0.93	0.97
Color	Light Yellow	Light Yellow
Penetration (100 g, 5 sec, 77°F)	1–3	1–4

Duroxon Waxes—Group H

These are synthetic waxes of high molecular weight containing carboxylic acids and their esters and such other oxygenated compounds as facilitate the manufacture of dry-bright polishes. They are high-melting, hard, have a dense conchoidal fracture, are dry to the touch, and are of light color.

It is also characteristic of the waxes of the H group that they can produce stable, opaque-to-transparent emulsions not only of anionic, but also of nonionic emulsifiers. Duroxon H-110 can be used to produce translucent anionic emulsions completely without any added fatty acids.

Besides its established use in dry-bright wax compositions, Duroxon H-111 has been favorably reported as a primary hard-wax component of carbon paper inks because it is an outstanding dispersant of pigments and dyes, and oil binder.

Duroxon H-120, the latest development in this series, offers advantages of leveling, dry-bright gloss, and compatibility with natural and synthetic resins.

Characteristic Properties	Duroxon		
	H-110	H-111	H-120
Solidification Point, rotating thermometer (ASTM D938–49) °F	175/183	181/189	190/194
Specific Gravity 20°C/4°C	0.98	0.97	0.97
Color	Light Amber	Light Yellow	Light Amber
Penetration (100 g, 5 sec, 77°F)	2–4	2–4	2–4
Acid Value	15–30	10–20	25–35
Saponification Value	60–75	40–60	60–70

Duroxon Waxes—Group R

These are hard waxes for colloidal solvent dispersions. Type R-21 is regularly available from a pilot-plant facility operated by Ruhrchemie A.G.; it is marketed by Krupp Kohlechemie GMBH through Dura Commodities.

Properties:	R-21
Solidification Pt., Rotating Therm, °C	90–95
Acid and Saponification Value	3–10
Penetration (100g/5 sec/77°F)	3–5

Duroxon Waxes J-324 Series

Multi-purpose, emulsifiable hardwaxes derived from the Fischer-Tropsch process, and recommended for the manufacture of wax emulsions, solvent gels, dispersions and aerosol sprays and foams. High-melting, hard, dry to the touch, opaque, conchoidal fracture.

Technical Data	*J-324*	*J-324H*	*J-324V*	*J-324VH*
Congeal. Pt., ASTM D-938, °C	88/90	88/90	90/93	90/93
Melt. Pt., ASTM D-127, °C [(1)]	108	108	106	106
Penetration, ASTM D-1321	1–2	1–2	0–1	0–1
Specif. Gravity @ 100°C	0.81	0.81	0.82	0.82
Flash Pt. Open Cup., °C	245	240	250	250
Fire Pt., °C	330	320	330	333
Acid Value	5–9	5–9	5–9	5–9
Saponification Value	20–30	20–30	20–30	20–30
Ester Value (Average)	15	15	15	15
Iodine Value	4	4	4	4
Hydroxyl Value	25	34	34	34
Refractive Index/90°C	1.4401	1.4405	1.4399	1.4400

[(1)] Approximate Value

Special Properties:

The Duroxon waxes of the "J-324" group are characterized by easy emulsifiability with anionic or nonionic systems. They also have a high solvent-binding capacity, and yield pastes of very hard, yet smooth consistency and excellent gloss effect.

Differences Between the Four Grades:

Duroxon J-324 is pale yellow, J-324H is ivory to white. Duroxon J-324V is light-tan colored, J-324VH is light ivory.

Duroxon J-324V and VH have been "stripped" of lower-melting oxidation products. These types are, therefore, nearly odorless and they are also somewhat harder than the other two grades. Also, the congealing point of J-324V and VH is somewhat higher than that of J-324 and J-324H. The extra processing explains the slight price difference.

RUHRWAX "A"

(Now called Veba Wax)

RUHRWAXES of the "A" Series are saturated, synthetic hydrocarbons, of hard, wax-like consistency. They are pure, chemically neutral, colorless, and high-melting. The special VEBA-CHEMIE modification of the Ziegler low-pressure polymerization of ethylene accounts for the unusual variety of properties which are available.

1. RUHRWAXES SERIES "A" are produced either with nearly perfect straight-chain, paraffinic, highly crystalline structures *or* highly branched-chain, micro-crystalline iso-paraffinic configurations.
2. RUHRWAXES SERIES "A" range in density from about 0.90 to about 0.96, depending upon the degree of branching.
3. RUHRWAXES SERIES "A" are available with molecular weights ranging from about 900 to 3,500 and higher.

Thus, the RUHRWAXES SERIES "A" fill the gap between Fischer-Tropsch waxes and petroleum waxes (paraffins and micros) on one hand, and the low to intermediate molecular-weight polyolefins (polyethylene, polypropylene etc.) on the other.

Combining a well defined high melting point with low melt viscosity, hardness and compatibility, the RUHRWAXES SERIES "A" are ideal building blocks for a great number of industrial compounds such as water-resistant coatings, release agents, lubricants, construction products, adhesives and special wax compounds.

TECHNICAL DATA

Ruhrwax Type:	*A-409*	*A-613*	*A-616*	*A-415*	*A-217*	*A-116*	*A-227*	*A-235*
Molecular Weight osmometric, approx.	900	1,300	1,600	1,500	1,700	1,600	2,700	3,500
Congealing Point, ° F ASTM D-938	220-226	226-234	228-235	216-225	220-226	216-223	198-205	196-203
Drop Point, °F ASTM D-566	230-248	239-257	244-262	234-252	244-262	237-255	216-230	216-230
Penetration at 77° F ASTM D-1321	1	1	1	1-2	2-4	7-10	2-3	1-2
Ball Pressure Hardness								
10 sec. kilos/sq. cm	350-400	450-500	500-550	350-400	180-230	80-120	180-220	220-270
60 sec. kilos/sq. cm	280-330	400-450	400-450	300-350	120-180	60-100	150-180	180-220
Density at 68°F. approx.	0.94	0.96	0.96	0.94	0.92	0.91	0.92	0.92
Viscosity cps at 300°F.	20-30	40-60	110-130	70-90	110-130	60-80	210-230	350-380
n-Paraffin Content as approx. % addable to Urea	90	90	90	70	30	10	30	30

The structure and composition of RUHRWAXES were determined by customary methods of wax analysis such as fractional extraction of paraffin waxes and characterization of the fractions by means of physical data, molecular weight and ability to form urea adducts. The values obtained according to the above experiments could be verified by applying modern methods such as IR-analysis, dilatometric and, above all, calorimetric measurements.

For example, the iso-paraffin content increases approximately as follows:

A-613 → A-409 → A-616 → A-415 →
→ A-217 → A-227 → A-235 → A-116

The RUHRWAXES A-613, A-409 and A-616 consist almost entirely of straightchain hydrocarbons. A-415 contains approx. 30% of iso-paraffins.

The numbering system for the RUHRWAXES SERES "A" was fixed in such a way that the first digit indicates the second digit of the density value after the decimal point. The succeeding two digits represent the molecular weight divided by hundred.

The difference between the individual types of the RUHRWAX A-SERIES can be seen most strikingly in a graph obtained from a differential thermal analysis apparatus or, better still, from a differential calorimeter (DSC 1 of Perkin-Elmer).

The following will be observed:

Highly crystalline material generally shows a high melt enthalpy which is concentrated in a relatively narrow and high peak, provided there is also a narrow molecular weight distribution. On the other hand, low-crystalline material with broad molecular weight distribution has only low melt enthalpy spread over a very broad and low-level band (example: Ruhrwax A-116)

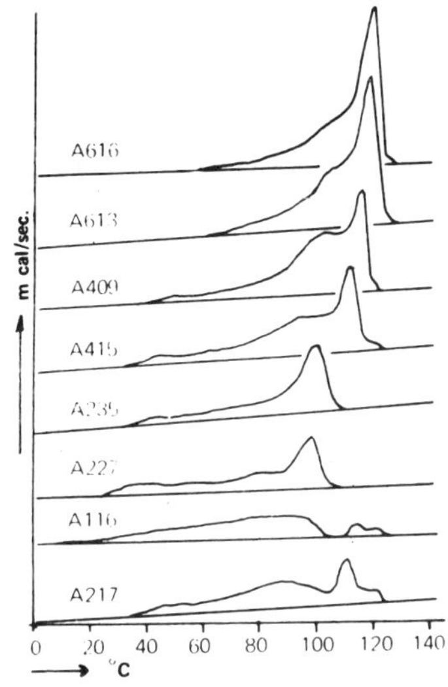

FIG. 11:1 **Melt Diagrams of RUHRWAXES SERIES "A" (Perkin-Elmer DSC-1)**

Solubilities

At elevated temperatures RUHRWAXES SERIES "A" will form transparent solutions in most customary hydrocarbon-type solvents. The following table gives the cloud points for all "A" Types in a 10% solution of different solvents:

	A-409	*A-613*	*A-415*	*A-616*	*A-217*	*A-116*	*A-227*	*A-235*
Mineral Spirits	190°F	196°F	196°F	198°F	187°F	187°F	154°F	156°F
Toluene	176°F	183°F	176°F	180°F	174°F	174°F	142°F	140°F
Xylene	176°F	181°F	180°F	180°F	176°F	176°F	142°F	144°F
Gum Turpentine	183°F	187°F	189°F	190°F	180°F	180°F	145°F	145°F
Butyl Acetate	219°F	223°F	225°F	225°F	221°F	219°F	223°F	223°F
Isophorone S 63	219°F	228°F	225°F	223°F	225°F	225°F	199°F	199°F

At 2 to 9°F above these temperatures the solutions are completely transparent. At room temperature the RUHRWAXES TYPE "A" are almost insoluble in most of the solvents. The gel-like paste, formed by cooling such wax solutions, is excellent for use in varnishes and printing inks.

Compatibilities

RUHRWAXES SERIES "A" are compatible with all mineral and petroleum waxes, most natural waxes, polyethylene and many other resins. The high melting point requires high temperatures for blending. More thorough agitation over an adequately long period of time is recommended in order to assure a uniformly blended compound, because of the slightly higher viscosity of the RUHRWAXES SERIES "A" when compared with lower melting petroleum or Fischer-Trosch waxes.

Influence on Congealing Point and Hardness of Blends

Having a high congealing point, the "A" Types also show very strong melting point boosting properties for other paraffins and waxes. The melting points of blends of RUHRWAX A-616 with fully refined paraffin 126/129°F are shown in the following graph.

Similar to the properties raising the melting point, the hardening effect of RUHRWAXES TYPE "A" is considerable and may already be produced by adding small quantities. The following graph shows how strong this effect is:

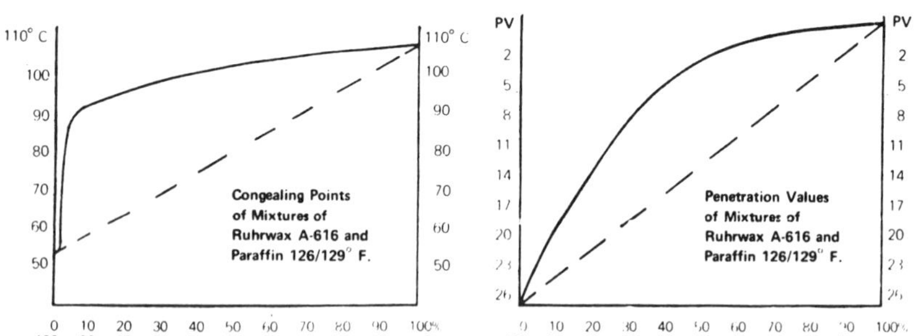

FIG. 11:2

Viscosity

RUHRWAXES of the "A" Series have a higher amount of branched paraffin than the Fischer-Tropsch waxes. Therefore their viscosity is somewhat higher than that of the latter, although it is low for products with their high melting point and hardness. The "A" types raise the melting point and the hardness of blends without noticeably increasing the viscosity of the melt. This low viscosity of melted RUHRWAXES of the "A" Series is of special advantage in paper coating compounds, hot melts, etc.

Electrical Properties

Ruhrwax	*Dielectric Constant at 125°C. 1kc*	*Loss Factor tang.δ at 125°C. 1kc*	*Specific Volume Resistivity at 125°C and 300 V*
A-409	2.0	1.0×10^{-3}	3×10^{12}
A-613	2.0	0.5×10^{-3}	3×10^{12}
A-616	2.1	1.0×10^{-3}	3×10^{12}
A-415	2.0	0.5×10^{-3}	6×10^{11}
A-217	2.1	2.0×10^{-3}	1×10^{12}
A-116	2.1	1.0×10^{-3}	6×10^{11}
A-227	2.0	0.3×10^{-3}	2×10^{13}
A-235	2.1	0.4×10^{-3}	2×10^{13}

Efton D Super

Melting Point (ASTM D-127)	210–215°F
Congealing Point (ASTM D-938)	195–200°F
Acid Value	8–12
Saponification Value	16–24
Hydroxyl Value	9–10
Iodine Value	Abt. 1
Specific Gravity/100°C	0.795
Refractive Index (90°C)	1.4392
Penetration (ASTM D-1321) @ 77°F	1–2
Flash Point, C.O.C. °F	548
Fire Point, C.O.C. °F	654
Color	Less than 0.5 N.P.A. (white)

EFTON D SUPER resembles the unoxidized Fischer-Tropsch wax with respect to hardness, color, color stability, melting point and refractive index. It is, however, polar, whereas the non-oxidized Fischer-Tropsch wax is non-polar. Because of its polarity, the Efton D Super is far more compatible with anionic soaps. By combining Efton D Super with suitable emulsifiable waxes and soaps, it is possible to produce ultra-fine particle size emulsions which dry with a high gloss. Suitable starting formulations for very light colored bright-drying floor finishes and shoe polishes are available.

Efton UC Super

Efton UC Super is a white, emulsion-grade, oxygenated Fischer-Tropsch hydrocarbon hardwax. Compared with Efton UC (regular), it features greater hardness because it is produced from specially solvent extracted and fractionated raw material.

Technical Data:

Melting Point (ASTM D-127)	220–230°F
Congealing Point (ASTM D-938)	190–197°F
Acid Value	25–32
Saponification Value	45–55
Penetration (ASTM D-1321) @ 77°F	2–3
Color	Less than 1 N.P.A. (White)

Special Properties:

Efton UC Super, containing acids and esters of high molecular weight, is readily saponifiable with organic or inorganic agents, such as triethanolamine, morpholine, alkali carbonates or hydroxides. In most emulsion formulations, Efton UC Super is directly interchangeable with Oxidized Microcrystalline Waxes of petroleum origin, taking into account differences in acid values. Generally, the use of Efton UC Super results in more reliable emulsification, finer particle size, harder, more scuff-resistant film and, because of Efton white color, less discoloration. Combined with vegetable hardwaxes, like Carnauba, or Duroxon waxes, Efton UC Super plasticizes the polish film and enhances non-skid properties, although in that respect it is recommended to add softer, plastic waxes also. Combined with emulsifiable polyethylene for use in polymer floor finishes, Efton UC Super reduces melt viscosity, improves process reliability, hardens the polish film and makes it more resistant to scratching, scuffing, heel-marking and powdering.

Efton UC Super is capable of forming soaps with alkali hydroxides and alkali carbonates which are excellent emulsifying agents and carrying agents for paraffin, microcrystalline, unoxidized FT and other waxes which by themselves are not saponifiable. Efton UC Super is also suitable for dispersing organic and inorganic pigments.

In the field of lubricants, the addition of Efton UC Super in a saponified grease raises the dropping point and improves the consistency of the grease.

In carbon paper, Efton UC Super is an excellent additive for Carnauba wax and improves the wetting out properties for pigments and dyestuffs.

The name of all *Ruhrwaxes Type "A"* is now

Veba-Waxes Type "A"

VEBA-WAX A-560

Softening Point (Ring & Ball)	130°
Penetration @ 25°C (77°F)	less than 1
Density @ 20°C (68°F)	.97
Molecular Weight	5,000
Viscosity @ 150°C (302°F), cps.	20,000
Form of Supply	Powder
Color	White

Veba-Wax A-560 is an especially effective lubricant for glass-clear PVC bottle blowing compounds, and conforms to FDA requirements. It is also a high-melting additive in hot-melt adhesives.

Vebafine

Synthetic Hardwaxes and Polywaxes in Ultra-Fine, Spherical Powder Form. Pigment-Like Slip, Anti-Rub and Flatting Agents for All Types of Printing Inks, Lacquers and Coatings

VEBAFINE spray-micronized waxes and polyethylenes are specially formulated for the compounder of inks, coatings and lacquers who prefers a high-performance slip or flatting additive free of any extraneous vehicle component without sacrificing ease of "stir-in" incorporation.

VEBAFINE FT-300 is made from a highly extracted synthetic Fischer-Tropsch wax. In the extraction process all but the hardest, high-slip components and any residual oil is removed from the basic raw material. This has made FT-300 in flake and powder form the preferred slip agent in printing inks of all types for many years. Many of the best compounds on the market today contain FT-300 alone or in combination with polyethylene.

Shanco W-1121 Wax

SHANCO W-1121 is a pure hydrocarbon wax of Fischer-Tropsch type.

Properties

Melting Point (Air Drop Method)	223–226°F
Acid Number	0
Color—Molten—Gardner	1–2
Saponification Number	0
Penetration 100 grams, 77°F, 5 sec.	0.5–1.5

SHANCO W-1121 is a very white wax having excellent hardness and low penetration. It exhibits excellent compatibility with other natural and synthetic waxes.

SHANCO W-1121 Wax is recommended for use in solvent type liquid and paste polishes, textile and paper finishes, release agents, lubricating compounds and printing inks.

Shanco W-1122 Wax

SHANCO W-1122 is a very light colored, very high melting, oxidized and emulsifiable Fischer-Tropsch Wax.

Properties

Melting Point—Air Drop Method	211–215°F
Acid Number	14–15
Color—Gardner	5–6
Saponification Number	28–32
Penetration—100 grams, 77°F, 5 sec.	1–3

SHANCO W-1122 is one of a series of SHANCO oxidized Fischer-Tropsch waxes produced to close tolerances for use in the compounding of emulsion, solvent dispersion and blended compositions. It is intermediate in acid number and degree of emulsifiability between SHANCO W-1130 and W-1119 Waxes.

The outstanding attributes of SHANCO W-1122 are high melting point, low penetration and excellent water resistance coupled with emulsification properties. These characteristics suggest its use as a desirable component in a number of wax-based products. Such items include all varieties of emulsion, liquid and paste polishes and finishes and specialized compositions for paper and textile applications.

SHANCO W-1122 Wax is compatible with all commonly used waxes, polyethylene and wax compatible resins.

Shanco W-1123 Wax

SHANCO W-1123 is a very pale, oxidized Fischer-Tropsch Wax.

Properties

Melting Point—Air Drop Method	205–209
Acid Number	32–34
Color—Gardner	6–7
Saponification Number	52–57
Penetration—100 grams, 77°F, 5 sec.	4–6

SHANCO W-1123 is a high acid, oxidized Fischer-Tropsch Wax having very pale color. It is outstanding in its ease of emulsification and ability to produce high translucent emulsions having good film gloss and clarity.

The wax has a high degree of compatibility with other waxes and with wax compatible resins. It has the property of exerting a solvent action on other components in emulsions. This results in highly uniform emulsions of excellent appearance and film characteristics.

SHANCO W-1123 may be used as the sole wax ingredient or blended with hard waxes or with SHANCO Wax Compatible Resins. Resins such as SHANCO 300 and 320 tend to enhance gloss and contribute to film hardness.

Emulsions formulated with SHANCO W-1123 can be used in palest of emulsions including polymer types. It conveys good buffing properties to high wax compositions and adds plasticity and scuff resistance to polymer emulsions.

Shanco W-1130 Wax

SHANCO W-1130 is a very light colored, vary high melting oxidized Fischer-Tropsch Wax.

Properties

Melting Point—Air Drop Method	212–215°
Acid Number	10–12
Color—Gardner	5–6
Saponification Number	20–25
Penetration—100 grams, 77°F, 5 sec.	1–2.5

Shanco W-1131 Wax

SHANCO W-1131 is a pale colored, high melting, oxidized microcrystalline wax.

Properties

Melting Point—Air Drop Method	183–185°F
Acid Number	5–7
Color—Gardner	5
Saponification Number	5–10
Penetration—100 grams, 77°F, 5 sec.	2–3

Shanco W-1119 Wax

SHANCO W-1119 is a very light colored, high melting, oxidized Fischer-Tropsch Wax of medium acid number.

Properties

Melting Point—Air Drop Method	209–213°
Acid Number	18–22
Color—Gardner	5–6
Saponification Number	32–37
Penetration—100 grams, 77°F, 5 sec	2–4

SHANCO W-1119 is a very hard wax produced under controlled conditions to close tolerances. It is emulsifiable and this property plus its excellent color and hardness makes it a desirable component in emulsion polishes and wax finishes.

The use of relatively small amounts of SHANCO W-1119 Wax with softer emulsifiable waxes materially enhances film hardness and resistance to dirt pickup. These benefits are achieved without deterioration of color in dried films.

An important use for SHANCO W-1119 Wax is with emulsifiable polyethylene on a partial replacement basis in amounts ranging from 10 to 30%. The SHANCO W-1119 Wax produces greater uniformity in emulsions by decreasing the molten viscosity of polyethylene at emulsification temperatures. This allows for greater ease of emulsification.

The hardness of the wax is evidenced in the dried film by improved hardness and resistance to dirt pickup. No change in basic composition or procedure is generally required when combining SHANCO W-1119 Wax with polyethylene.

Paraflint

Paraflint is a hard, high melting point hydrocarbon wax. It is essentially a mixture of saturated, straight-chain paraffin hydrocarbons, having an average molecular weight of approximately 750, corresponding to approximately 50 to 55 carbon atoms. Thus, with the exception of the longer chain length, Paraflint is similar in molecular structure to conventional paraffin waxes derived from petroleum. It also resembles them in appearance, being white in color, brittle, and, when melted, of low viscosity.

Paraflint differs from conventional paraffin waxes primarily in its high temperature properties and crystal structure, the crystal size being much finer and more like petroleum microcrystalline waxes. The melting point is much higher, 214°F (as compared to 150°F for the upper limit of conventional paraffin waxes), and the hardness is much greater, particularly at elevated temperatures, where it is comparable to carnauba and candelilla. Resistance to scuff and blocking is much superior, and the coefficient of friction is substantially lower, than in conventional paraffins.

Paraflint is synthesized by the familiar Fischer-Tropsch Process. Low-cost coal is burned in the presence of oxygen and steam to produce hydrogen and carbon monoxide, which are then reacted in the presence of various catalysts to make a wide variety of hydrocarbons, depending upon conditions. The wax produced is unusually low in cost.

Properties of Paraflint

Melting Point (ASTM)	214°F
Congealing Point (ASTM)	205°F
Needle Penetration at 77°F (ASTM)	1.5
Oil Content (MIBK)	1.0%
Average Molecular Weight	750 (approx.)
Acid, Saponification, Bromine Numbers	Nil
Ash Content	less than .01%
Color	White
Viscosity at 250°F	9.5 centipoises
Specific Gravity at 77°F	0.945
Dielectric Strength at 60 cps	760 v/mil
Dielectric Constant	2.33

Paraflint as Modifier

Paraflint has excellent compatibility with Polyethylene. Recent tests on formulations containing up to 20% Paraflint indicate that improvement in tensile yield strength, stiffness, gloss and permeability can be obtained. Paraflint raises the melt index of Polyethylene, enabling faster molding cycles without flow stresses, and permits lower molding temperatures, thus reducing thermal breakdown.

In paper coatings utilizing paraffin and/or microcrystalline waxes, Paraflint improves scuff, blocking, grease and dust resistance, hardness, tensile strength, and gloss stability. It lowers friction (better slip of coated surface in contact with itself or processing machine), and decreases tack. Higher percentages (up to 100% Paraflint) provide coatings or impregnating agents with unusual high temperature characteristics—especially suitable for hot drink containers or hot packaging of food products—particularly where greater grease resistance is required. From one-half to three percent of Paraflint in paraffin wax is an excellent hardening, stiffening, and opacifying agent for candles, crayons, and the like. Paraflint also can be utilized as a wick hardening agent and as a substitute or extender for carnauba in lacquer coating formulations for candles.

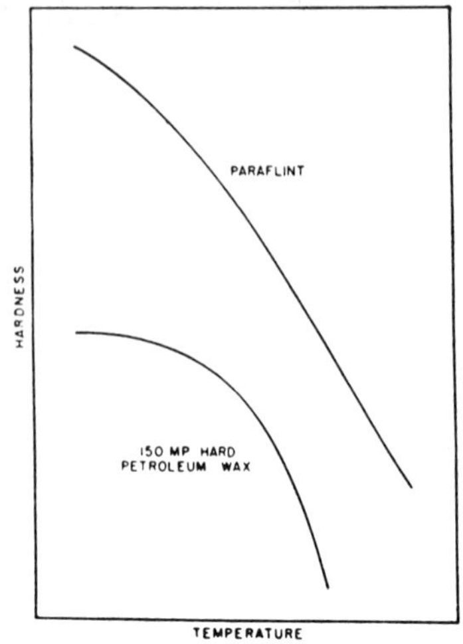

FIG. 11:3. TEMPERATURE–HARDNESS CHARACTERISTICS OF PARAFLINT

Polymekon[1]

Polymekon is derived from high melting point microcrystalline petroleum waxes. The chemical process transforms the original wax to Polymekon which has the characteristics of a resin and unique properties.

Specifications of Polymekon

Softening Point	195–200°F
Penetration 100/77/5	0–3
Color—Yellow N.P.A.	3–4
Melting Point (Drop Method)	200–260°F

Polymekon thickens progressively on heating, and, after prolonged heating, loses its solubility in aliphatic and aromatic solvents.

To prevent thermo-thickening by overheating, it is advisable to add Polymekon to molten waxes.

[1] Warwick Wax Div.

ROCSOL (OXIDIZED WAX)

Products	*Setting point (bulb)*	*Acid Value (mg. KOH/ gm)*	*Saponification value (mg. KOH/ gm)*	*Penetration (100g/5 sec/ 77°F)*	*Color*
Hard types					
Rocsol Wax CFD	176°F (80°C)	28–34	55–65	5–8	Light tan
Rocsol Wax COA	195–200°F (90–94°C)	15–19	50–55	1–3	3–5 NPA
Rocsol Wax COB	195–200°F (90–95°C)	22–25	55–65	2–3	3–7 NPA
Rocsol Wax COC	195–200°F (90–95°C)	27–30	55–65	2–3	3–7 NPA
Rocsol Wax CHE	195–200°F (90–95°C)	5–10	15–25	3	Tan
Soft types					
Rocsol Wax B	165–170°F (75–77°C)	27–30	55–65	15	5–7 NPA
Self-emulsifiable type					
Rocsol Wax CF	190–195°F (87–90°C)	20–25	40–50	6	3–5 NPA

Generally speaking the lower acid value waxes give white opaque emulsions, whereas very small particle size, which in fact may become almost colloidal, can be obtained with grades having higher acid values.

Rocsol Wax COC is similar to Rocsol Wax COB, but has an acid value of 27–30, which will give greater transparency when used in amine soap stabilized emulsions.

POLYMER WAXES

Some polyolefins and their derivatives have wax-like properties.

Vebafine

VEBAFINE A-616 is made from a very rough, very hard polyolefin. Due to its molecular weight and partly paraffinic structure, combined with a high melting point which even surpasses that of Fischer-Tropsch waxes, the A-616 has unusual rub resistance combined with slip.

Technical Data

	Vebafine FT-300	*Vebafine A-616*
Molecular Weight	abt. 700	abt. 1,600
Density @ 68°F	abt. 0.945	abt. 0.96
Penetration Hardness/77°F	0.3–0.9	0.5–1.0
Color	Snow-White	Snow-White
Melting Point, °F	221–225	244–262
Congealing Point, °F	205–212	228–235
Acid and Saponification V.	Nil	Nil
Ash	Nil	Nil
Particle Size: abt. 85%	10 microns or below	10 microns or below
abt. 15%	10–20 microns	10–20 microns

Electrical Properties

Typical electrical properties relative to SANWAX 151-P are shown below. (Measured under 20°C, 60% R.H. conditions.)

Dielectric constant (ϵ)* 2.27
Dielectric loss tangent ($\tan\delta$)* 4.3×10^{-4}
Volume resistivity (Ω cm)** Approx. 3×10^{17}

* 1 megacycle/sec.
** 500 volts D.C.

Viscosity of SANWAX-Paraffin Wax Blend

Relationship between the blend ratio (SANWAX to paraffin wax) and the viscosity of the blend is shown in Figure 11:4. Compared with the sharp increase in viscosity of a blend of paraffin wax with vinyl acetate/ethylene copolymer, the viscosity of SANWAX-paraffin wax blend shows only a small increase as SANWAX is increasingly added.

SANWAX

Non-Emulsifiable Low m.w. Polyethylene

Item	Grade			
	Sanwax 131-P	Sanwax 151-P	Sanwax 161-P	Sanwax 171-P
Appearance	White powder	White powder	White powder	White powder
Color *	100	80	30	60
Viscosity, cps. (140°C)	1000	290	4300	180
Softening point, °C **	108	107	111	105
Penetration hardness *** (100g/5 sec./25°C)	3.5	4	2	4.5
Acid value	Nil	Nil	Nil	Nil
Density (20°C)****	0.93	0.93	0.93	0.93
Average molecular weight	3500	2000	5000	1500

* APHA (melted)
** ASTM E 28–58T
*** ASTM D 1321–61T
**** ASTM D 792–60T

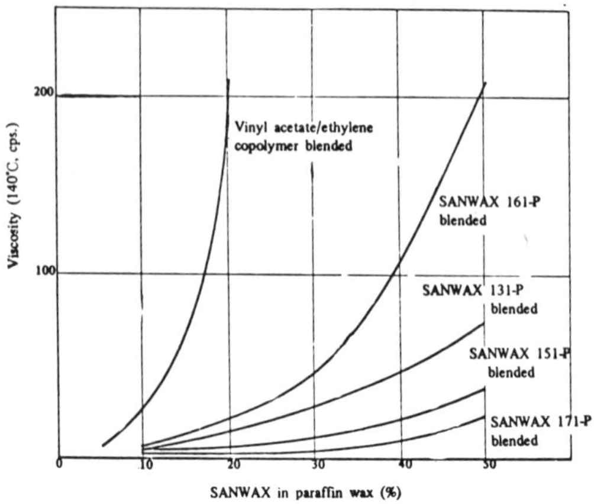

FIG. 11:4. VISCOSITY OF SANWAX-PARAFFIN WAX BLEND

Measuring Method:

The viscosity of the sample (SANWAX-paraffin wax blend) was measured using a Brookfield Viscometer. The sample was set in a oil bath (140 °C) in order to melt it and fix its temperature level at 140 °C. Then the viscosity was determined.

Cloud Point of SANWAX-Paraffin Wax Blend

Fig. 11:5 gives the cloud point of SANWAX-paraffin wax blend. SANWAX SERIES products possess an excellent compatibility with paraffin wax.

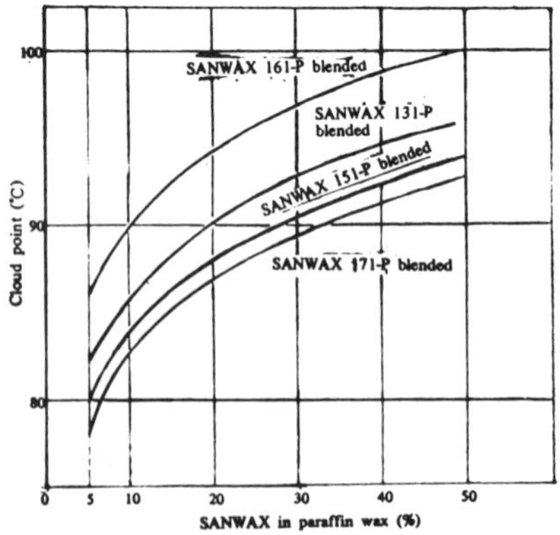

FIG. 11:5. CLOUD POINT OF SANWAX-PARAFFIN WAX BLEND.

Measuring Method:

Approx. 15g of the sample was taken into a test tube and heated until it became molten and clear. Then it was removed from the heater and air-cooled while being constantly stirred with a thermometer. The temperature of the thermometer at the instant the sample turned cloudy, was recorded as the cloud point.

Change in Hardness of SANWAX-Paraffin Wax Blend with Change in Temperature

Blending of SANWAX with paraffin wax is effective to prevent change in wax hardness caused by temperature fluctuation which is otherwise unavoidable with paraffin wax alone. This literally means SANWAX can be used as a blocking inhibitor of paraffin wax in summer when applied to paper processing. Figure 11:6 indicates typical examples of hardness vs. temperature relationship when SANWAX 151-P is added to paraffin wax.

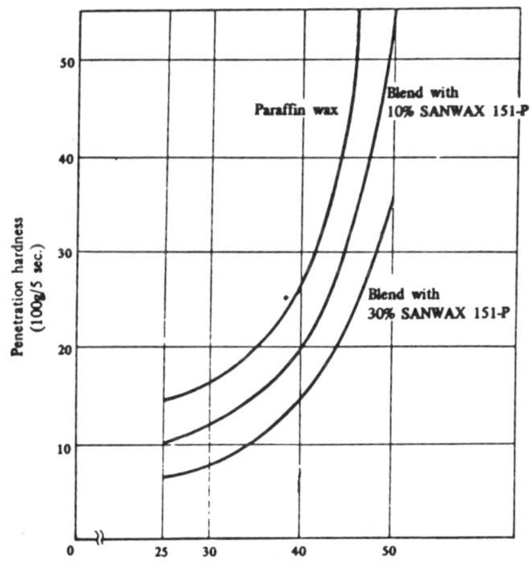

FIG. 11:6. CHANGE IN HARDNESS OF SANWAX 151-P—PARAFFIN WAX BLEND WITH CHANGE IN TEMPERATURE

Hardness of SANWAX-Paraffin Wax Blend

As shown in Figure 11:7, addition of SANWAX to paraffin wax increases hardness of the wax. Therefore, when the blend is applied to the process of paper coating, the increased surface hardness of the wax-coated paper can be achieved. (The hardness is expressed in terms of penetration hardness. The smaller the value, the higher the hardness.)

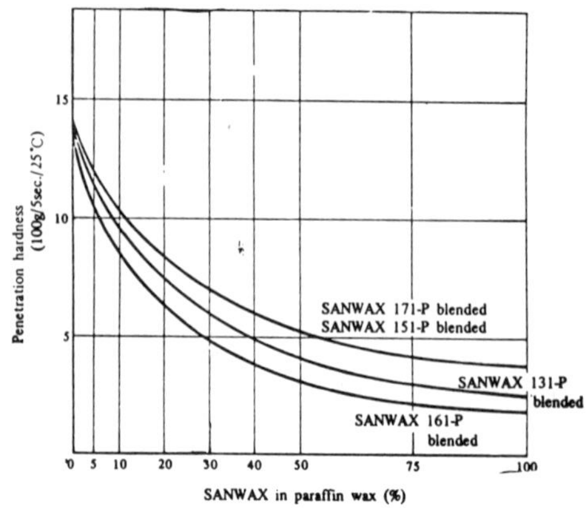

FIG. 11:7. HARDNESS OF SANWAX-PARAFFIN WAX BLEND

Measuring Method:

The penetration hardness was determined according to ASTM D 1321-61T using a penetrometer: the sample was melted, poured into a test container and air-cooled. It was conditioned at 25°C in a water bath and then the penetration hardness was measured by the use of a penetrometer. The penetration hardness is the depth, in tenths of a millimeter, to which the loaded needle (100g) is allowed to penetrate into the sample for five seconds at 25°C.

Softening Point of SANWAX-Paraffin Wax Blend

Figure 11:8 illustrates relationship between the blend ratio and the softening point of SANWAX-paraffin wax blend. As clearly shown here, the softening point of the blend becomes higher as the amount of SANWAX is increased. This characteristic is helpful to prevent softening and sticking of paraffin wax in summer.

Measuring Method:

The softening point was determined according to ASTM E 28-58T (ring and ball method): it was expressed as the temperature at which the sample held within a prescribed size ring was forced downward to touch the lower plate (the distance between the ring and the plate was 25.4 mm) under the weight of a standard steel ball positioned in the center of the sample, as the sample was heated at the rate of 5°C/min. in a glycerin bath.

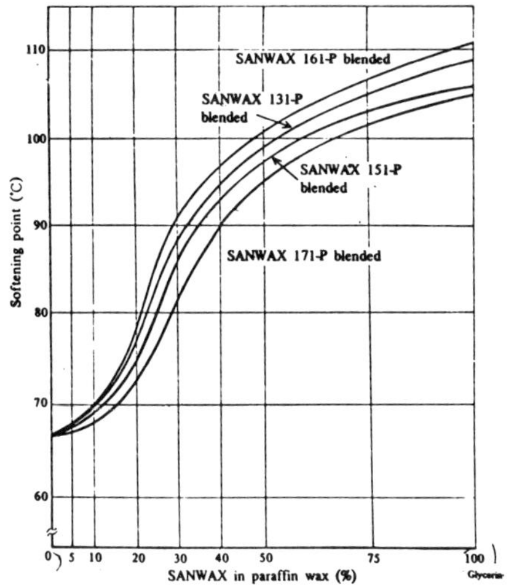

FIG. 11:8. SOFTENING POINT OF SANWAX-PARAFFIN WAX BLEND

SANWAX E-200

EMULSIFIABLE LOW-MOLECULAR-WEIGHT POLYETHYLENE

Properties:

Apearance	White pellet
Color (Hazen)	200
Softening point (°C)*	103.5
Penetration hardness ** (100, 5 sec., 25°C)	6
Viscosity (cps., 140°C)	230
Acid number	20

* ASTM : E28–58T
** ASTM : D1321–61T

Epolene (Polyethylene)

	Emulsifiable Types							Nonemulsifiable Types										
	E-10	E-11	E-12	E-14	E-15	E-43	E-45	N-10	N-11	N-12	N-14	N-45	C-10	C-13	C-14	C-15	C-16	C-17
Ring and Ball Softening Point, °C	106	106	107	104	100	157	114	111	108	113	106	118	104	110	90†	102	106	83†
Penetration Hardness, 100g/5 sec/25°C, tenths of mm	3	3	1	4	7	0-1	1	2	2	1	3	0.1	4	—	—	3	4	—
Density, 25°C	0.942	0.941	0.955	0.939	0.925	0.934	0.964	0.925	0.921	0.938	0.920	0.947	0.906	0.913	0.918	0.906	0.908	0.917
Acid Number	15	15	16	16	16	47	16	<0.05	<0.05	<0.05	<0.05	<0.05	<0.05	<0.05	<0.05	<0.05	5‡	<0.05
Brookfield Viscosity, cp*																		
125°C (257°F)	1200	450	390	320	520	—	—	1800	480	430	230	450	—	—	—	—	—	—
150°C (302°F)	—	—	—	—	—	—	160	—	—	—	—	—	9400	—	—	4300	12,000	—
190°C (374°F)	—	—	—	—	—	370	—	—	—	—	—	—	—	§ $4x10^4$	§ $5x10^6$	—	—	§ $4x10^5$
Color, Gardner Scale	2	2	1	2	2	11	1	1	1	1	1	1	1	1	1	1	1	1
Molecular Weight, approximate	3,000	2,200	2,300	1,800	3,400	—	2,100	3,000	2,200	2,300	1,800	2,100	8,000	12,000	23,000	4,000	8,000	19,000

*Spindle #3; 6rpm †Vicat softening point ‡Saponification number §Approximate

Uniwax 2050

Metal-Crosslinked, Self Emulsifiable Polyethylene

Plastiflow POP

	Modified partially oxidized polyethylene
Form	Fine, free flowing powder
Color	White
Specific Gravity, 24°C	0.96
Acid Number	65
Softening Point, °C	116
Viscosity at 120°C, cps	180

Appearance	Pale Yellow Flakes
Congealing Point, ASTM D-938	163–174°F
Drop Melting Point, ASTM D-566	181–191°F
Acid Value	16–20
Penetration @ 77°F, ASTM D-1321	5–6
Viscosity @ 250°F	100–150 cps.

"Bareco" Polywax E-2020

"Bareco" Polywax E-2020 is an oxidized, high density, low molecular weight polyethylene which emulsifies easily under pressure to produce a stable, fine particle sized, light colored emulsion.

Property	*Test Methods*	*Units*	*Typical Value*	*Mfg. Limit*
Melting Point	ASTM D–127	°F	242	235/245
Needle Penetration @ 77°F	ASTM D-1327	0.1 mm	1	2 Max.
Density		grams/cc	.99	
Viscosity, Brookfield @ 300°F	ASTM D-2669	Centipoise	38	
Color	ASTM D-1500		0.5	1·Max.
Acid Number			21	19/24
Saponification Number			38	32/44

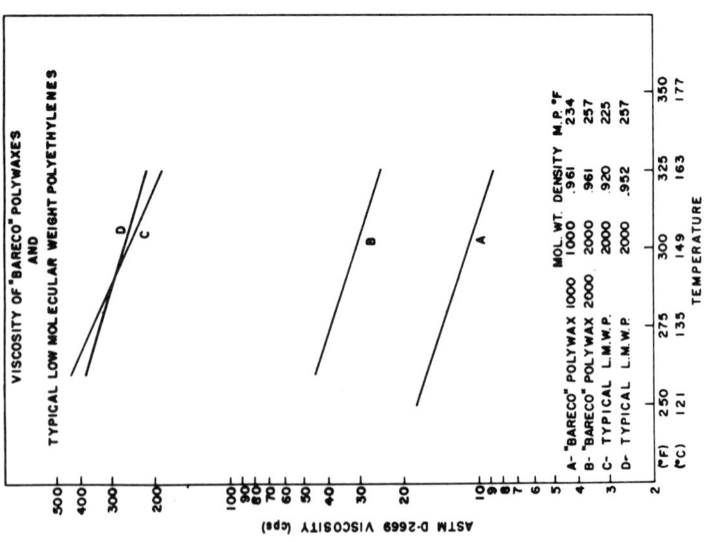

Fig. 11:9

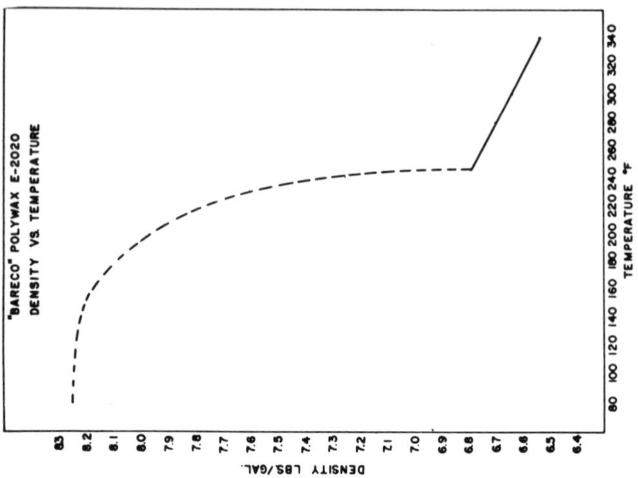

Fig. 11:10

Density of "Bareco" Polywax E-2020

Physical Properties/Specifications

Property	Test Method	Units	"Bareco" Polywax 1000 Typical Value	"Bareco" Polywax 1000 Spec	"Bareco" Polywax 2000 Typical Value	"Bareco" Polywax 2000 Spec
Molecular Weight	Vapor Pressure		1000		2000	
Molecular Weight Distribution	Mw/Mn (Note 1)		1.17		1.3	
Density		grams/cc	0.96		0.96	
Melting Point	ASTM D-127	°F	235	230/235	257	255/260
Viscosity, Saybolt @ 300°F	ASTM D-88	SUS	74	80 max.	196	300 max.
Viscosity, Brookfield @ 300°F	ASTM D-2669	Centipoise	11		34	
Needle Penetration @ 77°F	ASTM D-1321	0.1 mm	1.0	2 max.	0.5	1 max.
Needle Penetration @ 90°F	ASTM D-1321	0.1 mm	1.1		0.9	
Needle Penetration @ 100°F	ASTM D-1321	0.1 mm	1.5		1.1	
Needle Penetration @ 110°F	ASTM D-1321	0.1 mm	2.0		1.3	
Needle Penetration @ 120°F	ASTM D-1321	0.1 mm	2.8		1.6	
Needle Penetration @130°F	ASTM D-1321	0.1 mm	3.7		1.9	
Needle Penetration @140°F	ASTM D-1321	0.1 mm	5.0	7 max.	2.4	3 max.
Needle Penetration @ 150°F	ASTM D-1321	0.1 mm	7.0		2.9	
Needle Penetration @ 160°F	ASTM D-1321	0.1 mm	9.5		3.5	
Color Saybolt	ASTM D-156	Color	16	5 min	16	5 min.
Color	ASTM D-1500	Color	0		0	
Acid Number	ASTM D-664	mg KOH/g wax	0		0	
Saponification Number	ASTM D-94	mg KOH/g wax	0		0	
Odor	ASTM D-1833		0		0	

Note 1: $\frac{\text{Weight-Average Molecular Weight}}{\text{Number-Average Molecular Weight}}$

Effects of "Bareco" Polywax 1000 on Crude Scale Wax

"Bareco" Polywax 1000, %		—	2.5	5.
124/126 Crude Scale, %		100	97.5	95.
Cloud Point, °F		130.	190.	195.
Congealing Point °F	D- 938	123.	165.	184.
Melting Point, °F	D- 127	124.	167.	190.
Viscosity, SUS @210	D- 88	37.	38.	39.
Penetration @ 77	D-1321	35.	30.	27.
@ 90		112.	87.	72.
@100		145.	104.	92.
@110		255.	226.	192.
Blocking Temperature °F		86.	86.	88.
Pick Point °F		82.	82.	84.
% Oil Content		0.80	—	—
Melting Point, °F	D- 87	122.2	—	—
American Melting Point, °F		125.2	—	—

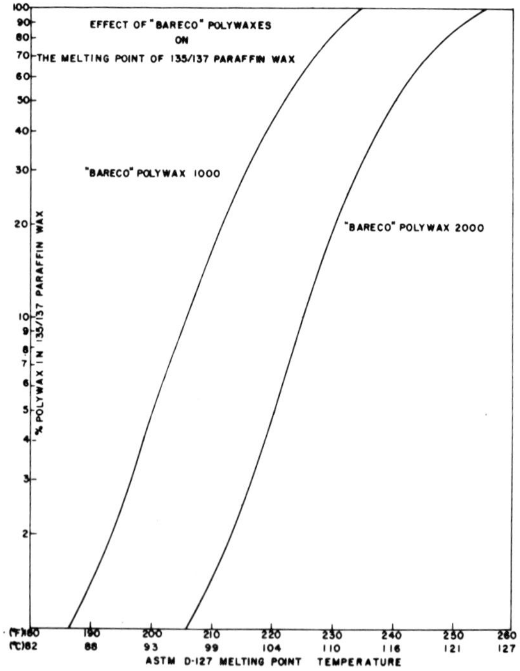

Fig. 11:11

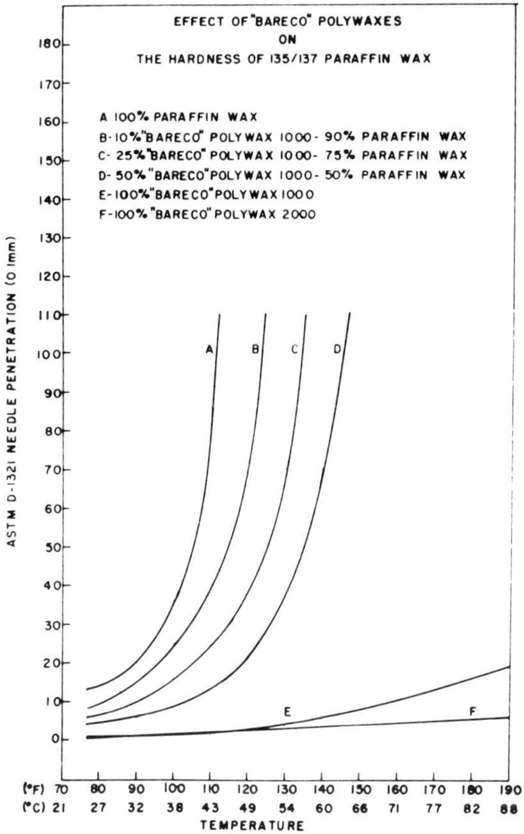

FIG. 11:12

Compatability of "Bareco" Polywaxes and Ethylene Vinyl Acetate Copolymers

"Bareco" Polywax 1000 exhibits complete compatibility in all ratios with all commercially available ethylene vinyl acetate (EVA) and most other copolymers.

"Bareco" Polywax 2000, because of its higher molecular weight, exhibits unusual compatibility with EVA Resins as outlined below. Based on observations of liquid phase separation or lack thereof on samples at 270°F after aging for 24 hours, the following comments can be made:

Two Component Blends

"Bareco" Polywax 2000 and EVA Resin

Only EVA Resins with a vinyl acetate content of 18% or less are compatible.

Three Component Blends

"Bareco" Polywax 2000, Partially Hydrogenated Rosin, and EVA Resin.

Only low molecular weight EVA resins (melt indexes above 125) exhibit compatibility. These resins are compatible only in concentrations of 50% or more EVA resin. Higher molecular weight EVA resins exhibit no compatibility in concentrations below 50% EVA resin, and limited compatibility in concentrations greater than 50%. In the latter case, the blends are initially compatible, but phase separation occurs on aging.

Three Component Blends

"Bareco" Polywax 2000, Paraffin Wax and EVA Resins.

Compatibility is excellent in all ratios with all commercially available EVA Resins with vinyl acetate contents of 18% to 28%.

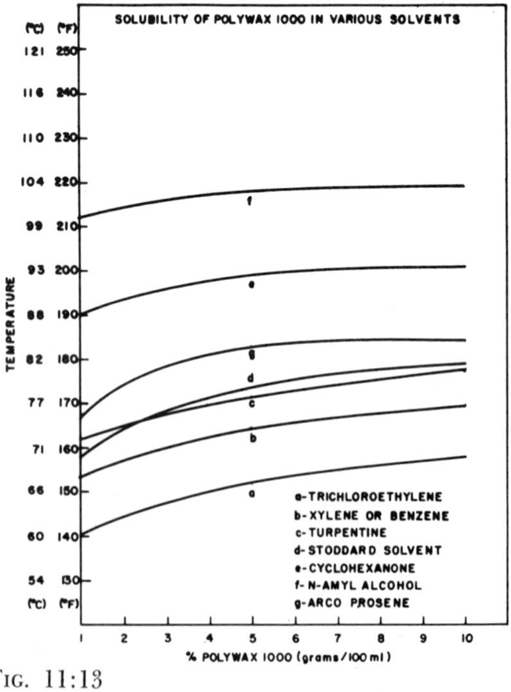

FIG. 11:13

Chapter 12

SYNTHETIC ANIMAL WAXES

SYNTHETIC BEESWAX

"B" Wax[1]

SPECIFICATIONS OF "B" WAX

Equals U.S.P.	Acid Value	20	Odor	Faint and Pleasant
	Saponification Number	93	Fracture	Dull and Plastic
	Ester Number	73	Emulsifying Properties	Excellent
	Ratio Number	3.6	Stability	Excellent
	Specific Gravity	.950	Colors	Extra White Bleached
	Homogeneity	Uniform		Light Refined Yellow

"B" Wax is used in many industries for cosmetics, salves, ointments, protective coatings, polishes, emulsions, soaps, greases, rubber, inks, textile and leather treating, etc.

The following formula produces a wax which is a suitable substitute for beeswax in certain applications. By changing the proportions of mineral oil, various degrees of plasticity may be obtained. By leaving out the Armid HT a soft, very tacky paste results.

W. W. Rosin	60%
Armid HT	30
Mineral oil (100 vis @ 100°F)	10

MODIFIED SPERMACETI WAXES

Hywax 122, 123

These are made from spermaceti or hydrogenated sperm oil by chemical reaction (esterification) to make them emulsifiable.

In addition to the fatty alcohol esters of fatty acids named above, Hywax 122 and 123 contain a small percentage of a chemically-reacted nitrogen compound, which acts as an emulsifier without causing the Hywax to be water soluble.

---

[1] International Wax Refining Co.

Chemical Properties	Hywax 122	Hywax 123
Melting Points	49–50°C	49–50°C
Unsaponifiable (Fatty Alcohols)	38–48%	38–48%
Acid No.	Nil	Nil
Iodine No.	5.0 max.	5.0 max
Saponification No.	95–110	95–110
Viscosity @ 210°F SSU	56	51
Fire Point	420°F	380°F
Flash Point	440°F	400°F
Color—Gardner	5 max	4 max
pH of 10% Emulsion	8–9.5	8–9.5

Average Composition of Esters

	Fatty Alcohols	*Fatty Acids*
C12	—	4.0
C14	1.5	9.5
C16	24.0	21.5
C18	17.5	14.5
C20	1.0	6.5

Physical Properties

Both Hywax 122 and Hywax 123 are excellent lubricants, which is typical of cetyl alcohol esters of fatty acids. Because of the shape of the molecules, a monolayer will deposit on metal, plastic, fiber, wood, lacquer, or any other surface. The portion which causes the stable emulsions will also act as a detergent and will clean surfaces. Hywax is composed of flat, graphite-like crystals which plate-out on a surface in a continuous film or invisible coating. Under slight pressure, these flat crystals form a film which is very strong. It is practically impossible to wipe a film af Hywax from a surface, but it can be removed by emulsification or by using detergents.

Both Hywax 122 and Hywax 123 are heat-stable and there is no noticeable change in color when heated to 165°C (329°F) for 30 minutes.

At concentrations of 40% or higher, Hywax 122 and Hywax 123 will form a white paste which can easily be diluted to stable emulsions.

Both Hywax 122 and Hywax 123 will dissolve and are quite soluble in mineral spirits, chlorinated solvents, and in white mineral oil. At levels below 15 or 20%, they will drop out of paraffin oils, but will be soluble above this level.

At room temperature, Hywax 122 and Hywax 123 are light, crystalline solids. The slight odor is not objectionable and easy to mask, if necessary.

Both Hywax 122 and Hywax 123 are non-toxic and will not oxidize, turn rancid, or change even if stored for an indefinite period of time.

These products are non-corrosive, have anti-rust qualities, and will not attack metal, wood, or other surfaces. After continuous contact, rubber will swell. The waxes are not compatible with plastics, such as vinyl or polyethylene, but will act as calender lubricants and slicking agents.

Hywax 120

Hywax 120 is a hydrogenated sperm oil product. The major components are fatty alcohol esters (C16 Cetyl and C18 Stearyl) of fatty acids (C14 Myristic, C16 Palmitic, and C18 Stearic) with a small amount of glycerides (glycerine esters of fatty acids). It is soluble in organic solvents above its melting point but insoluble in alcohol. Being low in free fatty acid it is non-corrosive, and it is compatible with most all fats and waxes.

Specifications	Composition		
	Fatty Acids	Chain Length	Alcohols
Melting Point 115 to 125°F (46–52°C)			
Color F.A.C. 5.0	5.0	%C12	1.0
Free Fatty Acid (as oleic) 2.0% Maximum	10.0	%C14	25.0
Acid Number 4.0 Maximum			
Saponification Number 124–135	20.0	%C16	17.0
Iodine Number 7.0 Maximum	15.0	%C18	
Unsaponifiable (Fatty Alcohols) 40 to 45%	7.0	%C20	
Flash Point 485°F			

Carolate[1]

Carolate is a mixture of spermaceti-amide and cetyl palmitic alkylol-amide.

It is ivory-colored, soft and amorphous, of bland taste and odor; it is used in cosmetics.

Properties

Sp. Gr. 15°C 0.99	pH (1% sol) 9.3–10.0
M.P. 40–45°C	Acid No. 0–5
Active Amide 45–55%	Alkali No. 48–64
Soap Content 5% max	Unsaponifiable 34–39%
Titratible Amine 12% max	

Synthetic Spermwax (Emerwax 4266-D)

Specifications:

Saponification Value	105–115
Hydroxyl Value	10.0 Max.
Acid Value	2.0 Max.
Iodine Value	1.0 Max.
Color	2 G Max.
Melting Point (Capillary Tube)	41–51°

[1] Robeco, Inc.

Toxicity and Irritation

Preliminary Skin and Draize Eye Irritation studies (rabbits) yielded a value of 0.0, indicative of an extremely low irritative potential. Acute oral toxicity studies (rats) yielded a LD_{50} of 19.9 cc/kg.

Cetina

Fatty Ester Alkylolamide

The most desirable properties and structure of natural Spermaceti and its fatty alcohols combined in an emulsifiable synthetic form.

White colored, bland taste and odor, soft amorphous non-ionic flake wax.

Specifications:

Melting Point	44–45°C
ph 1%	9.0–9.5
Acid Value	3.5
Alkali	18.0
Saponification Value	105

Toxicological Studies:

Primary skin application on the rabbit of 0.5 grams for 72 hours produced no erythema, eschar nor edema.

LD_{50} in the mouse, by oral administration, up to 20 grams per kilo of 20% suspension produced no deaths.

Mucous membrane test of 1 gram for 7 days in rabbit eye produced no conjunctivae.

Spermwax

(Synthetic Spermaceti)

Wax ester which near duplicates the properties of Natural Spermaceti.

White, somewhat translucent flake wax with a faint odor, bland mild taste, free from rancidity.

Specifications:

Acid Value	5 Max.
Saponification Value	109–119
Melting Point	43.5–46.5°C
Iodine Value	1 Max.

SPERMWAX is non-irritating to skin and mucous membrane.

MODIFIED LANOLIN WAXES

Solulan

The Solulans are ethylene oxide ethers of lanolin alcohols.

Properties of Solulans No. 16

Acid No.	1.5
Saponification No.	10.0
Hydroxyl No.	60.0
Iodine No.	15.0
Appearance	Soft waxy solid

Soluble in water and alcohol.

Hydrolan is hydrogenated lanolin.[2]

Constants

Melting Point °C	48°–50°
Specific gravity 60°C	0.858–0.867
Ash	0.05 Max.
Refractive Index @ 50°C	1.464–1.467
Viscosity Saybolt 210°F	46–50
Moisture loss on drying	0.3 max.
Acid value	0.2 max.
Saponification value	6 max.
Iodine value[1]	12–15
Unsaponifiables %	94–98
Acetyl value of unsaponifiables	110–130
Chlorides	0.00
Soluble oxidizable impurities	0.00
Petrolatum	0.00

Hartolan[3]

Hartolan is a mixture of wool wax alcohols obtained by the total saponification of wool fat followed by the extraction and purification of the unsaponifiable portion. It is rich in cholesterol and other nonionic surface-active agents useful as emulsifying agents for producing water-in-oil emulsions.

Hartolan is available in two grades, ordinary and super-refined. The super-refined grade represents wool wax alcohols refined to a degree not previously regarded as technically possible.

The approximate percentage is as follows:

n-Octadecanol and *n*-docasanol	3
Cholesterol	20
Cholesterol–3, 5-di-7-one	1
Dihydrocholesterol	5
Lanosterol	10
Dihydrolagnosterol	10

[2] Amer. Lanolin Corp.
[3] Croda, Inc.

Agnosterol	1
Dihydroagnosterol	4
Unidentified straight-chain alcohols	15
Unidentified diols	5
Unidentified components	26

Of the straight-chain alcohols present, ten branched-chain ones have been identified, comprising dextrorotary ante-*iso* alcohols of odd carbon number, C_{17} to C_{37}, and *iso*-alcohols of even carbon number, C_{20} to C_{26}.

Properties

Ordinary Hartolan is an amber-colored wax-like material with a slight characteristic but not unpleasant odor. Its average analysis is as follows:

Ash Content	0.21% max
Acid Value	2.0 mg max
Acetyl Value	130–140 mg
Cholesterol	28% min
Melting Point °C	58°C min
Mean Molecular Weight	377 average
pH Value	5.7 average
Specific Optical Rotation	20/D–11.8 average
Saponification Value	12 max
Volatile Matter	0.5% max

Super-Hartolan has an attractive, pale yellow appearance with only a very slight odor.

Acid Value	0.24 mg max
Melting Point	70°C approx.
Cholesterol Content	30–35 %
Peroxide Value	2.5 %

Uses

Wool Wax Alcohols can advantageously be substituted for lanolin in a large number of toilet preparations, the product having the advantage of less odor, less drag and stickiness, and having a better color than would result from the use of lanolin. These advantages apply equally to pharmaceutical ointment bases. As the physical and chemical constants of wool wax alcohols are standardized and must conform with the standards laid down in the British Pharmacopoeia, emulsions and ointments prepared from different batches will have the same consistency and texture.

Hartolan is particularly valuable for the production of emulsion creams of the water-in-oil type. These creams are ideal for use on the skin of infants and as a protective preparation against cold and moisture. The protective properties are very similar to those of a layer of petroleum jelly but they are less greasy and of closer relationship to the skin excretions. The value of wool wax alcohols as an ingredient of an ointment base (ointment of wool alcohols) and also as an emulsion (hydrous ointment) is reflected in the inclusion of these preparations in pharmacopoeial ointment bases.

Hartolan is included in cosmetics for several reasons. It produces soft creams; it has an emollient effect on the skin, and gives a very smooth and satisfactory feeling.

Lanfrax[1]

Lanfrax is a lanolin wax of high melting point, made from U.S.P. lanolin by a crystallization from solvents. It is a very light, almost cream-colored waxy solid.

Typical Analysis

Free Fatty Acid (U.S.P.)	0.56%
Color (ASTM Union—when liquid at 150°F)	$2\frac{1}{2}$ max
Odor	Bland
Melting range (U.S.P.)	49–51°C
Saponification number	95–110
Hydroxyl number	25–32
Loss on drying (24 Hrs. at 220°F)	None
Ash (U.S.P.)	0.10% max
Iodine Number (U.S.P.)	25
Alkalinity (U.S.P.)	None
Chlorides (U.S.P.)	Absent
Ammonia (U.S.P.)	Absent
Glycerine (U.S.P.)	Absent
Water-Soluble Oxidizable Substances (U.S.P.)	Absent

Lanfrax-WS55

This is ethoxylated Lanfrax.

Typical Analysis

Melting range	49–51°C
Color (ASTM Union—liquid at 150°F)	3
Saponification number	12
Iodine number	10

Solubility

In water—complete; in ethanol—5%; in methanol—3%; in isopropanol—10%; in acetone—5%; in ethyl acetate—10%; in ethylene glycol—complete; in toluene—10%.

Insoluble in hexane, mineral oil and vegetable oils.

Lantox[2]

Lantox is a wax derived from lanolin, similar to Lanfrax, which is used in cosmetics and leather finishes.

Chemactant PFC

This is a barium soap of saturated hydroxy and branched-chain fatty acids of lanolin.

Properties

Form	Powder (waxy on compression), mill run
Color	Light tan
Melting point	±170°C

[1] Malmstrom Chem. Corp.

[2] Lanatex Products, Inc.

Softening point	±125°C
Moisture	Negligible
Composition	75% Lanolin fatty acids — 25% barium (approx.)
Water solubility	Insoluble; very hydrophobic
General solubility	Soluble in benzene, toluene, turpentine. Slightly soluble in mineral oil, kerosine, pine oil, butyl lactate, chloroform, dioxane, amyl acetate, dibutyl ether, di isobutylketone; Forms gels and dispersions in many of these. Insoluble in lower alcohols and ketones, hexane, Stoddard's solvent.

Waxolan

This is the hard waxy fraction of lanolin.

Lanocerin

(lanolin wax)

A hydrophilic, waxy ester fraction isolated from pure lanolin.

Specifications:

Color	ASTM 3 maximum
Melting point	41° to 51°C
Sapon. value	85 to 115
Iodine value (Hanus)	15 to 30
Acid value	2.5 maximum
Loss on drying	0.25 maximum
Residue on ignition	0.1% maximum

Chapter 13

MISCELLANEOUS SYNTHETIC WAXES

Albacer

(Hydrogenated castor oil & glycol borate)

Properties

Color and Form	white, hard wax
Luster	high
Melting Point	99–104°C. (flow point)

Soluble in: Hot hydrocarbons. The best solvent for Albacer is a mixture of 1 part ethylene dichloride, 6 parts toluene and 4 parts butyl acetate, which will dissolve up to 5% Albacer, most of it remaining in solution even at room temperature.

Uses: The addition of 5–10% Albacer to paraffin wax, for wax-coated paper, gives a finish which exhibits considerable "slip." This is of interest for wrapping sticky candies (caramels, etc.). The paraffin wax is heated to 100°C and the Albacer, in small pieces, is added slowly with stirring until dissolved. Albacer, plasticized with up to 40% castor oil, is used as a paper coating for packaging polybutene (Vistanex-type) synthetic rubbers to prevent adhesion.

When Albacer is melted, it changes to a heavy-bodied nongelatinous liquid which sets very slowly.

Albacer can be blended with certain synthetic resins, mineral and vegetable oils, and other waxes.

ATLASENE WAXES

Substitutes (in whole or in part) for carnauba wax in polishes and finishes requiring high glass and hardness.

Specifications	Atlasene 2020	Atlasene 2525	Atlasene 3030
Acid No.	22.76	28.38	34.00
Sap. Value	30.00	44.00	58.00
Penetration (100 g/5 sec/77°F)	1–2	2	4–5
Color	White	Off White	Pale Straw
Specific Gravity	0.959	0.959	0.961
Melting Point Range	200–216°F	200–214°F	182–210°F

BASF WAXES[1]

Constants	OP Wax BASF	O Wax BASF	A Wax BASF
Color	slightly yellowish	slightly yellowish	almost white, slightly translucent
Dropping Point (Ubbelohde)	approx. 100°C (212°F)	approx. 100°C (212°F)	—
Melting Point	—	—	103–105°C (217–221°F)
Consistency and Structure	amorphous, very hard, brittle	amorphous, very hard, brittle	fine crystalline, hard, very tough
Acid Number	10–15	10–15	0
Saponification Value	110–132	111–133	0
Unsaponifiable Matter	5–8%	5–8%	100%
Specific Gravity at 20°C (68°F)	1.03–1.04	1.03–1.04	0.92

Constants	SPO Wax BASF	E Wax BASF	ES Wax BASF
Color Constants	dark brown to black	slightly yellowish	slightly yellowish
Melting Point	92–98°C (197–208°F)	79–82°C (174–180°F)	79–82°C 174–180°F)
Consistency and Structure	fine-crystalline, very hard, brittle	crystalline, very hard, brittle	crystalline, very hard, brittle
Acid Number	14–22	17–25	17–25
Saponification Value	67–72	158–178	158–178
Unsaponifiable Matter	15–20%	5–8%	5–8%
Specific Gravity at 20°C (68°F)	1.01	1.01–1.02	1.01–1.02

Constants	S Wax BASF	L Wax BASF	V Wax BASF
Color	slightly yellowish	pale yellow	almost white
Melting Point	80–83°C (176–181°F)	80–83°C (176–181°F)	48–50°C (118–122°F)
Consistency and Structure	crystalline, very hard, brittle	crystalline, very hard, brittle	fine-crystalline, hard, brittle
Acid Number	142–157	115–135	0
Saponification Value	160–175	145–165	0
Unsaponifiable Matter	7–10%	7–10%	100%
Specific Gravity at 20°C (68°F)	1.01–1.02	1.01–1.02	0.93–0.94

CARDIS WAXES

Name of Wax	*M.P. ASTM D-127-30*	*Penetration 100g/77°/ 5 sec*	*Color N.P.A.*	*Acid Number*	*Saponification Value*	*Type*
Cardis Polymer # 10	212–216	0–1	1–1.5	8–10	25–30	Polymer
Cardis Polymer # 8 Ⓟ	205–210	1–2	0–2	7–10	24–28	Polymer
Albaplex	200–205	1–2	0–2	7–10	24–28	Plastomer
Cardis 935	210–214	2–3	1–1.5	16–18	38–42	Polymer

[1] Badische Anilin & Soda Fabrik AG.

CERAMID

Ceramid (Stearic Acid Monoethanolamide) is a hard, yellow wax, melting at 78°–81°C. Its solubilities in cold solvent are of particular interest.

Compatibilities

1 part Ceramid
1 part Acrawax C, asphalt, carnauba wax, cumar, ethyl cellulose, microcrystalline wax.
1 part Ceramid
9 parts Acrawax C, asphalt, carnauba wax, cumar, ethyl cellulose, microcrystalline wax, paraffin wax.

Solubilities

Solvent	Hot	Cold[1]
Water	I	0
Alcohol, absolute	VS	5.4
Alcohol, 96%	VS	1.4
Toluol	VS	3.8
Naphtha, VM & P	S	3.4
Acetone	VS	1.6
Ethyl Acetate	S	2.0
Turpentine	VS	1.6

[1] Grams dissolved wax remaining per 100 grams solvent after cooling from hot solution.

For physical properties see Table 13:1.

DIOLIN (OCTADECANEDIOL)

This white, small-lump wax powders with pressure.

Properties

Melting Point	65°C
Molecular Weight	286
Boiling Point	172–186°C at 0.5 mm
Specific Gravity	0.84 at M.P.
Hydroxyl Number	>350
Iodine Number	<4
Saponification Number	<5
Acid Number	<2
Insoluble in Water	

Table 13:1 Properties of Some Synthetic Waxes

Wax	*Color*	*Luster*	*Melting Point, °C*	*Flash Point, °C*	*Specific Gravity at 20° C*	*Acid No.*
Acrawax	tan	high	94–97	230	1.04 (24)	—
Acrawax B	light brown	good	83–86	235	0.97	<2
Acrawax C	tan	high	140–143	285	0.97 (25)	<10
Albacer	white	high	99–104 (flow point)	285–295	1.05	—
Ceramid	yellow	good	78–81	260	0.98	<12
Flexo Wax C	tan	dull	69–73	257	0.82	—
Flexo Wax C Lt.	cream	dull	62–64	230–240	0.82	—
Glyco Wax S932	white	good	60–64	328	0.96	—
Rilan Wax	cream	hard	84–85	—	0.98–0.99	25

	SOLUBILITY IN					
Wax	*Water*	*Alcohol*	*Toluol*	*Naphtha*	*Mineral Oil*	*Vegetable Oil*
Acrawax	IC	SH	SH, GC	PSH	PSH	SH
Acrawax B	I	SH	SH	SH, GC	SH	SH
Acrawax C	I	I	SH	SH	SH	SH
Albacer	I	I	SH	SH	SH	SH
Ceramid	I	SH	SH	SH	SH, GC	SH, GC
Flexo Wax C	I	I	SH	SH	SH	SH, GC
Flexo Wax C Lt.	I	I	SH	SH	SH	SH, GC
Glyco Wax S932	1	I	SH	SH	SH	SH
Rilan Wax			S	S		

Key: S—Soluble; D—Dispersible; H—Hot; I—Insoluble; G—Gel; C—Cold; P—Partially.

GLYCO WAXES

Flexo Wax C

Flexo Wax C is a non-crystalline hydrocarbon wax with high adhesive properties, melting at 66°–74°C. Because of its amorphous character it is more waterproof than paraffin wax. It has a cold flow of 63°–65°C. In thin layers on paper and cloth, it will not flow even at much higher temperatures.

Properties

Color and Form	tan, amorphous wax
Luster	dull
Melting Point (Ball and Ring)	69–73°C
Flash Point	495°F or 257°C
Fire Point	565°F
A.P.I. Gravity, 60°C.	36.5
Viscosity 210°F	76 sec
Penetration, A.S.T.M, Needle, 77°F.	33

Soluble in: Hot hydrocarbons, toluene.

Uses: As an adhesive for cellophane, cellulose acetate, etc.; self-sealing waterproof coatings; greaseproofing; in nitrocellulose lacquers as waterproofing and antiblock agent; protective coating for nickel alloys; for increasing slip in non-mar enamels; stop-off wax for differential plating; water-repellent coatings for fabrics.

Compatibilities: 9 parts Flexo Wax C, 1 part Acrawax B, Acrawax C, beeswax, carnauba wax, cumar, Diglycol Stearate S, Glyco Wax S-932, Opalwax, zinc stearate.

Flexo Wax C Light is a bleached Flexo Wax C. It is slightly softer, has a lower melting point (60–64°C) and is tackier than Flexo Wax C.

Glyco Wax S932 is an edible waxlike ester (tristearin) melting at 60–64°C. It has a flash point of 328°C and does not smoke or give off much volatile matter on heating to 260°C. It is a quick-setting wax with a softening point within 1°C of its melting point.

Compatibility: 1 part Glyco Wax S932 to 1 part of: Acrawax B, Acrawax C, Carnauba Wax, Ethyl Cellulose, Flexo Wax C, Japan Wax.

Uses: In greaseproof coatings for paper; plasticizer for chewing gum (4–8% based on the chicle in chewing gum); in solder fluxes.

OXAZOLINE WAXES[1]

The six oxazoline waxes now available have a wide variety of properties which warrant their evaluation in all applications where natural waxes are now used. As oxazoline waxes are synthetic compounds, numerous variations can be made to give products with different properties. For example, TS-254A has one free hydroxy group and TS-254AA has two hydroxy groups which can be further esterified.

Physical Properties

Product	*ES-254*	*TS-254*	*TS-254A*	*TS-254AA*	*TX-1*	*TS-970*
Molecular weight (approx.)[2]	723	968	656	398	1352	937
Melting point, °C[3]	37	63	50	93	160	74
Volume expansion at m.p., %[4]	8.9	7.7	9.2	9.2	—	9.1
Penetration,[5] mm/100 g load	1.3	0.2	0.3	0.4	—	0.1
Acid number (approx.)	9.0	6.8	2.3	1.5	12.0	6.1
Nitrogen, % by wt	2.3	—	2.3	3.9	1.5	1.6
Solubility by wt, (approx.)						
In: Methanol	30%	2%	8%	11%	Ins.	1%
Butanol	50	7	22	13	49%	4
Toluene	49	16	16	7	52	16
Xylene	49	gel	gel	18	53	13
Stoddard Solvent	49	12	17	4	52	5
Skellysolve B	52	10	12	3	51	9
Skellysolve C	49	11	gel	6	52	9
Petroleum ether	55	11	13	7	Ins.	8
Carbon tetrachloride	49	16	gel	14	47	gel
Butyl acetate	49	11	13	9	55	6
Methyl isobutyl ketone	49	10	10	8	Ins.	5
Turpentine	24	21	21	10	10	10
Oil retention at 40% oil	Fair	Excellent	Excellent	—	—	Excellent

[2] Neutral Equivalent—RAM-30T.
[3] By Melting Point Block.
[4] Change of State, Solid to Liquid.
[5] ASTM D5–25.

The oxazoline structure is unique among waxes, both natural and synthetic. From it comes high adhesion, good wetting on hard surfaces, good lubricity, and solubility in organic liquids. These waxes are compatible with most natural and synthetic waxes. For example, they are completely

[1] Commercial Solvents Corp.

compatible with carnauba, candelilla, microcrystalline and polyethylene (dispersible type) waxes, and Durez Resin 225, in 1 : 1, 1 : 3 and 3 : 1 ratios.

The color of the waxes varies from cream to light tan except for Wax TX-1, which is brown, translucent, and rosin-like in appearance.

Toxicity: Preliminary tests indicate that these products are harmless. Because of difficulty of administration, it has not been possible to test acute oral toxicity at levels above 2.5 g/kg, and no deaths occurred in laboratory animals at that figure. Skin irritation tests have all been negative.

These materials would not be expected to exert any appreciable physiological influence because they are quite inert, and related materials have been shown to be practically harmless. Also, the raw materials from which they are made are all of a very low order of toxicity.

Table 13:2

PROPERTIES OF SOME BRITISH SYNTHETIC WAXES[1]

Product	*Melting Point °C*	*Spec. Gravity*	*Acid Value*	*Ash %*	*Color*
Abril X	68–74	0.99	7–20	0	Brown
Abril 1	79–83	1.02	6–16	0	Stone
Abril E	78–82	0.97	20–26	0	White
Abril S	75–78	1.02	below 3	0	White
Abril 10DS	138–142	0.99	below 5	0	White
Abril OS	45–47	—	below 3	0	Light Brown
Abril BPB	160–166	1.02	7–19	0	Grey
Abril PDS	160–166	0.94	5–20	0	Grey
Abril 93	68–72	1.00	89–96	below 0.5	Brown
Abril R	38–41	0.98	below 2	0	Brown
Cerabrit—BD.701	74	1.01	62–74	trace	Light Brown
Cerabrit—BD.706	76	1.01	42–54	trace	Brown
Sintra E	69–72	0.98	30–40	trace	Light Brown
Sintra SE	65–67	0.98	15–25	trace	Light Brown

EXPERIMENTAL SYNTHETIC WAXES

Most synthetic waxes are reaction products of stearic acid, its derivatives, or other high molecular weight fatty acids, e.g., myristic, palmitic, and lauric acids. It is the latter radical that gives the finished product waxy properties.

The following table indicates the melting points and penetration values of synthetic waxes made in the writer's laboratory. Unless otherwise stated molecular amounts of each reactant have been used. Where no penetration value is given it indicates a very soft product.

[1] Abril Industrial Waxes, Ltd.

Table 13:3

MELTING POINTS AND PENETRATIONS OF EXPERIMENTAL SYNTHETIC WAXES

Reactants	*M.P.°C*	*Penetration*
Ethylenediamine (1) + Stearic Acid; commercial (2)	143	10
+ Stearic Acid 100% (2)	143.5	10
+ 12-Dihydroxystearic Acid (2)	138	9
+ Methyl Stearate (2)	124–6	—
+ Ethyl Dihydroxystearate (2)	130–40	—
+ Stearic Acid (1) + Isobutyric Acid (1)	135–41	—
+ 2-Ethylhexoic Acid (2)	132–37	73
+ Pelargonic Acid (2)	136–7	—
+ Behonic Acid (2)	130–6	12
+ *n*-Dodecyl-β-Alanine (2)	47	17
+ Hydroxydecanoic Acid (1) + Stearic Acid (1)	126–8	8
+ *ter*-Butylbenzoic Acid (1) + Stearic Acid (1)	110	17
+ Hydrofol 30 Special (2)	134	13
+ Myristic Acid (2)	145	15
+ Lauric (2)	149–50	15
+ Isopropyl Benzoate (1) + Stearic Acid (1)	126–7	23
+ Ethylbenzene Sulfonate (1) + Stearic Acid (1)	112–5	32
+ Tolylstearic Acid (2)	129–30	22
+ Diethylphthalate (1) + Stearic Acid (1)	131–2	4
+ Hydrazine (0.15) + Stearic Acid (2)	137–8	13
+ Glycollic Acid, Polymerized (1) + Stearic Acid (1)	117–24	78
+ Hydrofol 200 Glycerides (2)	140–1	8
+ Methyl Formate (1) + Stearic Acid (1)	115–28	5–8
+ Octadecyl Isocyanate (2)	120–6	12
+ Tristearin (2)	143	12–17
1, 3-Diamino-2-Propanol (1) + Stearic Acid (2)	114	8
Triethylenetetramine (1) + Opalwax (1)	59–70	—
Hexanolamine (1) + Stearic Acid (1)	75–76	—
Morpholine (1) + Opalwax (1)	54–57	—
Polyglycol Amine 190 (1) + Stearic Acid (1)	87	13
Polyglycol Amine 190 (1) + Hydrofol 80 (1)[1]	53	—
Polyglycol Amine 275 (1) + Stearic Acid (1)	60–2	18
Polyglycol Amine 275 (1) + Hydrofol 80 (1)	46–7	—
p-aminophenol (1) + Stearic Acid (2)	119-21	—
p-aminobiphenyl (1) + Stearic Acid (2)	135-7	30
o-aminobiphenyl (1) + Stearic Acid (2)	45–50	10
p-phenylenediamine (1) + Hydrofol Acid AB (2)	170–2	10
p-phenylenediamine (1) + Stearic Acid (2)	166–8	9–10
m-phenylenediamine (1) + Stearic Acid (2)	113	9–10
Benzidine (1) + Stearic Acid (2)	223–8	12.5
Benzidine (1) + Stearoyl Chloride (2)	234–5	12.5
Benzidine Dihydrochloride (1) + Hydrofol Acid AB (2)	221–4	12.5
Dichlorobenzidine (1) + Stearoyl Chloride (2)	156–8	14
o-tolidine (1) + Stearoyl Chloride (2)	187–8	10

[1] Hydrofol Acid AB consists of Arachidic Acid, 30; Stearic Acid, 30; Behenic Acid, 25; Palmitic Acid, 13; Myristic Acid, 2.

o-tolidine (1) + Hydrofol Acid AB (2)	173–5	10
p, *p*′-Diaminodiphenylether (1) + Palmitic Acid (2)	180–2	10
p, *p*′-Diaminodiphenyl Ether (1) + Hydrofol Acid Ab (2)	178–80	10
p, *p*′-Diaminodiphenylmethane (1) + Stearoyl Chloride (2)	162–3	10
p,*p*′-Diaminodiphenylmethane (1) + Hydrofol Acid AB (2)	159-60	10
p, *p*′-Diaminodiphenylmethane (1) + Lauric Acid (2)	170–1	10
3, 3′-Iminobispropylamine (1) + Stearic Acid (2)	94–6	10–12
Cyclohexylamine (1) + Stearic Acid (2)	78–80	4
Aniline (1) + Coconut Oil Fatty Acids (2)	64–67	7
Aniline (1) + Stearic Acid (2)	84–6	8
Propylenediamine (1) + Stearic Acid (2)	55	—
Monoethanolamine (1) + 12-Hydroxystearic Acid (1)	100	18
+ Stearic Acid (2)	79–81	10
+ Lauric Acid (2)	73–4	14
+ Behonic Acid (2)	93.6	8
+ Lard Flakes (2)	90	25
+ Tristearin (2)	60–76	24–40
+ Opalwax (2)	×88–96	8.7
Diethanolamine (1) + Opal wax (2)	54–7	8.5
+ Tristearin (2)	51–2	—
Stearyl Alcohol (1) + Sebacic Acid (1)	65	10
+ Opalwax (1)	60	—
Octadecylamine (1) + Stearic Acid 1)	50–60	24
+ Phthalic Anhydride (1)	50–60	23
Octadecylamine (2) + Succinic Acid (1)	114–5	14
+ Diethyl Oxalate (1)	115	26
+ Salicylic Acid (1)	53–5	—
Octadecylamine Monoethylamine } (1) + Citric Acid (1)	80–108	—
Cyclohexylamine (1) + Stearic Acid (1) + Formaldehyde (1)	72–5	27
Dodecanamide (1) + Formaldehyde (1)	127	—
Hexadecanamide (1) + Formaldehyde (1)	132	—
Hydrazine (1) + Stearic Acid (2)	132	7

METALLIC SOAPS

Metallic soaps have some waxy properties and are described here. The metals are of the alkaline earth and heavy metal groups. The fatty acids include stearic, palmitic, myristic, lauric and octoic acids.

Aluminum Soaps

Stearates

	Monostearate USP	Distearates				Tri-stearate	Octoate	Laurate	Palmitate
	High-Gel #505	*Extreme High-Gel #209-B*	*Medium High-Gel #77*	*Medium High-Gel #303*	*Medium High-Gel #26-SW*	*Low Gel #404*	*Low Temperature Gels*		
Total Ash, %	16.0	10.6	10.0	9.7	8.9	7.2	15.0	13.0	9.5
Washed Ash, %	15.5	9.6	9.5	9.2	8.4	6.7	14.5	11.8	9.0
Water Soluble Salts, %	0.5	1.0	0.5	0.5	0.5	0.5	0.5	1.2	0.5
Free Fatty Acid, %	4.0	6.0	3.3	6.5	10.0	26.0	10.0	8.5	5.0
Moisture, %	0.5	0.5	0.5	0.5	0.5	0.5	1.5	0.5	0.5
Softening Point (Mean) °C	170	160	158	150	145	115	275	170	120
Specific Gravity	1.01	1.01	1.01	1.01	1.009	1.01	1.02	1.045	1.01
Apparent Density gm/ml	0.26	0.40	0.25	0.20	0.26	0.28	0.20	0.21	0.21
Fineness % through 200 mesh	95	95	95	95	95	95	95	95	95

Solubility:

Soluble in Benzene, Toluene, Xylene, Carbon Tetrachloride, Vegetable Oils, Animal Oils, Mineral Oils, Waxes and Turpentine. Insoluble in Water, Ketones, Esters and Lower Alcohols.

Barium Stearate *

	#159	#160
Total Ash, % *	28.5	28.0
Washed Ash, %	—	—
Water Soluble Salts, %	none	0.5
Free Fatty Acid, %	1.0	1.0 ma.
Moisture, %	0.2	0.2
Softening Point (Mean) °C	Degrades	Degrades
Specific Gravity	1.23	1.23
Apparent Density gm/ml	0.54	0.22
Fineness % through 200 mesh	90	98

* Synthetic Products Co.

Solubility:

Soluble in Benzene, Toluene, Xylene, Carbon Tetrachloride, Vegetable Oils, Animal Oils, Mineral Oils, Waxes and Turpentine. Insoluble in Water, Ketones, Esters and Lower Alcohols.

Cadmium Stearate

Total Ash, %	19.3
Washed Ash, %	18.8
Water Soluble Salts, %	0.5
Free Fatty Acid, %	1.0
Moisture, %	1.0
Softening Point (Mean) °C	104
Specific Gravity	1.20
Apparent Density gm/ml	0.20
Fineness % through 325 mesh	98

Solubility:

Slightly soluble in Benzene, Toluene, Xylene, Carbon Tetrachloride, Animal Oils, Vegetable Oils, Mineral Oils, Waxes and Turpentine.

Insoluble in Water, Esters, Ketones and Lower Alcohols.

Lead Stearate

Moisture %	0.15
Water Soluble Material, %	0.15
Alkalinity, %	0.0
Melting Point, °C	108
Specific Gravity	1.35

Solubility:

Slightly soluble in Benzene, Toluene, Xylene, Carbon Tetrachloride, Animal Oils, Vegetable Oils, Mineral Oils, Waxes and Turpentine.

Insoluble in Water, Esters, Ketones and Lower Alcohols.

Calcium Stearate

	Type XF	*Type 24–46*	*Type 15–36*	*Type SW*	*Food Grade*	*Wettable*	*Dispersion (25% solids) (*)*	*Calcium Migralube*
Total Ash, %	9.9	9.5	9.5	9.7	9.5	12.0	9.5	8.6
Water Soluble Salts, %	0.5 max.	0.5 max.	0.2	0.3	0.5	3.0	0.5 max.	none
Free Fatty Acid, %	1.0 max.	1.0 max.	1.0	1.0	1.0	1.0	1.0 max.	15
Moisture, %	1.5	2.0	2.8	2.8	2.0	2.0	—	0.5
Softening Point (Mean) °C	150	150	128	150	150	150	• 150	100
Specific Gravity	1.04	1.04	1.04	1.04	1.04	1.05	1.04	1.03
Apparent Density gm/ml	0.1	0.20	0.24	0.17	0.18	0.18	—	0.6
Fineness	99% thru 325 mesh	100% thru 200 mesh	100% thru 200 mesh	95% thru 325 mesh	100% thru 200 mesh	100% thru 200 mesh	99.5% thru 200 mesh	90% thru 200 mesh

(*) Analysis of Solids

Solubility:

Mildly soluble in Benzene, Toluene, Xylene, Carbon Tetrachloride, Vegetable Oils, Animal Oils, Mineral Oils, Waxes and Turpentine. Insoluble in Water, Esters, Ketones and Lower Alcohols.

Lithium Soaps

	Stearate	*12-Hydroxy Stearate*
Lithium, %	2.4	2.2
Free Fatty Acid, %	0.5	0.5
Moisture, %	0.2	0.5
Softening Point (Mean) °C	210	200
Specific Gravity	1.02	1.01
Apparent Density gm/ml	0.10	0.10
Fineness, % through 200 mesh	95	95

Solubility:

Slightly soluble in Benzene, Toluene, Xylene, Carbon Tetrachloride, Animal Oils, Vegetable Oils, Mineral Oils, Waxes and Turpentine.

Insoluble in Water, Esters, Ketones and Lower Alcohols.

Magnesium Stearate

	U.S.P. High Bulk
Total Ash, %	8.5
Washed Ash, %	8.0
Water Soluble Salts, %	0.5
Free Fatty Acids, %	0.5
Moisture, %	3.0
Softening Point (Mean) °C	154
Specific Gravity	1.04
Apparent Density gm/ml	0.12
Fineness, % through 325 mesh	99.5

Solubility:

Slightly soluble in Benzene, Toluene, Xylene, Carbon Tetrachloride, Animal Oils, Vegetable Oils, Mineral Oils, Waxes and Turpentine.

Insoluble in Water, Esters, Ketones and Lower Alcohols.

Zinc Stearate

Total Ash, %	14.2
Washed Ash, %	13.7
Water Soluble Salts, %	0.5
Free Fatty Acid, %	0.5
Moisture, %	0.5
Softening Point (Mean) °C	120
Specific Gravity	1.10
Apparent Density, gm/ml	0.18
Fineness—% Passing thru mesh	99.5% thru 325 mesh

(*) Solids Analysis

Solubility:

Mildly soluble in Benzene, Toluene, Xylene, Carbon Tetrachloride, Vegetable Oils, Animal Oils, Mineral Oils, Waxes and Turpentine. Insoluble in Water, Esters, Ketones and Lower Alcohols.

ALPHABETIC LISTING OF TRADEMARK PRODUCTS

PRODUCT	MANUFACTURER
Abalyn	Hercules Inc.
Abril	Abril Industrial Waxes, Ltd.
Acrawax	Glyco Chemicals, Inc.
Adogen	A-D-M Co.
Albacer	Glyco Chemicals, Inc.
Albaplex	Warwick Chemical Co.
Aldo	Glyco Chemicals, Inc.
Alrosal	Geigy Industrial Chemicals
Amberol	Rohm & Haas Co.
Amo	Quaker State Oil Refining Corp.
Amsco	American Mineral Spirits Co.
Aristowax	Petroleum Products Refining & Producing Co.
Armeen	Armak Co.
Armid HT	Armak Co.
Armowax	Armak Co.
Arochlor	Monsanto Chemical Co.
Atlantic	Atlantic Refining Co.
Atlasene Wax	Atlas Refining Co.
B Wax	International Refining Co.
Barnsdall Wax	Bareco Wax Div., Petrolite Corp.
Be Square	Bareco Wax Div., Petrolite Corp.
Canfield	Canfield Oil Co.
Carbowax	Union Carbide Corp.
Cardipol	Warwick Wax Division
Cardis	Warwick Chemical Co.
Carboset	B. F. Goodrich Chem. Co.
Carlisle Wax	Cinn. Millacron
Carolate	Robeco Chemicals, Inc.
Castor Wax	Baker Castor Oil Co.
Cellosolve	Union Carbide Corp.
Cerabrit	Abril Industrial Waxes, Ltd.
Ceroflex	Cornelius Products Co.

Cetina	Robeco Chemicals Inc.
Chlorowax	Diamond Shamrock
Circasol Wax	Sun Oil Co.
Citroflex	Chas. Pfizer & Co., Inc.
Crodamide	Croda Chem. Inc.
Cumar	Allied Chemical Corp.
Dallawax Black	Dallas Laboratories
Delvet	Diamond Shamrock
Diablo	Diamond Shamrock
Diolin	E. I. duPont de Nemours & Co., Inc.
Duplicane	E. I. duPont de Nemours & Co., Inc.
Durox	Dura Commodities Corp.
Durez	Hooker Chemical Corp.
Durmont	Dura Chemical Co.
Durexon	Dura Commodities Corp.
Dutrex	Shell Chemical Co.
Effon	Dura Commodities Corp.
Elastex	Allied Chemical Corp.
Elvax	E. I. duPont de Nemours & Co., Inc.
Epolene	Eastman Chemical Products, Inc.
Epon	Shell Chemical Co.
Esso	Exxon Co.
Estawax	Petrolite Corp.
FC-128	3M Mfg. Co.
Firwax	Weyerhaeuser Timber Co.
Flexal	Union Carbide Corp.
Flexalyn	Hercules Inc.
Flexo Wax	Glyco Chemicals, Inc.
Flexricin	Baker Castor Oil Co.
Fortex	Warwick Chemical Co.
Fybrene	Sonneborn Chemical & Refining Corp.
Gersthofen Wax	Wax & Rosin Products, Inc.
Glokem	Glovers Chemicals, Ltd.
Glycomel	Glyco Chemicals, Inc.
Glycosperse	Glyco Chemicals, Inc.
Glyco Wax	Glyco Chemicals, Inc
Halowax	Union Carbide Corp.
Hartolan	Croda, Inc.
Helix	Newport Industries Co.
Hoechst	Amer. Hoechst Corp.
Hydrolan	American Lanolin Corp.

Hyonic	Nopco Chemical Co.
Hywax	Werner G. Smith, Co., Inc.
Igepal	GAF Corp.
Kaopolite	Kaopolite Inc.
Kemamide	Humko Products
Kemstrene	Humko Products
KP 140	FMC Corp.
Kyrax	Air Reduction Chemical & Carbide Co.
Lacer	Dura Commodities Corp.
Lanfrax	Malmstrom Chemical Corp.
Lanette	E. I. duPont de Nemours & Co., Inc.
Lanocerin	Robinson Wagner Co.
Lantax	Lanotex Products, Inc.
Lewisol	Hercules Inc.
Lucite	E. I. duPont de Nemours & Co., Inc.
Mekon	Warwick Chemical Co.
Michelene	M. Michel & Co., Inc.
Mobilad	Socony Mobil Oil Co., Inc.
Mobilwax	Socony Mobil Oil Co., Inc.
Monamine	MONA Industries, Inc.
Multiwax	Sonneborn Chemical & Refining Corp.
Mylar	E. I. duPont de Nemours & Co., Inc.
Myvacet	Eastman Chemical Products
Myverol	Eastman Chemical Products
Neo Cryl	Polyvinyl Chemicals
Neofat	Armak Co.
Neoprene	E. I. duPont de Nemours & Co., Inc.
Ninol	Stepan Chem. Co.
Nopcogen	Nopco Chemical Co.
Norcowax	Neatsfoot Oil Refineries Corp.
Nuba	Neville Chemical Co.
OP Wax	BASF Inc.
Opalwax	E. I. duPont de Nemours & Co., Inc.
Oxazoline Wax	Commercial Solvents Corp.
Palawax	Croda, Inc.
Paraflint RG	Moore & Munger
Paraplex	Rohm & Haas Co.
Paricin	Baker Castor Oil Co.
Pearl	Standard Oil Co. of California
Pentalyn	Hercules Inc.

Pentawaxes	Tenneco Co.
Permaline	Refined Products Corp.
Petrolite	Bareco Wax Div., Petrolite Corp.
Petronauba	Bareco Wax Div., Petrolite Corp.
Petrosene	Sonneborn Chemical & Refining Corp.
Piccolastic	Pennsylvania Industrial Chemical Corp.
Piccolyte	Pennsylvania Industrial Chemical Corp.
Piccopole	Pennsylvania Industrial Chemical Corp.
Plastiflow	N. L. Industries
Pliolite	Goodyear Tire & Rubber Co.
Pluramine	Kearny Mfg. Co., Inc.
Polyco	Polyresins Ltd.
Polymekon	Warwick Chemical Co.
Poly Tergent	Olin Corp.
Polywax	Bareco Wax Division
Protowax	Prov Tech
Renex	ICI America Inc.
Resyn	National Starch & Chemical Corp.
Rilan Wax	Deutsche Hydrierwerke A.-G.
Rocsol	Fine Dyestuffs Chem. Ltd.
Ross Wax	Frank B. Ross Co., Inc.
Santicizer	Monsanto Chemical Co.
Santowax	Monsanto Chemical Co.
Santo-Resin	Monsanto Chemical Co.
Sanwax	Sanyo Chem. Industries Ltd.
Seekay Wax	Imperial Chemical Industries Ltd.
Shanco	Shanco Plastics
Shellwax	Shell Chemical Co.
Sintra	Abril Industrial Waxes, Ltd.
Smithol	Werner G. Smith Co.
Socony Paraffin Wax	Standard Oil Co.
Solulan	American Lanolin Corp.
Sovawaxes	Socony-Vacuum Oil Co.
Sovasol	Socony Vacuum Oil Co.
Spermwax	Robeco Chemicals Inc.
Star Wax	Bareco Division
Staybelite	Hercules Inc.
Superba Wax	Sonneborn Chemical & Refining Corp.
Superflex	Quaker State Oil Refining Corp.
Tergitol	Union Carbide Corp.
Triton	Rohm & Haas Co.
Ubatol	UBS Chemical Co.

Vebafine	Dura Commodities Corp.
Veba Wax	Dura Commodities Corp.
Vinylite	Union Carbide Corp.
Vistac	Cinn. Millacron Co.
Vistanex	Enjay Chemical Co.
Vydax	E. I. duPont de Nemours
Warco	Warwick Chemical Co.
Warcosine	Warwick Chemical Co.
Z-52 Wax	Allied Asphalt & Mineral Corp.
Zinar	Enjay Chemical Co.
Zinsser's Insulating Wax	Wm. Zinnsser Co., Inc.

index